Critical Problems in Physics

PRINCETON SERIES IN PHYSICS

Edited by Sam B. Treiman (published since 1976)

Studies in Mathematical Physics: Essays in Honor of Valentine Bargmann *edited by Elliott H. Lieb, B. Simon, and A. S. Wightman*

Convexity in the Theory of Lattice Gasses *by Robert B. Israel*

Works on the Foundations of Statistical Physics *by N. S. Krylov*

Surprises in Theoretical Physics *by Rudolf Peierls*

The Large-Scale Structure of the Universe *by P.J.E. Peebles*

Statistical Physics and the Atomic Theory of Matter, From Boyle and Newton to Landau and Onsager *by Stephen G. Brush*

Quantum Theory and Measurement *edited by John Archibald Wheeler and Wojciech Hubert Zurek*

Current Algebra and Anomalies *by Sam B. Treiman, Roman Jackiw, Bruno Zumino, and Edward Witten*

Quantum Fluctuations *by E. Nelson*

Spin Glasses and Other Frustrated Systems *by Debashish Chowdhury* (*Spin Glasses and Other Frustrated Systems* is published in co-operation with World Scientific Publishing Co. Pte. Ltd., Singapore.)

Weak Interactions in Nuclei *by Barry R. Holstein*

Large-Scale Motions in the Universe: A Vatican Study Week *edited by Vera C. Rubin and George V. Coyne, S.J.*

Instabilities and Fronts in Extended Systems *by Pierre Collet and Jean-Pierre Eckmann*

More Surprises in Theoretical Physics *by Rudolf Peierls*

From Perturbative to Constructive Renormalization *by Vincent Rivasseau*

Supersymmetry and Supergravity (2nd ed.) *by Julius Wess and Jonathan Bagger*

Maxwell's Demon: Entropy, Information, Computing *edited by Harvey S. Leff and Andrew F. Rex*

Introduction to Algebraic and Constructive Quantum Field Theory *by John C. Baez, Irving E. Segal, and Zhengfang Zhou*

Principles of Physical Cosmology *by P.J.E. Peebles*

Scattering in Quantum Field Theories: The Axiomatic and Constructive Approaches *by Daniel Iagolnitzer*

QED and the Men Who Made It: Dyson, Feynman, Schwinger, and Tomonaga *by Silvan S. Schweber*

The Interpretation of Quantum Mechanics *by Roland Omnès*

Gravitation and Inertia *by Ignazio Ciufolini and John Archibald Wheeler*

The Dawning of Gauge Theory *by Lochlainn O'Raifeartaigh*

Critical Problems in Physics *edited by Val L. Fitch, Daniel R. Marlow, and Margit A. E. Dementi*

Critical Problems in Physics

Proceedings of a Conference Celebrating the
250th Anniversary of Princeton University
Princeton, New Jersey
October 31, November 1, November 2, 1996

VAL L. FITCH
DANIEL R. MARLOW
MARGIT A. E. DEMENTI, EDITORS

Princeton Series in Physics

PRINCETON UNIVERSITY PRESS · PRINCETON, NEW JERSEY

Note: The numbering is hexidecimal.

01 Christian Bula, 02 Phyllis Morrison, 03 Philip Morrison, 04 Rubby Sherr, 05 George Charpak, 06 Harry Swinney, 07 Robert Pound, 08 Maurice Goldhaber, 09 John Wheeler, 0A James Cronin, 0B Bram Pais, 0C Frank Wilczek, 0D Joseph Taylor, 0E Will Happer, 0F Bruce Hillman, 10 Nick Samios, 11 Sam Ting, 12 Richard Taylor, 13 Jerry Gollub, 14 Jim Peebles, 15 Val Fitch, 16 Leon Lederman, 17 Rosanna Cester, 18 Henry Kendall, 19 Pierre Piroue, 1A Thibault Damour, 1B Alvin Tollestrup, 1C David Gross, 1D Vernon Hughes, 1E Sidney Drell, 1F Sam Treiman, 20 Erich Vogt, 21 Philip Anderson, 22 George Reynolds, 23 Anthony Hewish, 24 Claire Max, 25 Larry Sulak, 26 Karl Berkleman, 27 Nick Khuri, 28 Steven Adler, 29 Per Bak, 2A Val Telegdi, 2B Alfred Mann, 2C John Hopfield, 2D James Langer, 2E Sudip Chakravarty, 2F TV. Ramakrishnan, 30 Steven Shenker, 31 Phuan Ong, 32 Stewart Smith, 33 Albert Young, 34 David Clarke, 35 unidentified, 36 Zhengdong Cheng, 37 Yi Xiao, 38 Malvin Ruderman, 39 Paul Steinhardt, 3A Rick Balsano, 3B unidentified, 3C John Krommes, 3D Larus Thorlacius, 3E Chiara Nappi, 3F Robert Jaffe, 40 Howard Berg, 41 Claudio Teitelboim, 42 Herman Verlinde, 43 Tom Rosenbaum, 44 J. Giordmaine, 45 Andrei Baranga, 46 Kirk MacDonald, 47 Paul Chaiken, 48 Boris Altshuler, 49 Yitzhak Sharon, 4A Jeffrey Harvey, 4B Robert Naumann, 4C Donald Priour, 4D Kathryn McGrath, 4E Paul Shocklee, 4F Michael Chertkov, 50 A. Millard, 51 Huang Tang, 52 Mark Chen, 53 Valery Kiryukhin, 54 Anastasia Ruzmakia, 55 Patricia McBride, 56 Raghava Chari, 57 Kun Yang, 58 Duncan Haldane, 59 Anthony Zee, 5A G. Baskaran, 5B Frank Sciulli, 5C Eric Verlinde, 5D Ranier Weiss, 5E Lydia Sohn, 5F Elihu Abrahams, 60 Laulik Parikh, 61 Peter Meyers, 62 Mark Strovink, 63 Tom McIlrath, 64 Steven Block, 65 John Krommes, 66 Stephen Appelt, 67 Igor Klebanov, 68 Alexandre Polyakov, 69 Victor Yachot, 6A Giorgio Parisi, 6B Ravindra Bhatt, 6C Konstantin Savvidi, 6D Daniel Stein, 6E Donald Perkins, 6F Morten Krogh, 70 Shiraz Minwalla, 71 Amulya Madhav, 72 Eric Prebys, 73 Bernhard Keimer, 74 Kip Thorne, 75 Chris Quigg, 76 Jerry Sullivan, 77 David Nice, 78 Edward Groth, 79 Joseph Weingartner, 7A Gerald Brown, 7B Arthur Wightman, 7C M. Kiessling, 7D Gordon Cates, 7E Kapeeleshwar Krishana, 7F Dean Jens, 80 Jay Shrauner, 81 Peter Duesing, 82 Mark Tate, 83 Sol Gruner, 84 Piers Coleman, 85 Alberto Guijosa, 86 Oyvind Tafjord, 87 Kentaro Nagamine, 88 Christopher Erickson, 89 Kominis Ioannis, 8A Mikhail Romalis, 8B Keir Neuman

Published by Princeton University Press, 41 William Street,
Princeton, New Jersey 08540
In the United Kingdom: Princeton University Press, Chichester, West Sussex

Library of Congress Cataloging-in-Publication Data

Critical problems in physics : proceedings of a conference celebrating the 250th anniversary of Princeton University, Princeton, New Jersey / Val L. Fitch, Daniel R. Marlow, Margit A. E. Dementi, editors.
p. cm. — (Princeton series in physics)
Includes bibliographical references.
ISBN 0-691-05785-0
ISBN 0-691- 05784-2 (pbk.)
1. Physics—Congresses. 2. Astrophysics—Congresses. 3. Princeton University—Congresses. I. Fitch, Val L., 1923– . II. Marlow, Daniel R., 1954– . III. Dementi, Margit Ann Elisabeth, 1963– . IV. Series.
QC1.C765 1997
530—dc21 97-23916

The publisher would like to acknowledge the editors of this volume for providing the camera-ready copy from which this book was printed

Princeton University Press books are printed on acid-free paper and meet the guidelines for permanence and durability of the Committee on Production Guidelines for Book Longevity of the Council on Library Resources

http://pup.princeton.edu

Printed in the United States of America

1 3 5 7 9 10 8 6 4 2

1 3 5 7 9 10 8 6 4 2 (pbk.)

The Organizing Committee

PHILIP ANDERSON

VAL FITCH, CHAIRMAN

DAVID GROSS

DANIEL MARLOW

STEWART SMITH

JOSEPH TAYLOR

Contents

PREFACE

This volume contains the Proceedings of the Physics Conference held October 31, November 1 and 2, 1996 in celebration of Princeton University's 250th year. It comprises fourteen lectures and transcripts of the discussions which followed. The theme of the conference was "Critical Problems in Physics." Fifty years earlier, in celebration of Princeton's Bicentennial in September of 1946, a conference was held devoted to two subjects: "The Future of Nuclear Science" and "Physical Science and Human Values." At that time much of the new physics had originated from war-time research and was still classified. It would be another nine months before the result of Lamb and Retherford would be announced. The pion and its sequential decay to muon then electron with accompanying neutrinos was yet to be discovered, and it would be more than a year before the transistor was invented. The world of physics was on the threshold of momentous new discoveries. It was also on the threshold of a transition from the basically cloistered research activity of pre-World War II to big science. Correspondingly, of the three days of conference, only one was devoted to physics. Reflecting the awareness of the physicists to the new situation and attendant social responsibilities, wide-ranging discussions of government support of research, of education, and of "physical science and human values" occupied the other two days. It is remarkable to what extent the problems in this broad social context have remained the same. Professor Treiman summarizes his impressions of this earlier conference in the introductory chapter of this volume.

In contrast with the situation of 1947, one has to be impressed with the diversity of subjects that, today, we classify as physics. Indeed, it was not possible in three days to discuss all of the topics that currently compete for our attention. Therefore, it was not intended that the title imply "all" critical problems, and the organizing committee had some difficulty selecting topics to be included in the program. There are a number of notable omissions. An example is CP noninvariance. In addition to being on the experimental program of existing accelerator laboratories, there are now three major facilities being constructed throughout the world for the express purpose of elucidating the question. This topic was not covered because recently there have been several conferences and workshops devoted to the subject. For similar reasons the topic of the quark-gluon plasma was omitted. Elegant new experiments based on the laser cooling of trapped particles were likewise passed over. Finally, the societal problems which occupied two-thirds of the conference in 1947 were not discussed.

Many people in the Department of Physics contributed significantly to the organization and implementation of the conference agenda. Of special note are the graduate students who elected to become rapporteurs of the various sessions. Their names adjoin the discussions after each paper. The editors of this volume are especially grateful for the splendid cooperation of the contributing authors. They also thank Elaine D. Remillard for producing "camera-ready copy" with good humor and great skill.

Val L. Fitch
Daniel R. Marlow
Margit A.E. Dementi

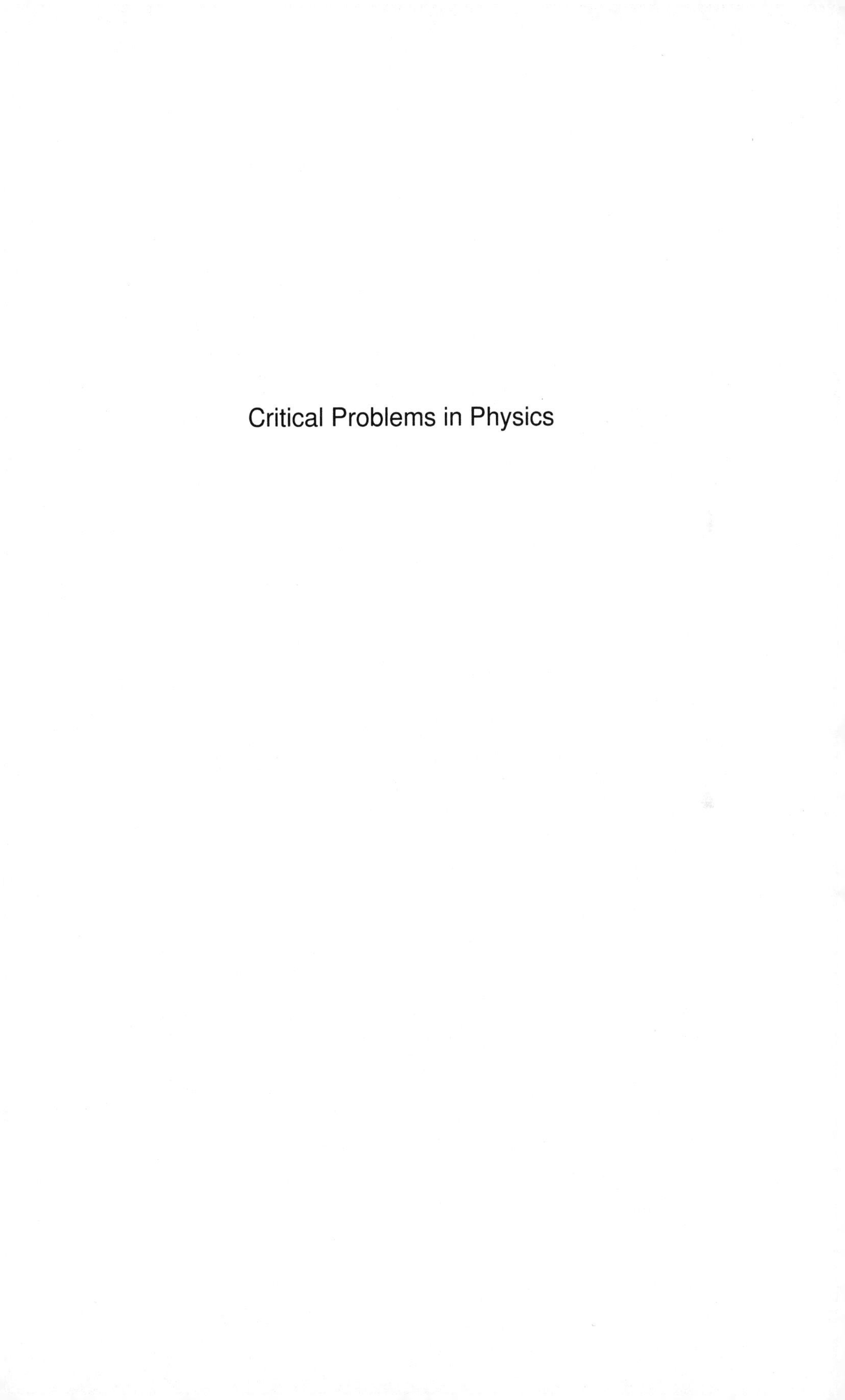

Critical Problems in Physics

CHAPTER 1

THE 1946 CONFERENCE: THEN AND NOW

SAM TREIMAN

Department of Physics
Princeton University, Princeton, NJ

Princeton celebrates its birthday every 50 years. The undergraduates and many of the old alums jump at the chance to party all year long. The professors hold learned conferences. In 1946 the physics department hosted such a conference; and I've been ordered by Val Fitch to say a few words about it. This assignment is all the more interesting because I was *not there*. I was scarcely born! But not to worry, said Val. There will be a few in the audience who *were* there. And anyhow, he had unearthed some records of the event.

Before I got around to looking at that material, I indulged myself in a bit of idle speculation about what those distinguished conference participants might have talked about. The year 1946, remember, was a transition time between two distinct eras in physics, eras separated by a great war. The scientists were for the most part only just returning to their normal pursuits, but in a changed climate. They were now very much in the public eye. A Cold War was getting started. Material support for science, physics in particular, was on the rise; students were flocking to the field; vast new projects were being envisaged. A new half-century of great scientific progress was getting started. Indeed, the historic Shelter Island conference, which ushered in the triumph of quantum electrodynamics, took place within a year. The first two V particles had been captured in the cosmic rays just before that. Did the bicentennial participants foresee these things or the flood of new particles that began to show up not long after? Did they anticipate the transistor, the laser, the giant particle accelerators, the Josephson junction, the BCS breakthrough in superconductivity, the ideas of non-Abelian gauge theory? Could they have imagined that we'd one day pick up microwave signals from the early universe? Could they even remotely have imagined quarks, gluons, the Standard Model, all that stuff? The more I pondered, the more I exulted in the glories of *our* 50 years in physics.

But then it occurred to me that those 1946 conferees might have chosen to look back at *their* 50 years. Well, 1896 was another great time of change. That's the year when the College of New Jersey changed its name to Princeton University. But other developments of almost equal significance set in. Within a ten year period: x-rays, radioactivity, the electron, special relativity, the first glimpse of the quantum; then later on, superconductivity, the discovery of cosmic rays, the Rutherford-Bohr atom, general relativity, quantum theory, the neutron, the positron, how the stars burn...

My exultation over *our* 50 years abated. I decided to abandon this competitive line of musing and instead turn to the actual record of the 1946 bicentennial conference, as much of it as could be found. And here I discovered an unexpected thing. Only one of the three days was devoted to physics shoptalk, to what Eugene Wigner described at the time as *technical physics*. There had to be at least some of that, Wigner implied, if you're going to bring a bunch of physicists together for any purpose whatsoever. But the bulk of the conference—the other two days—can best be described under the rubric "Physical Science and Human Values." That was in fact the title of the published conference proceedings, with a forward by Wigner. What were the main themes of those two days? Many interesting subjects were discussed, the American high school among them. But the main themes were: scientific freedom, secrecy, the lure and dangers of big money, the future role of large laboratories, relations among universities, government, and industry.

One can well understand this selection of topics. Many of the participants had been under wraps during the war, and their research, much of it, was *still* under wraps. Their work had been directed toward urgent, short-term goals. This, for many, was quite different from their expectations of a life in science and different from their pre-wartime experience. They had now tasted both the joys and irritations of large, directed, collaborative projects. They may have come across to the public as pointy-headed and difficult, but they were widely regarded as heroes and wizards all the same. There was the prospect of financial support for research on a scale that could not have been imagined before the war. They could begin to conceive vast new projects, both national and local to their universities.

I can, at most, convey only a small flavor of the conference, which I will attempt to do by means of a selection of snippets taken from the written record. First, from those technical sessions on the middle day:

Bob Wilson reported the latest results obtained at Berkeley on p-p scattering at 15 MeV. The published proceedings don't quite say it, but I think I detect between the lines a murmur that must have gone up in the audience at the first mention of this ultra-high energy. Ernest Lawrence talked about the coming 300 MeV machine and his hope that pi mesons would show up there. He then went on to speculate about still bigger machines, 10 GeV someday, he said! Gasps of disbelief in the audience! In the general discussion that ensued, Lawrence made the prophetic observation that big money would be needed not just for big accelerators, but for the big arrays of detecting equipment that would surely be required. Enrico Fermi talked about the future of the "chain reacting uranium pile." Let me quote: "I believe that the sensible scientific opinion is that the pile as an instrument for serving physics may not be as spectacular an instrument as in medicine, chemistry, and biology. But," he went on, "it will surely be good for basic neutron physics and radioactivity studies."

Here is Glenn Seaborg on a related matter: "It is not at all out of the question that the greatest gains to humanity from the atomic energy development project will result from the widespread use of tracers . . . rather than the harnessing of the power itself."

Paul Dirac's talk focused on the shortcomings of quantum electrodynamics and quantum field theory generally—the familiar worries about infinities, and so on. He talked some about his own inconclusive attempts to surmount the difficulties. Feynman, who was remarkably well-behaved throughout the conference as nearly as I can tell, spoke up in the discussion here: "We need an intuitive leap at the mathematical formalism," he said. Was he alerting the audience to forthcoming developments?

As for the non-technical portion of the conference, that got started with a talk by Isidor Rabi on the relation of research in universities to government and commercial laboratories. One has always been able to count on Rabi for memorable quotes. Here are a few: "University research is to a certain degree a part of the educational process and is parasitic on teaching." A few sentences later on: "The commercial laboratory is parasitic on commerce and industry." Then, on science in government labs: "As public servants, the scientists in these laboratories have to justify to every penny the expenditure of public funds. The ancient American suspicion that public officials are rascals who seek to drain the taxpayer's pocket serves to provide a sense of original sin in all their activities."

The ensuing discussion was quite lively indeed! Some in the audience were shocked at Rabi's characterization of research as parasitic. But Ed Condon spoke up in support of parasitism. "I like the idea of what I will call bootleg research, simply because it's one way of insuring freedom." As to the need to justify public expenditures for science, said Condon, "I do not deprecate that . . . . Numbers in the government are so big that we are likely to forget the decimal point." Oswald Veblen, too, saw merit in Rabi's position. It is better to do research as a by-product of something else, he said. Nevertheless, he argued against a subordinate relationship to teaching. He extolled the European system where, he said, research comes first—or maybe, as I might have said, where teaching comes second. There followed a broader discussion on the subject of how to accept government financial support for university research and yet retain independence; and more generally, how to cope with the risks of big money. These were running themes throughout much of the conference. Karl Compton, for example, referring to the development of quantum theory in Europe in the 1920s wondered "whether great theoretical advances were not sometimes stimulated by inability to get one's hands on an apparatus and the consequent necessity of simply thinking." And here is George Kistiakowsky: "It seems to me that the danger is not in the government offering large sums of money to the colleges, but in the eagerness with which a great many scientists accept the money and sacrifice their freedom."

All of this was of course very high-minded. But listen to what that wily chemist Harold Urey had to say: "If only we do not have unification of the armed services; if we have an army and navy which are separate and keep the traditional enmity between these two services, we may be in a position to escape domination by playing one off against the other. And if it should be possible to get a national science foundation established and to produce rivalry between that foundation and both the services, perhaps we should still be able to retain our independence."

Between the hand-ringing and the high-minded, there was extended discussion of the principles that ought to be recommended and adopted for the management of large, national laboratories and for the distribution of federal funds to the universities. Ilene Curie-Joliot described the organizational arrangements that were being adopted in France. P.M.S. Blackett reported on the English scene. In an extensive presentation, Lee DuBridge laid out—prophetically and accurately, I think—the policies that ought to govern the large basic research laboratories that were forming or being contemplated in the U.S.: the management by university consortia with scientific decisions and leadership in the hands of the scientists, strong in-house staffs, ample provision for accessibility by university scientists to laboratory facilities, and so on.

Secrecy and declassification were also lively topics at the conference. Richard Tolman, who was head of a governmental declassification committee, described the scheme being followed, deplored its slow pace, but regarded it as generally successful. Feynman spoke up here. He acknowledged the case for secrecy of technical know-how and gadgetry but regretted that too much of basic science was being retained in the secret recesses. "There isn't any secret science," he said. "Science is very easy. There are the same atoms all over the world." Fermi spoke up too. The real trouble, he said, was that he and his fellow scientists had been too lazy to write up and publish the seventy to eighty percent of their war-time work that *could* be published. Tolman argued forcefully against secret research in the universities. In general, there was little stridency in this extended and interesting discussion of secrecy. Given the times, there was general acceptance of a need for some degree of secrecy, but there was impatience at its excesses and idiocies.

In the midst of all the proceedings, the conference was treated to a philosophical disquisition on the relation of the natural sciences to philosophy and thence to economic and political system preferences. The speaker, a well-known Yale philosopher, argued that social and economic beliefs are firmly rooted in philosophical outlooks on the natural sciences. I will lift out a single snippet. "To look at the relation of a person to society from a scientific viewpoint, [consider that] a person is first of all a thermodynamic system which has not attained the state of maximum entropy. This means that he requires to take energy in from the outside world to survive.... It is from this that you derive your conception of the state." (He is referring there to political state, not thermodynamic state). And so on. In the ensuing discussion Karl Darrow offered the following comment: "This way of expressing the situation seems to imply that the necessity for eating was not realized until after the second law of thermodynamics had been formulated."

The conference returned to the subject of freedom in science in a session whose centerpiece was a thoughtful analysis presented by Michael Polanyi. It was, of course, not the freedom of scientists to pursue their own research whims on their own time and with their own resources that was at issue. At issue, implicitly, was freedom within a framework of approbation and material support by society. "Freedom of scientific research," Polanyi said, "is in harmony with intensely personal impulses . . . . An individualist philosophy would regard these personal impulses as justification for freedom in science." "But I find such a view rather superficial," he said. "For, clearly, not every strong personal impulse can claim respect, and it remains therefore to be shown why those of the scientist should be respected."

He then went on to argue that scientific freedom serves social efficiency by coordinating the efforts of individual scientists. One normally thinks of coordination as a process that imposes *restraints* on the discretionary power of individuals. But the nub of his argument was that coordination in science is achieved by the opposite method of releasing individual impulses. He illustrated this with the example of a very large jigsaw puzzle. It would be wasteful to give each player a separate set to work on in isolation. It would greatly speed up the solution if all players were allowed to work freely on the same set, each taking note of the results of the others.

Polanyi, of course, recognized the contrasts with science: Jigsaw puzzles have a definite solution, you know the objective and know when you're done. That's not the case with science, so further argumentation was needed—and supplied. But I will cease and desist here, except to say that in the course of his further analysis, Polanyi laid out some of the responsibilities that go with freedom in science: among these, the responsibility to prevent frauds, cranks, and "habitual blunderers" from gaining ground in science. "Habitual blunderers"—what a magnificent phrase!

The bicentennial conference was, to me, surprising in its focus and quite remarkable for that. Here were the world's scientific luminaries, in between terms, so to speak. Not dwelling much on their own great accomplishments, not many war stories told—whether cold or hot, restrained in their attempts at predicting the *substance* of future scientific developments. Rather, the conference was a series of wonderfully thoughtful conversations about the great changes that were looming in the climate for science, and how to cope and flourish with honor.

Let me conclude by saying that in those first moments of conjuring up the bicentennial conference, I inevitably mused also about the Princeton Tricentennial

Conference scheduled for October, 2046. What will the big themes be? I will leave it to you to make your own guesses. Here are just two of mine:

1. A major topic at that 2046 conference will be the pesky background of Higgs particles that clutter up the search for excited resonances in gravitino-hadron scattering. The experiments will already have gotten underway at the big new Japanese accelerator, the *Stupendous Supersymmetry Contraption.* An international lawsuit over the acronym SSC will be nearing resolution.

2. Here in the U.S., support for science will be on the rise again, after a long drought following the abolition of the departments of energy, commerce, education, the National Science Foundation, and other liberal, left-wing agencies. Funds will be pouring out of a new Department of Creationism. Many of the conference sessions in 2046 will be devoted to the formulation of compelling connections between creationist thought and physics. Our physics colleagues will have no trouble, I predict, in figuring out the angles.

1.1 DISCUSSION

Session Chair: A.J. Stewart Smith

SMITH: It might be appropriate to begin with remarks from those who were at the 1946 conference.

REYNOLDS: It was a heady experience for the young guys who were there. I can't avoid one personal reminiscence. Don Hamilton and I had to transcribe from wire recorders, not tape recorders or high-fidelity audio systems. In particular, we spent hours over Niels Bohr's remarks; we got all the "and so"s and a couple of "buts"; as for the rest... Sam did a remarkable job of summarizing. It was an exciting time because we could come from behind the walls of secrecy and begin to think about what we could could do now that we were free.

MORRISON: I was immediately struck by what Sam said about the tactical nature of the discussions. Except for the personal encounters with people, especially those we had heard about but never seen—Dirac, for example—everything that happened at that conference escapes my mind except memories of a significant encounter of perhaps twenty minutes in an anteroom off a library, as I recall, among four old friends from Los Alamos: Robert Oppenheimer, Robert Wilson, myself, and David Hawkins, who had been a top administrator at Los Alamos and who wrote a famous history of that project. [Editor's note: Oppenheimer had been their leader during the war.] Whatever it was about, there appeared to be agreement in the discussion, and then Oppenheimer, always very busy, suddenly left, leaving us three juniors more or less breathless. Wilson remarked ruefully, "Well, we almost stood up to him this time." This interaction is embedded, not without its tragic side, in the history of American politics, the U. S.-Soviet Union encounter and relations for the next ten years.

PAIS: I'm grateful to Sam for clarifying it for me. My memory was that it was extremely uninteresting. I was arrogant and wanted to learn new physics, and there wasn't much. Also, I'd like to tell a personal experience. You must understand it was my second week in the U.S. As a good European, at 4 pm I wanted a cup of tea. I went to Nassau Street and sat down in a restaurant. There was another man sitting next to me with a cup of tea; he turned and said, "How was the game?" I replied, "Which game?" to which he cryptically responded, "The Series." I inquired politely, "What's that?" He hurriedly put down his nickel tip and ran out of the restaurant.

WIGHTMAN: There was a fair amount of discussion about high-school education. Everything that was said then is relevant to 1996. No progress whatsoever has been made.

GOLLUB: Your summary seems to imply that there was little or no discussion of the role of science (or physics) in the development of armaments or the use of nuclear weapons. Is this correct, and, if so, did it arise from exhaustion, or were these subjects perhaps taboo in some respect?

TREIMAN: I don't know the answer. My guess is that they were exhausted; there was also still a question of secrecy. There was an interesting discussion at one point about whether it was OK to have an international meeting on the subject of nuclear physics. There was a sense of inhibition, and, in the modern context, a surprising lack of stridency. They simply accepted secrecy, but complained about its inefficiencies and excesses. The answer is that there was very little war talk at all. There was, however, some Cold War talk—of the nature of our system vs. their system. But you have to remember the time. The Russian invited to comment said that although we may have our differences, about science we think the same.

CRONIN: Another Russian who was there was Vavilov. When you go to Russia, they'll say it should be Vavilov radiation, not Shrenker. Does anyone have any memory of what this physicist said; how he participated in this conference?

UNIDENTIFIED: I think the Vavilov who was here was perhaps Sergei Vavilov's son. The photo identifies someone who appears to be 25-30 years of age; Sergei, the President of the Russian Academy of Science, was close to 70 at that time. This could be his son, who was a physicist.

DRELL: I have two comments. We have now stopped testing bombs. As of September 10th, the Comprehensive Test Ban Treaty was endorsed by 158 nations in the United Nations. A majority of nations in the world have now signed a treaty to stop all testing. This progress should not be forgotten. On a local level, one wonders. As I walked down Nassau Street remembering my undergraduate days, I was shocked. There was an Einstein Bagels and Landau clothing store. What has happened to Princeton?

GRUNER: Was there a sense that physicists should be the stewards for all science?

TREIMAN: I'm overwhelmed by the question. There was no sense of that, although it is true that at the time physicists were cocks of the walk, as I think chemists were after WWI. Physicists may have been regarded as pointy-headed and difficult, as I said, but they were heroes. Science in general has benefited from society's esteem for physicists. That esteem has waxed and waned. But the atmosphere then was surprisingly modest. They were not bragging about the past. These were the giants of their time, and they weren't pretending that they could foresee the future technically; rather they were thinking about their changed relationship to society.

NAPPI: Although many famous physicists attended the 200th conference, they appear to have played a minor role in it. In particular, they do not appear to have taken an active role in what was the major topic of the conference, i.e., discussion of the relationship between physics and society. Was there in 1946 already a separation between the physicists who do physics and those who concern themselves with its impact on society?

TREIMAN: We have our own opinions. Scientists who do science tend to worry about society as they age. The theme of the conference was science and society. In those days there were still things once couldn't talk about.

SAMIOS: We're notoriously bad at predicting the future (except for your Higgs prediction); they were probably wise not to predict, but to focus elsewhere.

POUND: The question arose as to why the talks in 1946 concerned the relationship between science and government, rather than physics itself. I have been speaking in several places about the 50th anniversary of NMR in 1995-6. For that I developed a statistic that explains the low level of new physics in 1946. The *Physical Review* published about 250 pages in one month in January 1940; 54 pages in January 1946, and approximately 7400 pages in January 1995, which raises the question of whether to laugh or cry.

ANDERSON: A footnote: The Vavilov who attended this conference was undoubtedly the young one. He was one of the first to visit the U.S. after the Stalin era ended. I met him at Bell Labs in 1959 or so where he seemed to be the "commissar" for a delegation of solid state physicists. (An uncle, incidentally, was imprisoned by Stalin.)

CLARKE: Given that two-thirds of the 1946 conference was devoted to science and society, was any thought given to having a talk at this conference on this issue? Or was there not a suitable speaker to be found?

TREIMAN: It was considered.

KENDALL: Nowadays if one speaks of science and society one thinks of technology for human use. Was technology discussed?

TREIMAN: Yes. Nothing too memorable. Heads of labs were in attendance, and they insisted vehemently that basic science was as important to them as to universities. There was also a discussion of patent policy.

GOLDHABER: I fear that by 2046 you can hold the conference in a smaller room with narrower seats, since the physicists will be thinner.

CHAPTER 2

NONEQUILIBRIUM PHYSICS AND THE ORIGINS OF COMPLEXITY IN NATURE

J.S. LANGER

Department of Physics
University of California, Santa Barbara, CA

ABSTRACT

Complex spatial or temporal patterns emerge when simple systems are driven from equilibrium in ways that cause them to undergo instabilities. Dendritic (i.e., snowflake-like) crystal growth is an especially clear example. We now have a well-tested and remarkably detailed theory of how dendrites form and grow. We understand the role of intrinsically small effects such as crystalline anisotropy, and we even can trace the growth of dendritic sidebranches all the way back to their origins as microscopic thermal fluctuations. This extreme sensitivity to perturbations explains many long-standing puzzles about snowflakes. A similar sensitivity is emerging in the physics of dynamic fracture, a phenomenon which is similar in some ways to dendritic growth but which now seems to be governed by entirely different mechanical and mathematical principles. Extreme sensitivity to perturbations and system parameters is likely to be a characteristic feature of pattern-forming systems in general, including those that occur in biology. If so, we shall not succeed in reducing the physics of complexity to a small number of universality classes. On the contrary, we shall have to be prepared for a wide variety of challenges and surprises in this field.

2.1 INTRODUCTION

I cannot remember a time when I have felt more optimistic about the intellectual vitality of physics than I do now. I am cautiously optimistic about the remarkable developments in particle theory and cosmology, but those are too far from my areas of expertise for me to say much about them in public. What impresses me more is that we are on the verge of deep new understanding of a wide range of every-day phenomena, most of which appear to be "complex" in some sense of that word. For the first time in history, we have the tools—the experimental apparatus and the computational and conceptual capabilities—to answer questions of a kind that have always caught the imaginations of thoughtful human beings.

I want especially to emphasize my conviction that the physics of complex, every-day phenomena will be an extraordinarily rich and multi-faceted part of the science of the next century. It may, however, have to be pursued in a manner that is different from that of physics in the twentieth century. It will be strongly interdisciplinary, of course. We physicists shall have to work in close collaboration with many kinds of scientists and engineers, often in areas where we have not been the pioneers. That will be a major challenge for us and our institutions.

It is even more important, I think to recognize that we shall have to modify our innate urge to speculate about unifying principles at very early stages of our research projects, often long before we have dug deeply into the details of specific phenomena. Here I differ from some of my colleagues. For reasons that I shall try to explain, I suspect that complex systems will not fall usefully into a small set of universality classes. On the contrary, I think that our glimpses so far into the physics of complex systems indicate that this world is larger than we had expected, and that we may be much closer to the beginning than to the end of this chapter in the history of science.

About twenty-five years ago, P.W. Anderson wrote an essay entitled "More Is Different" [1]. He meant that systems with large numbers of degrees of freedom often behave in novel and surprising ways. In his words: "at each [increasing] level of complexity new properties appear, and the understanding of the new behaviors requires research which ... is as fundamental in its nature as any other." He was right, as usual, and quite early in arguing this point. I think, however, that even Anderson may have underestimated how many fundamentally new directions for investigation would emerge, and how profoundly that diversity of challenges may change the nature of research in physics.

Biology, of course, comes first on my list of new areas of inquiry for physicists in the next century. Life on earth may in some sense be the most complex phenomenon in the universe—a sobering thought with profound philosophical as well as scientific implications. I think that we must place the understanding of life, and human life in particular, at the top of the list of our most compelling intellectual challenges. Unfortunately, I am no more competent to talk about biology than I am to say anything sensible about elementary particles. However, the subject that I shall address, nonequilibrium physics and the origins of complexity in nature, moves us in the direction of thinking about the physics of biological systems.

My favorite way to illustrate the impact of new developments in nonequilibrium physics is to talk about snowflakes and dendritic crystal growth. This is a classic problem in the theory of pattern formation, one with a long history that is relevant to the issues I want to raise. I also shall make some brief remarks about the fracture of solids and earthquake dynamics, primarily to emphasize the scope and diversity of research in this field even within the conventional bounds of non-biological materials physics.

2.2 KEPLER'S SNOWFLAKES

In 1611, Johannes Kepler published a monograph called "The Six-Cornered Snowflake" [2]. Kepler wrote it as a "New Year's Gift" for one of his patrons in Prague. It was re-published in 1966 with commentary by the historian and philosopher Lancelot Law Whyte, who pointed out that it marked the first recorded instance in which morphogenesis had been treated as a subject for scientific rather than theological discussion.

In his essay, Kepler wondered how such beautifully complex and symmetric structures can emerge, literally, out of thin air. It would be easy to understand, he thought, if these were animate objects that had souls and God-given purposes for being, and which therefore possessed an innate force for growth and development. But no such option seemed available. He went on to speculate about natural geometric ways to generate six-fold symmetries and, long before he had any reason to believe in the existence of atoms, he drew pictures of hexagonal close-packed arrays of spheres. But he concluded that this was a problem that would have to wait for future generations of scientists; he did not have the tools to solve it in his own time. (I wonder whether my patrons, the DOE or the NSF, would be satisfied with a progress report in which I admitted that I couldn't solve the problems that I had posed!)

2.3 THOREAU'S SNOWFLAKES

About two and a half centuries after Kepler (on January 5, 1856), the American essayist and naturalist Henry David Thoreau described in his journal the snowflakes that he observed while walking in the woods, and then remarked: "How full of the creative genius is the air in which these are generated! I should hardly admire them more if real stars fell and lodged on my coat. Nature is full of genius, full of the divinity; so that not a snowflake escapes its fashioning hand" [3]. Like Kepler, Thoreau spoke of the "genius" of nature, and he could not resist using the term "divinity."

Kepler's and Thoreau's questions, and ours to this day, are very deep. We still ask how snowflakes form. What causes their six-fold symmetry? Why the tree-like structure? Why do their decorative features—sidebranches, tip splittings, bulges, and the like—occur at roughly the same places on each of the six branches? Do the growing parts of these branches somehow communicate with each other? And why does it seem that no two snowflakes are ever exactly alike?

2.4 Metallurgical Microstructures

By the 1960's, when Whyte was writing his comments about Kepler, the crystallographers and the atomic and solid-state physicists had figured out, at least in principle, why ice crystals have intrinsic hexagonal symmetry. At about the same time, however, the metallurgists were realizing that crystallography is only a small part of the story. At best, crystallography can predict the *equilibrium* form of the material which, in the case of ice, is a compact crystal with hexagonal symmetry, but with no branches or sidebranches or any of the complex structure of snowflakes. Crystallography, plus equilibrium thermodynamics and statistical mechanics, provides no clue about how to predict and control the microstructures of cast alloys, or welds, or any of the vast number of examples in nature where complex patterns emerge spontaneously in relatively simple materials.

A solidified metallic alloy, when etched and examined through a microscope, often looks like a forest of overdeveloped snowflakes with generations upon generations of sidebranches. During solidification, the molten material crystallizes in much the same way that water vapor crystallizes in the atmosphere, and the resulting "dendritic" or tree-like structure—the so-called "microstructure" of the alloy—is what determines its mechanical and electrical properties. To predict microstructures, metallurgists need to know the growth rates and characteristic length scales that occur during the transient processes of solidification. They need especially to know how those speeds and lengths are determined by constitutive properties of the materials such as thermal conductivity, surface tension, etc., and also by growth conditions such as the temperature and composition of the melt.

2.5 Solidification as a Free-Boundary Problem

The dendrite problem is easiest to describe if we restrict our attention to the case of solidification of a pure fluid held at a temperature slightly below its freezing point [4]. We must also, at this point, turn our attention to systems that are intrinsically simpler than real snowflakes made out of water molecules. Most of the interesting metallurgical materials are much easier than water for theoretical purposes.

Imagine a piece of solid completely immersed in its undercooled melt. Thermodynamics tells us that the free energy of the system decreases as the solid grows into the liquid. For this to happen, the latent heat generated during growth must be carried away from the solidification front by some transport mechanism such as diffusion. The motion of the front is governed by the interplay between two simple and familiar processes—the irreversible diffusion of heat and the reversible work done in the formation of new surface area.

In mathematical language, we say that the shape of the emerging crystal is the solution of a free-boundary problem. That problem consists of a diffusion equation for the temperature field u,

$$\frac{\partial u}{\partial t} - D\nabla^2 u = 0; \tag{2.1}$$

the condition of heat conservation at the moving solidification front:

$$v_n = -D\left[\frac{\partial u}{\partial n}\right]_{\text{liquid}} + D\left[\frac{\partial u}{\partial n}\right]_{\text{solid}}; \tag{2.2}$$

and the Gibbs-Thomson condition for the temperature u_s at a curved interface:

$$u_{\text{interface}} = -d_0\kappa. \tag{2.3}$$

Here, the dimensionless field u is the difference between the local temperature and the bulk freezing temperature, measured in units of the ratio of the latent heat to the specific heat. At the boundaries of the system, very far from the solidification front, $u = u_\infty < 0$; where $|u_\infty|$ is the dimensionless undercooling that plays the role of the driving force for this nonequilibrium system. D is the thermal diffusion constant; n denotes distance along the normal directed outward from the solid; v_n is the normal growth velocity; d_0 is a "capillary length"—ordinarily of the order of Angstroms—that is proportional to the liquid-solid surface tension; and κ is the sum of the principal curvatures of the interface. It is important that the surface tension may depend on the orientation of the interface relative to the axes of symmetry of the crystal.

2.6 SHAPE INSTABILITIES AND SINGULAR PERTURBATIONS

The key to understanding how such a simply posed problem can have complex dendritic solutions is the observation that smooth shapes generated by these equations are intrinsically unstable. Such instabilities appear throughout physics—in the lightning-rod effect, for example, or in the stress concentrations at the tips of the cracks that I shall mention later. In the case of dendritic crystal growth, the slightest bump on the surface of an otherwise smooth solidification front concentrates the diffusion gradients and increases the heat flow from the surface, thus enhancing the local growth rate. The small bump grows out into a dendrite; the primary dendrite undergoes secondary instabilities that lead to sidebranches; and thus a complex pattern emerges whose shape is highly sensitive to the precise way in which these instabilities have occurred.

There is a natural length scale associated with this instability. A solidification front growing at speed v is unstable against deformations whose wavelengths are larger than

$$\lambda_s = 2\pi\sqrt{\frac{2Dd_0}{v}}. \tag{2.4}$$

Note that $\lambda_s/2\pi$ is the geometric mean of a macroscopic length, the diffusion length $2D/v$, and the microscopic length d_0. It provides a first order-of-magnitude estimate for the length scales observed in dendritic growth. In particular, ρ, the radius of curvature of the dendritic tip, scales like λ_s. Thus the ratio of λ_s to ρ, conventionally defined by the parameter

$$\sigma^* \equiv \left(\frac{\lambda_s}{2\pi\rho}\right)^2 = \frac{2Dd_0}{v\rho^2}, \tag{2.5}$$

has been found experimentally to be very nearly constant for a given material over a wide range of growth conditions.

The problem of obtaining a first-principles estimate of σ^* remained unsolved, however, until about ten years ago. The crucial observation was that the capillary length d_0, and therefore σ^* itself, is a singular perturbation whose presence, no matter how small, qualitatively changes the mathematical nature of the equations. For $d_0 = 0$, there is a continuous family of steady-state solutions. For any nonzero d_0, however, there is only at most a discrete set of solutions, one of which may (or may not) be a stable attractor for this dynamical system. Moreover, the attracting solution looks like a regular dendrite only if there is some crystalline anisotropy in the surface tension. The parameter σ^* determines growth rates and tip radii in much the same way we had guessed from stability theory, but it is a function of the anisotropy, and it vanishes when the anisotropy strength goes to zero. In the latter case, when the surface tension is isotropic, the patterns are irregular and even more complex. They may, in some limiting cases, become truly fractal. In short, complex dendritic patterns are governed by a weak correction—the anisotropy—to a weak but singular perturbation—the surface tension.

2.7 THERMAL FLUCTUATIONS AND SIDEBRANCHES

There are two more recent developments in the dendrite story whose implications are relevant to the issues I want to raise here. The first has to do with the origin of sidebranches. Consider a small perturbation at the dendrite tip, caused perhaps by an impurity in the melt or a thermal fluctuation. The latest theories predict that such a perturbation grows like

$$\zeta(s) \approx \zeta(0)\, \exp\left[+\frac{\text{const.}}{\sqrt{\sigma^*}}\left(\frac{s}{\rho}\right)^b\right] \tag{2.6}$$

where $\zeta(s)$ is the size of the deviation from a smooth steady-state shape at a distance s from the tip of the dendrite. The exponent b is $1/4$ in an approximation in which the dendrite is cylindrically symmetric [5]. When realistic axial anisotropy is taken into account, the relevant exponent is $b = 2/5$ [6]. This equation is typical of the results that emerge from the singular perturbation theory. Note the special role of small values of σ^* in this formula. It is impossible to approximate ζ by a series expansion in powers of σ^*. The bump that is formed from this deformation grows out at a fixed position in the laboratory frame of reference; it becomes a sidebranch.

A natural question to ask is: How strong are the initial perturbations $\zeta(0)$ that produce the sidebranches seen in the real world? Might thermal fluctuations be strong enough? We can answer this question by using formulas like (2.6) to predict the way in which the first sidebranches appear behind the tip. We now know, thanks to recent work along these lines by Brener, Bilgram, and their colleagues [6, 7, 8, 9] that sidebranches in the most carefully studied three-dimensional dendritic systems are indeed formed by selective amplification of thermal fluctuations. I find it quite remarkable that this pattern-forming process is so sensitive that we can trace macroscopic structures all the way back to their origins as fluctuations on the scale of molecular motions.

2.8 A FIELD-THEORETIC DESCRIPTION OF SOLIDIFICATION PATTERNS

The last part of the dendrite story that I want to mention is the so-called "phase-field model." The free-boundary problem described in (2.1)-(2.3) looks esoteric to most theoretical physicists who, like myself, grew up immersed in field theories. It happens, however, that it is easy and perhaps even useful to rewrite these equations in a field-theoretic language. To do this, add a source to the thermal diffusion equation (2.1):

$$\frac{\partial u}{\partial t} - D\nabla^2 u = -\alpha\,\frac{\partial \phi}{\partial t}. \tag{2.7}$$

Here, α is a constant that is proportional to the latent heat, and ϕ is a "phase field"—i.e., a local order parameter—that tells us, by being near one or the other of two equilibrium values that we shall fix for it, whether the system is in its liquid or solid phase. According to (2.7), heat is generated or absorbed at a moving liquid-solid interface when, at such a point, ϕ is changing from one of its equilibrium values to the other.

Equation (2.7) must be supplemented by an equation of motion for ϕ, which I shall write in a specially simple but suggestive form:

$$-\frac{1}{\Gamma}\,\frac{\partial \phi}{\partial t} = -\nabla^2\phi - \mu^2\phi + \lambda\,\phi^3 + \beta\,u. \tag{2.8}$$

This is the familiar time-dependent Ginzburg-Landau equation with an explicit minus sign in front of $\mu^2\phi$, indicating that we are in the symmetry-broken low-temperature state of the system. The two stable values of ϕ are approximately $\pm\mu/\sqrt{\lambda}$. As long as μ is large, so that the interface between the two phases is thin, Eqs. (2.7) and (2.8) reduce to Eqs. (2.1)-(2.3), in this case with a nonzero but purely isotropic surface tension. The parameters α, β, Γ, λ, and μ can be evaluated in terms of the parameters that appear in the earlier equations.

The phase-field model has been studied extensively in recent years [10]. It even has been used to confirm the singular-perturbation theory of dendritic pattern selection and to compute the parameter σ^* defined in (2.5) [11]. Although (2.8) is a very stiff partial differential equation for large μ, the combination of (2.7) and (2.8) lends itself to numerical analysis more easily than the equivalent free-boundary problem. My main reason for mentioning this model, however, is to point out that a familiar-looking field theory of this kind may produce patterns of unbounded complexity. These patterns are exquisitely sensitive to small perturbations, for example, the addition of crystalline anisotropy or external noise. They are also "chaotic," in the technical meaning of the word, in their sensitivity to initial conditions. The fact that such complex behavior can occur here must be kept in mind when studying similar equations in circumstances where reliable experimental or numerical tests are not available.

2.9 The Symmetry and Diversity of Snowflakes

Recent theoretical and experimental developments finally have provided answers to our age-old questions about snowflakes. We now understand that they nucleate as tiny crystals in a supersaturated mixture of water vapor and air, and that the intrinsic hexagonal symmetry of ice guides the subsequent morphological instability and produces six nearly identical arms. The temperature and humidity of the air that surrounds each growing snowflake are very nearly constant on the relevant length scales of millimeters or less; thus all the arms respond in much the same way to their surroundings, and the precision and sensitivity of the controlling mechanisms make it unnecessary for there to be any direct communication between the distant parts of the emerging structure in order for them to grow in essentially identical ways.

All of this is happening in a turbulent atmosphere. The characteristic length scales of atmospheric turbulence, however, are meters, or kilometers–enormously larger than the sizes of the snowflakes themselves. As the snowflakes undergo advection through this environment, they encounter changing growth conditions. Sometimes their arms freeze quickly in regions that are particularly cold; at other times they grow slowly or even melt back as they encounter warmer regions. Our twentieth century understanding of turbulence tells us that the trajectories of neighboring snowflakes must diverge, that each flake must encounter its own unique sequence of growth conditions. In this technical sense, no two snowflakes are alike.

2.10 FRACTURE DYNAMICS

The dendrite problem seemed "mature" thirty years ago; but we know now that it was not. It did mature greatly in the last decade and, although there remain many important unsolved problems, it may have peaked as a center of intellectual excitement. To counter any notion that the field of nonequilibrium pattern formation as a whole is reaching maturity, I want to describe briefly one other supposedly mature area of research that is poised for major fundamental developments. This is the broad area of mechanical failure of solids, fracture in particular.

For obvious reasons, human beings must have become interested in fracture long before they started to ask philosophical questions about snowflakes. Similarly, the modern field of fracture mechanics—the study of when things break—was a well established specialty within engineering and applied mathematics before the dendrite problem had even been stated in precise terms. Remarkably, however, fracture *dynamics*—the study of how cracks move—is literally *terra incognita*, largely, I suppose, because the engineer's responsibility is to prevent such catastrophes before they start, but also because propagating fracture has been very hard to study either experimentally or theoretically until now.

I first started looking at dynamic fracture because I thought that a crack moving through a brittle solid should have much in common theoretically with a dendrite. The two phenomena look quite similar, at least superficially. Both are nonequilibrium, finger-like structures that move under the influence of external forces. The elastic stress concentration at the tip of a crack occurs for much the same reason that the diffusion flux is concentrated near the tip of a dendrite. Both exhibit some kind of sidebranching behavior. The scientific problem, in both cases, is to predict the propagation speed, the characteristic length scales, and the stability of the patterns that are formed. I also became interested in the dynamics of fracture on earthquake faults, and I wondered whether this behavior is an example of "self-organized criticality" [12]. It is not; but it does come close. The underlying physics of fracture, on the other hand, turns out to be entirely different from the physics of dendrites [13]. My education in dendrite theory has been of very little use for studying fracture, but I have learned a great deal that is new to me.

One of the first things I learned about fracture is how many obvious and ostensibly simple questions remain unanswered. For example, we do not know in any really fundamental way how to predict whether failure will occur in a solid *via* brittle fracture or ductile yielding. There are elegant and, and least to some extent, successful theories involving the creation and motion of dislocations. But all the same phenomena occur in amorphous materials, where dislocations have no meaning. So there must be more to learn here. In fact, we still have no first-principles theory of the glassy state, and certainly have no fundamental understanding of failure mechanisms in glasses. The state of affairs regarding granular materials is even more wonderfully uncertain.

My recent research has focussed on the stability of dynamic fracture and on the question of how—sometimes—complex patterns are generated during the failure of solids. Here again, the open questions have to do with familiar phenomena. We generally think of fracture surfaces as being intrinsically rough, perhaps even fractal. But we also know that they can be very smooth as, for example, in the case of a slow crack in an automobile windshield, or a sharp cleavage surface on a prehistoric stone tool. What is the nature of the instability that determines whether a crack will produce a smooth or a rough surface? Is this the same instability that causes an earthquake fault zone to be an intricate, interconnected network of fractures that extends for tens of kilometers both laterally and vertically in the neighborhood of a main fault [14]? Is this geometric complexity the reason for the broad, Gutenberg-Richter distribution of earthquake magnitudes, or does the chaotic dynamics of individual fault segments also play a role [15]?

I think I now know at least the broad outlines of answers to some of these questions. Recent experiments by Fineberg et al. [16, 17] have made it clear that bending instabilities at high speeds are intrinsic features of fracture; they are not just artifacts of special experimental conditions. My recent work with Ching and Nakanishi [18] leads me to be fairly certain that this instability is a general property of stress fields near the tips of moving cracks. The "lightning-rod" effect does play a crucial role here in concentrating stresses near the crack tip, but the crucial ingredient is the "relativistic" response of an elastic solid when the crack tip is moving at an appreciable fraction of the sound speed. There is no analog of such an effect in dendritic crystal growth. So far as I can see, the fracture problem and the dendrite problem are in two entirely different universality classes.

If I am correct in these assertions, then fracture dynamics is at roughly the same state as was solidification theory thirty years ago. We know (maybe) what triggers complex behavior in fracture, but little or nothing about mechanisms that may control the underlying instability. Our new results imply that, just as in the dendrite problem, the dynamics is strongly sensitive to the specifics of the models, in this case to the detailed mechanisms of cohesion and decohesion at crack tips or stick-slip friction on earthquake faults. There are even hints that some of these mechanisms may be singular perturbations whose presence makes qualitative changes in the system. But this is a subject for a another occasion.

2.11 THE PHYSICS OF COMPLEXITY

I hope that it is clear by now where I have been going in this discussion. Complex patterns are produced during crystal growth or fracture because systems are driven from thermal or mechanical equilibrium in ways that cause them to become unstable. Departure from equilibrium, external driving forces, and instability are important, fundamental elements of our understanding of these classes of phenomena. But those general elements tell us only a small part of what we need to know. A deep scientific understanding requires that we find out what mechanisms control the instabilities and that we learn in detail how the interplay between those mechanisms and the driving forces produces the patterns that we see. Because those mechanisms are acting in intrinsically unstable situations, dynamical systems may be strongly sensitive to them. Their effects may be highly specific. The patterns in some extreme cases may be completely regular, or fractal, or fully chaotic; but those are just special limits. Between those limits, the range of possibilities is enormous, often bewilderingly so. Whether we like it or not, that is where we must look to understand the real world.

What, then, are the prospects for a new "science of complexity"? Quite good, I think, if our measure of goodness is the variety and scope of the intellectual challenges that we shall encounter. Note that my scale of goodness is different from the traditions of twentieth century physics, which insist on grand unifications and underlying simplicity. I am perfectly comfortable with the idea that we may never find a set of unifying principles to guide us, in the manner of Gibbsian equilibrium statistical mechanics, to solutions of problems in the nonequilibrium physics of complex systems. We seem to be discovering far too large a variety of intrinsically different behaviors for such a synthesis to be possible. Nor, as I have said, do I think that we shall discover that all complex systems can be sorted into some small set of universality classes. That is why I find it hard to be enthusiastic about modelistic searches for generic behavior, usually computational experiments, that are not tightly coupled to real physical situations.

This is not to say, however, that I am pessimistic about the discovery of new basic principles. Just the opposite—I am sure that they will appear, and that they will do so in ways that will surprise us. We already have one example of a new idea in "self-organized criticality" [12]. Whether or not "SOC" matures into a precise predictive mode of analysis, I believe that it will persist as a provocative way of thinking about many complex systems. Other such insights, perhaps pertaining to more specific classes of phenomena, or to phenomena of a kind that we have never seen before, may be just around the corner. In a field so rich with new experimental and observational opportunities, and with the huge world of biological systems just beginning to be explored from this point of view, it seems obvious that new unifying concepts will emerge as well as a vast amount of new physics.

ACKNOWLEDGMENTS

The research on which this paper is based was supported by DOE Grant No. DE-FG03-84ER45108 and NSF Grant No. PHY-9407194. I specially want to thank J.M. Carlson and S.A. Langer for constructive critiques of early versions of this manuscript, and Elizabeth Witherell for pointing me toward Thoreau.

This lecture was presented at both the Princeton 250th Anniversary Conference on Critical Problems in Physics, October 31, 1996 and at the 11th Nishinomiya-Yukawa Memorial Symposium on Physics in the Twenty-First Century, November 7, 1996.

REFERENCES

[1] P.W. Anderson, Science **177** 393 (1972).

[2] J. Kepler, *The Six-Cornered Snowflake*, translated by C. Hardie, with accompanying essays by B.J. Mason and L.L. Whyte (Clarendon Press, Oxford, 1966) [originally published as *De Nive Sexangula* (Godfrey Tampach, Frankfurt am Main, 1611)].

[3] Henry D. Thoreau, *The Journal of Henry David Thoreau*, 14 volumes, edited by Bradford Torrey and Francis Allen (Houghton Mifflin, Boston, 1906; Peregrine Smith, Salt Lake City, 1984) Volume 8, 87-88.

[4] I can make no attempt here to cite detailed references to the literature in solidification physics. A few reviews that may be useful are the following: J.S. Langer, Rev. Mod. Phys. **52** 1 (1980); J.S. Langer, in *Chance and Matter*, proceedings of the Les Houches Summer School, Session XLVI, edited by J. Souletie, J. Vannimenus, and R. Stora (North Holland, Amsterdam, 1987), p. 629; D. Kessler, J. Koplik, and H. Levine, Adv. Phys. **37** 255 (1988); P. Pelcé, *Dynamics of Curved Fronts* (Academic Press, New York, 1988).

[5] J.S. Langer, Phys. Rev. A **36** 3350 (1987).

[6] E. Brener and D. Temkin, Phys. Rev. E **51** 351 (1995).

[7] M. Ben Amar and E. Brener, Phys. Rev. Lett. **71** 589 (1993).

[8] E. Brener, Phys. Rev. Lett. **71** 3653 (1993).

[9] U. Bisang and J.H. Bilgram, Phys. Rev. Lett. **75** 3898 (1995).

[10] J.A. Warren and W.J. Boettinger, Acta Metall. et Mater. **43** 689 (1995).

[11] A. Karma and W.-J. Rappel, Phys. Rev. Lett. **77** 4050 (1996).

[12] P. Bak, C. Tang and K. Weisenfeld, Phys. Rev. Lett. **59** 381 (1987).

[13] My favorite routes into the fracture literature are *via* B. Lawn, *Fracture of Brittle Solids* (Cambridge University Press, New York, 1993) or L.B. Freund, *Dynamic Fracture Mechanics* (Cambridge University Press, New York, 1990).

[14] C. Scholz, *The Mechanics of Earthquakes and Faulting* (Cambridge University Press, New York, 1990).

[15] J.M. Carlson, J.S. Langer, and B.E. Shaw, Rev. Mod. Phys. **66** 657 (1994).

[16] J. Fineberg, S.P. Gross, M. Marder and H. Swinney, Phys. Rev. Lett. **67** 457 (1991); Phys. Rev. B **45** 5146 (1992).

[17] S.P. Gross, J. Fineberg, M. Marder, W.D. McCormick, and H. Swinney, Phys. Rev. Lett. **71** 3162 (1993).

[18] E.S.C. Ching, J.S. Langer and Hiizu Nakanishi, Phys. Rev. Lett. **76** 1087 (1996); Phys. Rev. E **53** 2864 (1996).

2.12 DISCUSSION

Session Chair: Albert Libchaber
Rapporteur: Katya Chashechkina

BAK: Your view is that in order to understand complexity one has to look into the details. Indeed, if you wish to understand the symmetry of snowflakes or why there are elephants, you must study specifics within the relevant domains of science. However, there is a great and general question that is rarely asked. Why are the things complex? Why, starting from the Big Bang, did the Universe take the path of more and more complexity instead of ending up as a simple gas of particles or a simple solid where every place looks like every other place? More specifically, after the Big Bang we had the formation of galaxies, black holes, neutron stars, pulsars. Then the solar system with planets emerged, the earth with its complicated geophysics, and eventually life. The question is not what specific complex phenomena occur, but why they occur. What are the principles that govern us? What are the dynamical mechanisms which lead to complex but not simple things?

LANGER: My response to Per Bak is that he is absolutely right. Of course, it is he who has taken the lead in trying to answer such questions. As I said in my talk, his theory of self-organized criticality—the conjecture that nonequilibrium systems with large numbers of degrees of freedom move spontaneously toward states of instability—is very fascinating. It gives us a thought-provoking hint about why complexity emerges so naturally in our world.

I think, however, that SOC is at best only a first step toward a scientific understanding of complexity. If we are going to take the next steps, for example, in finding a predictive theory of earthquakes, then we must dig into the details and be prepared for surprises. Most real complex systems are far from being pure cases of SOC, or fractality, or fully developed chaos, or whatever. The very fact that they may be operating close to limits of stability tells us that they are likely to be specially sensitive to small perturbations that break symmetries or introduce unexpectedly relevant length or time scales. That sensitivity can make these systems behave in remarkable ways, and make them very hard for theorists to understand.

PARISI: I liked very much the idea that we are solving long-standing problems which have fascinated mankind for centuries. I have two comments. First, complexity is not confined to off-equilibrium systems. Sometimes, equilibrium systems also have complexity, which should be taken into account. Secondly, I believe that we must stress the importance of universality, especially if we move towards biology. I understand that a detailed mechanism might be crucial for explaining why you have a certain kind of phenomenon. Therefore, the form of the detailed mechanism can determine the universality class to which the system belongs. And if these mechanisms are not under control, the study of universality class is mandatory. After determination of this class we can understand many important properties without knowing all the microscopic details. If we turn to biophysics or biology we should try to understand universality classes of biological systems. This universality thinking is not common for people doing biology. We badly need universality to organize the very large amount of data collected by biologists in such a form that it could be explained by physicists.

LANGER: Parisi's remark that there is complexity in equilibrium systems is well-taken. I am sure that he is thinking about the wonderfully complex equilibrium states that occur, for example, in systems with competing or frustrated interactions between the components. But that is not the kind of complexity that I am addressing here.

With regard to universality, I did not in any way mean to imply that we physicists should abandon our search for the fundamental principles that govern the behavior of various kinds of systems. On the contrary, as Parisi says, the essence of our job is to identify the truly relevant underlying ingredients of physical situations and, by doing so, to discover relationships between different kinds of phenomena.

I think that this approach is working well in both the dendrite and fracture problems that I used as examples in my talk. We have made good progress with the dendrites, and we now understand—at least in principle—the physics of a large class of dendritic pattern-forming systems. So we can do what Parisi suggests; we can look at some kinds of dendritic phenomena whose constituents are not completely under our control and understand in a useful way what is happening. Dynamic fracture, on the other hand, may look superficially similar to dendritic growth, but the mechanism that produces complex fracture patterns is intrinsically different from the one that produces dendrites. To develop predictive theories, we must recognize that these phenomena belong to separate universality classes. More general approaches don't seem (to me) to be very productive.

In short, I expect that the search for universality will continue to be an extremely important feature of the physicist's mode of operation. But I also expect that we shall encounter an indefinitely large number of universality classes in investigating the physics of complexity. We shall have to resist drawing conclusions from superficial similarities, and we shall have to be willing to do the hard work needed to be sure we understand what is happening in many different situations.

GOLLUB: You argued that a careful understanding of the role of instability can help us to understand the origin of complexity in nature, and you implied this paradigm can be extended to biology. I would like to ask you to what extent this is expected to work. For example, we know that large protein molecules are able, spontaneously, to form biologically functional and specific shapes in times of the order of a second, though statistical estimates would suggest astronomical time scales for this process. Can stability concepts help us to understand this or other examples of biological pattern formation? And if so, how conceptually can the specific microscopic molecular information, which is obviously central, be combined with a stability paradigm that is usually based on underlying homogeneity? These aspects seem to be mutually contradictory.

LANGER: I am certainly not the best person here to answer questions about biological complexity, but I shall try to make some remarks that may be useful.

The protein folding problem mentioned by Gollub may be a particularly tough example. So far as I know, nobody really understands how it works. Clearly these very large molecules do not simply explore their multidimensional configuration spaces in a random manner, which would indeed take astronomically long times. Perhaps they undergo a sequence of morphological instabilities driven by interactions on successively larger (or smaller) length scales. Perhaps only very special proteins fold efficiently, and then these are the ones that are biologically important. Perhaps other proteins act like folding machines to accelerate and guide these processes. Most likely none of these speculations are completely correct.

Gollub then asks whether there might be a basic inconsistency in trying to use stability theory to explain how information that is encoded in individual molecules generates complex biological patterns. As he points out, our stability analyses generally describe the emergence of structure in initially homogeneous systems. What are the molecules doing? Isn't there a major disconnect here?

This question goes right to the heart of any serious discussion about the future role of physics in biology. I think that a first step toward an answer is the observation that the information contained in the human genome is very many orders of magnitude less than what would be needed to construct a living human being starting with just the chemical ingredients. Nature must have a repertoire of physical mechanisms that it uses to implement the genetic code. In significant part, it is the job of the physicist to discover these mechanisms and understand how they work.

Instability of initially homogeneous distributions of molecules does seem to be one of nature's tricks. It may play a role, for example, in determining the shapes of cells and the ways in which cells become differentiated from each other during embryonic development. Those ideas, based on instabilities in reaction-diffusion equations, go all the way back to Turing. My tentative answer to Gollub's question, therefore, is that the molecularly encoded genetic information may primarily determine what chemicals are produced at which places and at what times, and that basic physical mechanisms such as instabilities may take over from there.

CALLEGARI: Could the cause of complexity be ascribed to the flow of energy through the system which makes the most disordered state unstable and forces the system into a more stable minimum with a spontaneous symmetry breaking?

LANGER: I think that Callegari is asking whether there might be a Gibbsian formulation of nonequilibrium statistical mechanics in which driven systems can be seen to flow toward a stable minimum of some analog of the free energy.

Statistical physicists have tried to find such a variational formulation for many years because, if it existed in a useful form, it might be a powerful tool for the solution of many kinds of problems. My guess—based on what I think are plausible but not entirely rigorous arguments—is that no such general principle exists, but that it may be quite profitable to look for new variational methods for solving nonequilibrium problems in special circumstances. I would be delighted to be proven wrong about the general principle.

GUBSER: Was the square snowflake (shown in connection with the phase-field model) the result of numerics on a square lattice? Was it "grown" from an initial seed?

LANGER: Probably the phase-field code was run using a square lattice in the case that I showed. However, the equations being solved there contained an explicit four-fold crystalline anisotropy, and that is what produced the square symmetry. The Eqs. (2.7, 2.8) that I wrote down in the lecture contained no such anisotropy. And yes, those patterns were grown on the computer by starting with a simple (circular) seed and letting it go unstable as it grows.

LIBCHABER: Let me make one remark. I do not see in all the questions here any convergence to biology. If you apply the laws of physics at the level of molecules, at room temperature, you end up with an information machine which is DNA and little motors of nanometer scale which are just proteins. How can we reconcile it with everything that we have heard? That stretches my mind.

CHU: I have two simple questions. Can you give a simple but nontrivial example of a system, starting from the first principles, say Maxwell's Equations, such that one can show that the thermodynamic equilibrium does not exist for it?

And my second question is related, for instance, to your analysis of the dendrite problem that follows a traditional method, namely we start from equilibrium, then go to stability, then to so-called transport. Now, in such an analysis, because the system is always driven, why does the throughput in the system never explicitly appear in the result?

LANGER: I do not understand the first question about an example in which, in a well-posed system, there is no state of thermal equilibrium.

CHU: The implication is that it is always nonequilibrium when it is driven.

LANGER: This is a question of definition. Usually, when we say that a system is being "driven," we mean that there is energy coming in somewhere and going out somewhere else, so the system is out of equilibrium with its surroundings. In such a system, the energy flux would be one measure of the driving force. For instance, in the dendrite problem the undercooling is the driving force. The corresponding energy flux has its source at the solid-liquid interface where latent heat is generated and its sink at the boundaries where the cooling takes place. In that sense, the throughput does appear explicitly in the result.

GOLDHABER: Could you test your ideas about the dendrite growth by adding an external force like an electric field and predicting what would happen?

LANGER: Just such experiments have been done by Herman Cummins and his coworkers at CCNY. He exposed a growing dendrite to a pulsed laser beam to see whether the response agreed with our theories. Basically it did. So we think we have most of the basic physics there pretty well under control, which is not to say that there aren't still some important outstanding problems related to dendrites.

LIBCHABER: You said that side branching was related to kT fluctuations. Do quantum fluctuations replace kT?

LANGER: I don't know whether quantum fluctuations could generate dendrites; I've never thought through that possibility. But the question whether kT is big enough to generate dendrites at low temperatures is a quantitative one that must be answered on a case-by-case basis. For example, Bilgram did his experiments with liquid Xenon at fairly low temperatures, and he saw thermal sidebranches in accord with the theory. People see sidebranching dendrites growing in liquid Helium.

Before I looked for quantum effects, I'd want to find out whether some dendrites, even in the absence of thermal fluctuations, might have oscillatory growth modes. That is, in situations where there are extra degrees of freedom such as composition variations, I think the growth mode at the dendritic tip could be a limit cycle instead of just a simple fixed point. Such an oscillation would generate a more coherent train of sidebranches than those that we see in the thermal cases. That would be very interesting.

CHAPTER 3

DYNAMICS, COMPUTATION, AND NEUROBIOLOGY

JOHN J. HOPFIELD

Beckman Institute
California Institute of Technology, Pasadena, CA

3.1 BIOLOGY AND PHYSICS

Biology has progressed enormously in last sixty years, with changes in the field and viewpoint which might be likened to that which have taken place in physics in a period twice that long, starting with Faraday or Henry. However, these two revolutions seem very different because of the qualitative difference in the scale of the essential information in the two sciences. The Rosetta Stone of condensed matter physics and chemistry is the periodic table, a few facts each about 100-odd elements. The equivalent Rosetta Stone for biology could be the genome of a simple bacterium, containing a million entries. No biologist will ever know the genome of *E. coli* in the same sense that a condensed matter physicist knows the periodic table.

At the same time, biology has, to date, displayed rather fewer unifying principles from which predictions can be made than has physics. Few today would go to a particular paper by Faraday for information. All the knowledge that he accumulated about electricity and magnetism has been codified into the unifying mathematical description of Maxwell's Equations. But many of the important biological facts have not been put into a unified framework from which they could be predicted, and many are intrinsically not predictable. Physics was at one time taken as a paradigm for what any science should be. In truth, there are many different kinds of science. Physics has tended to define itself as a science of general principles, in which a major goal is to compact the information on a diversity of experiments into a simple conceptual framework which (were it known in advance) would make most of the experimental results predictable. Physics is thus, by definition, an information-poor science. Biology is an information-rich science. A physicist can hope to explain "why there are protons" in terms of a predictive theory of everything. A biologist will never similarly be able to explain "why the particular 20 amino acids are used in protein synthesis" in a comparably predictive theory.

However, while the nature of theory and the generality of principles in biology is rather different from that in physics, there are some broad problems which underlie much current research in biology. To me, the three greatest problems might be posed as:

1. How did life–and we–arise, evolve?

2. How does a complex organism develop from a one-dimensional DNA sequence to a functioning three-dimensional organism?

3. How do we think?

What physics is involved in biology? Quantum mechanics lies behind biology, as it does behind all interesting phenomena of chemistry and molecules. However, with one minor exception (long-range electron transfers), the quantum mechanics of electrons is essentially irrelevant in biology. The electrons can be adiabaticly removed, and produce only a complicated potential function on which the nuclear motion takes place. Essentially all biochemistry can then be described by the molecular dynamics of nuclear motions. The most interesting part of that molecular dynamics is classical, because the important modes are "soft" and groups of atoms are heavy. Thus quantum mechanics is only trivially involved in biology. There is no macroscopic coherence, no quantum-mechanical subtlety. The temperature is too high, electrons are well-localized, groups of atoms are heavy, and zero-point motions of individual atoms are small.

There are those who have written otherwise on the subject of how the brain works, the focus of the present essay. Some of them are from past eras, where all interesting (i.e., unexplained) phenomena were assigned quantum-mechanical origins. Some authors have been biased by religious views, and some merely confused. Those who invoke quantum mechanics in biology in serious ways tend to fall into sinkholes of philosophical sophistry. The subject of quantum coherence in biology does not merit further serious words.

There is a degeneracy from which the molecules of a particular organism have been chosen (e.g., there are thousands of different forms of hemoglobin). There is a degeneracy of description of a particular protein in DNA, and of the order of exons in DNA. Biological systems have immensely more broken symmetry per cubic centimeter than other physical systems. For every broken symmetry, a fact must be known. This is what makes biology information-rich. In addition, the development and functioning of biological systems are examples of systems in which simple rules can generate complex structures and behaviors. In biology, single molecular events at the DNA level are routinely amplified into macroscopic observables in a cell, and single cell events into actions of a large organism. While this kind of amplification can occasionally be easily seen in physics (e.g., in crystallization of only one isomeric form of a crystal), it is relatively rare and special. The combination of immense broken symmetry, reproduction, amplification, and evolutionary selection dominate biology. Simple macroscopic physical systems are not dominated by such features.

Biology is a physical informational system. The duplication, proliferation, amplification, transformation, and selection of *information* dominate the behavior of biological systems. Physicists seldom think in terms of information processing (or else Shannon's theorem would take its place in the physics curriculum along with the second law of thermodynamics), further increasing the apparent gap between these sciences.

3.2 NEURAL COMPUTATION AND DYNAMICS

Batch-mode digital computation is the easiest computation to describe in physical terms. (We emphasize here the computer as a physical device, not the abstract logic of computation.) In this description, a digital computer consists of a large number N of storage registers, each one storing a single bit. A modern workstation has more than 10^8 such registers. The state of the machine can be described as an N-bit binary word, or N Ising spins. With each clock cycle the state of the machine changes, as the content of some storage registers change from 0 to 1 or vice versa. In doing a computation, the initial state of the machine is set by reading a fixed initialization condition, the program, and data. The state of the machine then evolves in time. The present state of the machine determines the next state according to a prescription which is implicitly built in by the designer of the machine. The user determines only the starting state, not the state transition map. The computation ends when the state of the computer stops changing, and the answer is read out from a subset of the storage registers. Thus the computation itself is carried out by the motion of a dynamical system from a starting state to a final state. The computations done by neurobiology are not batch mode; are not binary; and are not run by a system clock. But the basic idea of computation as being carried out by the motion of a dynamical system is immediately translatable into continuous time and continuous system variables. End states of digital systems translate into transiently fixed points or more general attractors of dynamical systems. Finally, batch mode computation can be generalized to a more continuing form of operation and dynamics.

The nerve cell, or neuron, provides the starting point for describing the dynamics for a set of neurobiology. The cell is an electrically closed compartment filled with a conducting fluid and surrounded by an insulating membrane. A typical behavior of the potential difference between the inside and the outside of a cell as a function of time is shown in Fig. 3.1. When the cell potential rises to a threshold level of about 65 millivolts, a short paroxysm of activity results, with a potential swing of about 100 millivolts over a couple of milliseconds, followed by a return to a lower quiescent potential. These events are called action potentials, and the cellular biophysics lying behind them was beautifully elucidated by Hodgkin and Huxley fifty years ago. We will simply represent the action potentials as delta-functions which occur whenever the cell potential rises above a threshold V_t, and result in a resetting of the cell potential to $V_t - \Delta$.

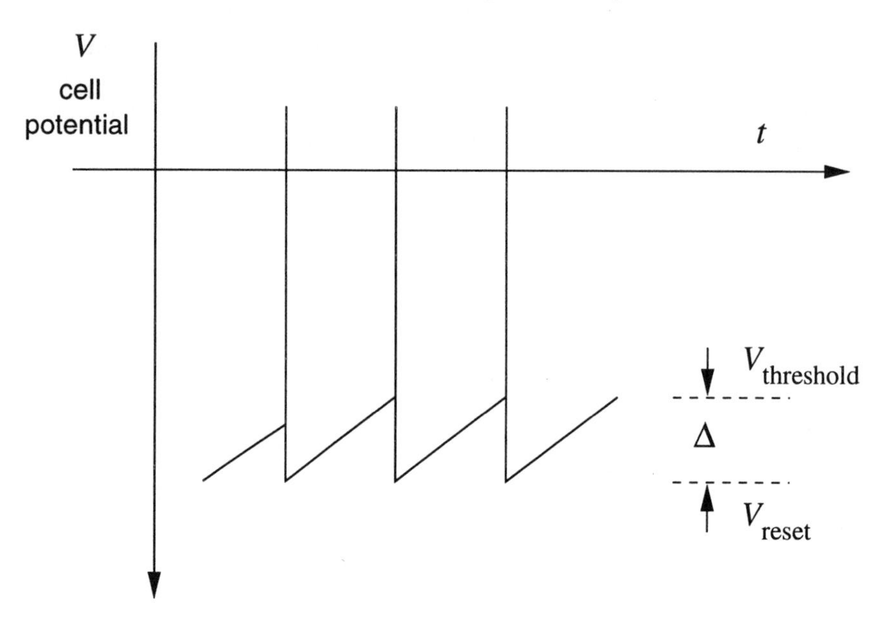

Figure 3.1: V as a function of t for fixed input current i. dV/dt = I/C except when $V_{threshold}$ is reached, when a brief action potential is generated and the potential is reset to V_{reset}.

Because the cell membrane is only about 50 Å thick, a neuron has appreciable electrical capacitance C. When a constant input current I is sent into a cell, the potential rises linearly with time until it reaches V_t. An action potential is then produced and the cell potential is reset. If the current to the cell is held fixed, it will again charge up to V_t after a time ΔC/I. The "firing rate" f of the cell would then be linear in I indicated in Fig. 3.2. In reality, there are many complicating factors. Action potentials last about a millisecond, so firing rates can never rise above about 1000 Hz. The leakage resistance of the membrane results in a minimum current necessary for the cell to fire at all. Noise is present. A more realistic neural model would have a firing rate as a function of input current more like the solid line of Fig 3.2. But the essence of the generation of action potentials is contained in this "integrate and fire" model of a neuron. The times when a cell produces an action potential are given in terms of the history of the current into a cell.

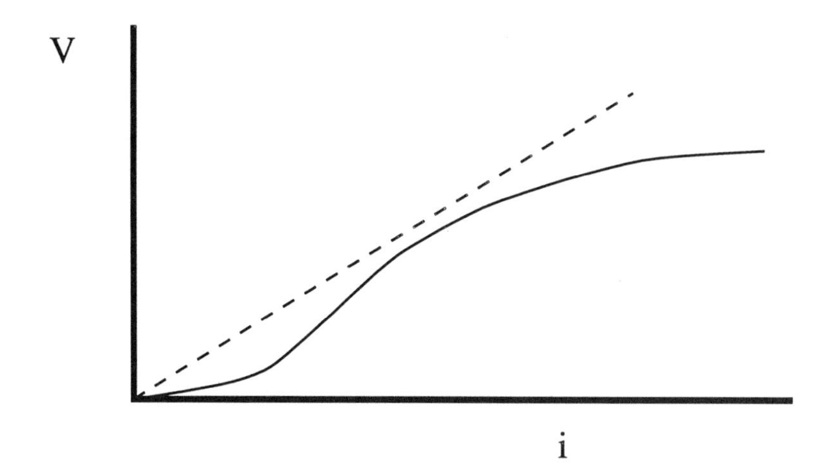

Figure 3.2: The firing rate vs current for an idealized integrate-and-fire neuron (dashed line) and for a more realistic case (solid line) including the effects of leakage currents and the finite kinetics of action potential biophysics.

The biological currents into cells arise from two processes. First, sensory cells such as cones in the eye (or hair cells in the ear) have cellular currents which are caused by light (or sound). Second, cells interact at synapses. When an action potential is produced in a pre-synaptic cell, one or more vesicles containing a neurotransmitter fuse with the cell membrane, dumping neurotransmitter onto the post-synaptic cell. The postsynaptic cell membrane contains protein molecules which open transmembrane conducting channels, specific to particular ions, when they bind neurotransmitter. As a result of this process, the electrical conductivity of the post-synaptic cell membrane is transiently turned on.

This transient conductivity results in a current into (or out of) the cell, depending on the ion which can best selectively flow through the membrane channel. Na^+ and K^+ ions are kept at non-equilibrium concentrations across a cell membrane by means of ion pumps connected to an energy source. When Na^+ channels open, a current flows into the cell. When K+ channels open, a current flows out of a cell. The form of the current into the post-synaptic cell i due to an action potential at time t_j^k in a presynaptic cell j may be approximated by

$$
\begin{aligned}
\mathrm{I}_{ij} &= 0 & t < t^k, \\
\mathrm{I}_{ij} &= T_{ij} \exp - (t - t_j^k)/\tau & t > t^k .
\end{aligned}
\tag{3.1}
$$

The magnitude and sign of T_{ij} depend on the particular synapse involved.

This completes the simplest description of the activity of an interacting set of neurons. Action potentials are generated as a result of the currents which flow;

and the synaptic component of the cellular current is generated in response to action potentials. We can define the instantaneous firing rate of cell j as

$$f_j(t) = \sum_k \delta(t - t^k j), \tag{3.2}$$

and find the total synaptic current I_i into cell i obeys

$$dI_i/dt = -I_i/\tau + \sum T_{ij} f_j(t). \tag{3.3}$$

This equation, though exact, is very complicated because the functions fj(t) are not smooth, and the times of the action potentials are only implicitly described by the "integrate-and-fire" description given. There is at this point a great divide between two possible views on approximations. In one view, action potentials are only statistically important. It takes many action potentials arriving within a time τ to make a post-synaptic cell fire an action potential. In this view, the functions f_j(t) can be replaced by smooth statistical averages, just as the granularity of electrons is replaced by continuous values of the charge on a capacitor in the usual description of circuit theory for engineering purposes. There are many places in neurobiology where this attitude is believed to be an excellent approximation. In such circumstances, these equations can be replaced by the more collective

$$dI_i/dt = -I_i/\tau + \sum T_{ij} f(I_j), \tag{3.4}$$

where f is a smooth sigmoid function of the type sketched in Fig. 3.2. This set of equations has complex behaviors because of the non-linearity of f(I).

A second equation of motion is also of paramount importance to neurobiology. The above equation describes a simple idealization of the way the activity in a network of neurons changes with time when the synaptic connections T_{ij} are fixed. However, the synapses themselves change with time, usually over a slower time scale than the change of activity (though whether the time scales are always separate is open to debate). Most long term memory is believed to reside in synaptic changes produced as a result of experience, i.e., as a result of the activity of the neurons themselves. The details of this synapse change algorithm are less well known, and the rest of this essay is based on a system with fixed synapses.

While the statistics of action potentials seem to be an adequate description of some aspects of neurobiological computation, others appear to involve in more detail the timing of action potentials. For example, humans and owls have the ability to identify the azimuthal location of a "click" or rustling sound source with a precision of about one degree on the basis of the time difference of the arrival of sound at the two ears. Detecting coincidences between action potentials initiated (indirectly) by the arrival of sound at the two ears is believed to be the mechanism. There are also suggestions that action potential synchronization plays a part in computations in the visual system and in olfaction.

A huge number of biophysical and physiological details have been omitted from the above model. Many neurons adapt to changes in input current, so that in responding to a step input current, they first fire rapidly, but then the firing rate slows. Different synapses have different time constants. The post-synaptic cell potential changes with time, so that the synaptic current cannot exactly be given in terms of action potentials alone. Vesicle release at synapses is only probabilistic. Action potentials propagate along axons at a finite rate, so a cell j actually sees a delayed version of events at cell i. The interior of the cell is not at all at the same potential, for the dendritic tree on which synapses are made by the axons from other cells has a finite resistivity. Action potentials under some circumstances propagate into the dendritic tree as well as along axons. The Ca^{++} concentration is not uniform within a cell. There are many peptide modulators of synapse efficacy and membrane properties, and these modulators are themselves distributed by specialized neurons. Synaptic responses are somewhat delayed. Synapses may change on the same time scale as the nominal computation is being carried out. In short, the model described is a gross oversimplification of real neurobiology. At the same time, it is capable of describing major features—the actions potentials, the detailed anatomy (implicitly contained in the description of T_{ij}), and the delayed dynamics. Many of the features which are not included could be readily incorporated in slightly more complicated mathematics if we knew when each was important.

One lesson learned from efforts to do analog computation in silicon hardware is the importance of device physics. A circuit is or is not effective depending in great part on whether details of the device physics can be used in the algorithm. If logarithms or exponentials are needed, the physics of a single transistor device can supply them. If time delays are needed, a surface wave acoustoelectric device may work well. If we are interested in Bessel functions, the exponential response of analog silicon transistors is not of much relevance, and the brute force approach of using silicon to make digital hardware may be the simplest approach. In any particular computation carried out by neurobiology, a few of the features of neurobiology and cellular biophysics may be of great help in carrying out the desired computation. It may not be possible to capitalize on many other features, which can then probably be subsumed in the parameters of a model in the right "equivalence class." Unfortunately, the larger tradition of neurobiology has been to collect all details without much emphasis on understanding the whys and wherefores of the computation involved.

3.3 STATISTICAL AND COLLECTIVE ASPECTS

A digital workstation works by detail. Each Sun SPARC 20 of a given model is exactly like its siblings. There is a 1:1 correspondence between transistors and their connections between two such machines. If some of the connections or transistors are omitted or fail, a machine will fail or be significantly different from its sibling in some respect.

A human has about 10^{11} neurons, with about 10^{14} connections between them. To describe the circuit diagram by brute force, each cell would need a 33-bit number tag, 1000x33 bits to describe the names of the 1000 other neurons to which it makes synapses, plus a few bits per synapse to describe the strength of a synapse. Altogether, such a brute-force listing of the circuit diagram would require about 3×10^{15} bits. Our DNA contains less than 10^{10} bits, of which less than 10^{9} are believed to be used in detail for coding. Thus our neural circuit cannot be brute-force coded into our DNA, though the DNA could implicitly contain a program which describes the circuitry.

The simple worm *C. elegans* has exactly 302 neurons and about 9000 synapses. The total information necessary to describe the circuit diagram is about 20 kilobits. The DNA of *C. elegans* contains about 10^{8} bits of information. It would be easy to directly encode the entire nematode circuit in its genes. In fact, while the nematode is not assembled with complete accuracy, a comparison of any two animals finds about 90% of the synapses in common.

This is not surprising—simple animals have identifiable neurons, the same from animal to animal, with determined (or very nearly so) connections. At the other end of the spectrum, mammals do not even have identifiable specific neurons. It is not possible to point to a particular cell in each of two animals and claim that they are the same. Nevertheless, *behavior* is specific. All cats move similarly when shown an erratically moving mouse-sized object, in spite of having in detail different neural circuitry. The regions of a brain are identifiably the same from cat to cat. The general connectivity patterns are the same. The brains of two different cats might be likened to two different samples of a glass, where no precise correspondence between the circumstance of two atoms in the different samples can be found. It is as though the brains from two animals of the same species are two frozen samples from an ensemble, all of which have the same macroscopic properties. Evidence indicates that we all see the Muller-Lyre illusion in Fig. 3.3 the same way, in spite of having brains which are in detail different. The laws of psychology and the predictabilities of behavior in higher animals of a single species with large brains seem to be collective phenomena. Only very simple animals are like the SPARC 20.

The same visual stimulus can be repeated over and over. Carefully done, the response of a retinal cone cell, or of a retinal ganglion cell sending signals off toward the brain, will be the same on each repetition. When cells are examined which are eight synapses further into the visual processing system, the cell behavior is quite variable from trial to trial, while averages over many trials can be stationary. However, the psychological percept ("what is seen") is often robustly invariant in spite of the variability of the response of the cells participating in it. The percept represents a collective aspect of the activity of many neurons, while the cellular responses are particular examples of an ensemble, all of whose

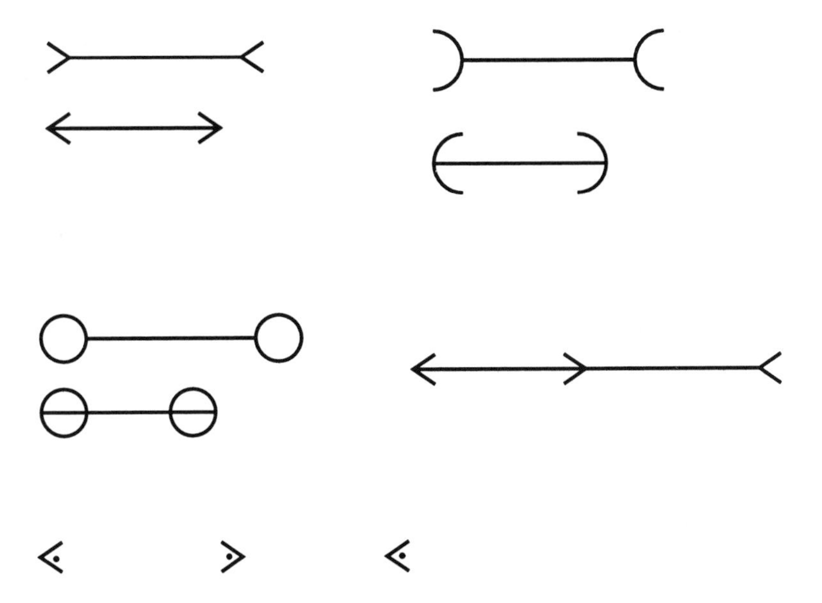

Figure 3.3: The classic Muller-Lyre illusion. We all see it in the same way, in spite of the detailed differences in our neural circuits.

members have the same collective aspect. The collective aspect of mammalian neurobiological computation is also seen in the robustness against loss of individual cells.

The relationship between the activity of neurons and psychology (or the laws of behavior) can be likened to the relationship between molecular motions and hydrodynamics. However, the hydrodynamics of simple fluids is dominated by local collisions and the conservation of energy and momentum. For complex fluids with long and varied molecules and structures, the details begin to appear in the collective properties. Similarly, many details of cellular and anatomical neurobiology will strongly affect the collective aspects of the neurodynamics.

Occasionally we can find cellular details which are directly related to psychological percepts. One astonishing example concerns monocular depth perception. When we see a movie of a rotating stick figure representing the edges of a cube or a molecular structure, we easily and automatically perceive the three-dimensional structure. Suppose the stick figure is in red, and on a green background. When the stick figure is bright and the background is dark, the three dimensional structure is perceived. If the intensity of the red and green are cleverly chosen, however, the three dimensional structure disappears, and we perceive only a two-dimensional form writhing in an unfathomable way. The origin of this effect is due to the importance for depth processing of cells which are red-green color-blind, and do not "see" the figure at all when the red and the green are equally luminant. Peculiarly, the situations which test most clearly how a brain works are often those which produce illusions, or "wrong" results. Time-order also seems to display effects of a microscopic cellular property. What is the letter in the alphabet which occurs after B? How long did it take you to find it? What is the letter before U? How long did it take you to find it, and how did you go about it? I believe the difference between forward and backward access in such a case is built into the synapse-change algorithm itself, and is also responsible for the difference seen in classical Pavlovian experiments between forward and backward conditioning. Biology seems to have built into our synaptic hardware a rule for behavior which is, if not *post hoc, ergo propter hoc*, at least *pre hoc, ergo non propter hoc*.

3.4 DEVELOPMENT AND ENVIRONMENT

A developmental program contained in DNA describes how cells replicate, and describes the biophysics of how the cells migrate and move as a result of the chemical environment and cell-cell contacts. In this regard, a similar description might be given for a kidney or a brain. However, development of the nervous system of higher animals has one important add-on not shared by other organ developmental programs, namely the major influence of the electrical activity of neurons during development. During development synapses are constantly being formed and being eliminated. The relationship between the electrical activity of the pre- and post-synaptic cells determines whether a synapse remains, and if so, how strong it is; it even determines if the cells themselves survive. Development and neural activity are irretrievably linked.

In early mammalian development, electrical signals represent noise in the sensory processing systems, or are internally generated. After birth, the developmental process continues, and synapses continue to change, but now there are also "meaningful" (or "patterned") signals from sensory cells exposed to a diverse environment. Cells in the primary visual cortex of a cat reared in darkness for six weeks have very different responses from those of a cat reared in a normal environment; cats reared while exposed to diffuse light but without real images of the world on their retinas are different from both.

At the level of perceiving lines or binocular vision, all normal humans are probably very similar, for the neural wiring for such "early processing" tasks is dominated by the hard-wiring rules of developmental biology and by early visual experiences which are broadly similar. At higher levels, it is much less clear how to describe what the signals in neurons are representing, and it is less clear that all humans represent higher levels of information in the same way. Bilingual people who learn both languages concurrently tend to have the languages physically intermixed in their brains, so that a stroke to a language area produces deficits in both languages. If, however, the languages are learned sequentially, it is much more likely that a stroke can effect one language but not the other.

Thus the question of how the brain is wired is simplified by the fact that to some extent, we all see the same things. We might say that the regularities of the environment get built into our brains. But in detail we are different, with different developmental and environmental histories, and with different representations of the world. This produces differences between us at higher processing levels, while at the level of early sensory processing, we all seem to be roughly equivalent.

3.5 EVOLUTION AS A SOURCE OF UNITY

Evolution produces strong cross-species unity in biology, and the structure and function of our brains is much like that of our first-cousin pygmy chimpanzees. Parts of our brains have structural and computational similarities to fish and reptiles. When looking in detail at structures in the olfactory system, a similarity can be found between humans and bees. But evolution can also produce related processing between different functions within the brain of one species.

A story from my time at Princeton might introduce the subject. A pair of theoretical students, one from mathematics, one from physics, were assigned to make the daily afternoon tea. The procedures were explained to them by the previous pair—get the kettle out of the cabinet, fill it with water, heat it on the burner, get the teapot out of the cabinet... The next day, the two new students started on their duties. The mathematician volunteered to do it the first day, and they could then alternate. Observing the kettle sitting on the table, he thought carefully, then picked up the kettle, opened the cabinet, put in the kettle, and closed the door. The physicist was aghast. "What the h— are you doing?" he asked. "Transforming the problem to one I know how to solve," replied the mathematician.

Enzymes for biochemical reactions are highly specific. One of the older biochemical pathways provides energy from sugars by a sequence of reactions. For the 6-carbon sugar fructose, the pathway in a typical bacterium begins

$$fructose \rightarrow fructose\,6\,phosphate \rightarrow$$

$$fructose\,1-6\,diphosphate \rightarrow 2\,triose\,phosphates,$$

where each of the arrows is catalyzed by a different enzyme. Suppose now that glucose, another 6-carbon sugar, became available in the environment. One evolutionary strategy would be to evolve a whole new glucose-enzyme cascade, duplicating at each step the function of the fructose-utilizing enzymes.

There is a second and equally conservative way evolution might find to utilize glucose. This alternative strategy would transform a glucose-based molecule to a fructose-based one at a very early step in the cascade. In *E. coli*, the glucose pathway actually begins

$$glucose \rightarrow glucose\,6\,phosphate \rightarrow fructose\,6\,phosphate \rightarrow ...$$

$$\text{glucose6P - fructose6P isomerase .}$$

The enzyme glucose6P-fructose6P isomerase *transforms the glucose problem into the fructose problem.* This solution by transformation can be simpler than evolving an entire new set of glycolytic enzymes.

It can be similarly argued that one of the ways that the brain evolves to "solve new problems" and produce new capabilities is by developing (through evolution) transformations which convert new problems to problems which could already be solved by older structures. (A concomitant duplication or enlargement of the older structure may also be needed. Such developmental duplications involve relatively small evolutionary steps, just as in molecular biology the duplication of a contiguous set of genes is a simple process.) If this is so, we can sensibly look for evidence of appropriate transformations being carried out, which would convert the new problem into an older one, even in the absence of understanding how the older problem is solved. The next few paragraphs show how a transformation, which can be readily carried out by the known biophysics of neurons and simple neural circuits, can convert a problem in understanding sequences of sounds into an older problem of olfaction.

The sense of smell, and its aquatic equivalent, is an ancient and ubiquitous form of remote sensing. Bacteria, slugs, insects, and vertebrates can all use olfaction to locate, identify, and move with respect to sources of nutrition, mates, and danger. Most natural odors consist of mixtures of several molecular species. At some particular strength a complex odor "b" can be described by the concentrations N_i^b of its constitutive molecular species i. If the stimulus intensity changes, each component increases (or decreases) by the same multiplicative factor. It is convenient to describe the stimulus as a product of two factors, an intensity λ and normalized components n_i^b as

$$\lambda = \sum_j N_j^b \,, \tag{3.5}$$

$$\mathrm{n}_i^b = N_i^b/\lambda \;\; or \;\; N_i^b = \lambda\, n_i^b \,.$$

The n_i^b are normalized, or relative concentrations of different molecules, and λ describes the overall odor intensity. Ideally, a given odor *quality* is described by the pattern of n_i^b, which does not change when the odor *intensity* λ changes. When a stimulus described by a set $\{\mathrm{N}_j^s\}$ is presented, an ideal odor quality detector answers "yes" to the question "is odor b present?" if and only if for some value of λ

$$\mathrm{N}_j^s \cong \lambda\, n_j^b \quad \text{for all j} \,. \tag{3.6}$$

Appropriate functional criteria for how stringent the approximate equality should be will depend on the biological context. This computation has been called *analog match.*

The identification of the color of a light could be described in similar terms. Ideally, an animal would have thousands of very specific light receptors, each in a different very narrow wavelength band. Ideally, the comparison of a new stimulus to a known color would be an analog match problem in thousands of dimensions. Our color vision, however, uses only three broadly-tuned receptors. The color comparison which our brains carry out still has the form of an analog match problem, but in only three variables.

Direct studies of vertebrate sensory cells exposed to simple molecular stimuli from behaviorally relevant objects indicate that each molecular species stimulates many different sensory cells, some strongly and others more weakly, while each cell is excited by many different molecular species. The pattern of relative excitation across the population of sensory cell classes determines the odor quality in the generalist olfactory system, whether the odorant is a single chemical species or a mixture. The compaction of the ideal problem (a receptor for each molecule type) down onto a set of, at most, a few hundred broadly responsive receptor cell types, is similar to the compaction done by the visual system onto three receptor cell classes. The generalist olfactory systems of higher animals apparently solve the computational problem

$$\begin{aligned} \lambda &= \sum_j M_j^b \\ \mathrm{m}_i^b &= M_i^b / \lambda \\ \mathrm{M}_j^s &\cong \lambda m_j^b \end{aligned} \tag{3.7}$$

exactly like that described earlier, except that the index j now refers to the effective channels of input, rather than to molecular types, and the components of M refer to the activity of the various input channels.

The archetypical sequence recognition problem is to identify known syllables or words in connected speech, where the beginnings and ends of words are not necessarily delineated. Speech is a stereotyped sound signal whose power spectrum changes relatively slowly compared to the acoustic frequencies present. The speech sound in a small interval (typically 20 milliseconds) is often classified as belonging to a discrete set of possibilities A, B, C, D, etc., so that a speech signal might be represented by a sequence in time as

$$\ldots\ldots\ldots\ldots\ldots CCAAAAAADDDDDDDDDDDBBBBBEEE\ldots\ldots\ldots\ldots\ldots$$

A computational system which can reliably recognize known words in connected speech faces three difficulties. First, there are often no delineators at the beginnings and ends of words. Second, the sound of a given syllable in a particular word is not fixed, even for a single speaker. Third, the duration of a syllable or word is not invariant across speakers or for a given speaker. This third difficulty is termed the *time warp problem*, and creates a particularly hard recognition problem in the presence of the other two difficulties. Without time warp, template matching would be an adequate solution to the problem.

The initial stage of a "neural" word recognition system, whether biological or artificial, might begin with "units" which recognize particular features A, B, C, etc. A particular feature detector has an output only while its feature is present. Syllable lengths vary considerably in normal speech. The following letter strings illustrate the problem created by this fact.

AAABBBCCCCCCCCCDDDDDDEEEFFF
AABBCCCCCCDDDDEEFF

These two sequences represent time-warped versions of the same "word." Clearly a single rigid template cannot match both strings well, no matter how the template is fashioned or positioned with respect to each of these word strings.

There is an alternate and novel way to represent a symbol string. A string of symbols which have occurred at times earlier than the present time t

$$AAABBBCCCCCCCCCDDDDDDEEEFFF ------- < t >$$

can be represented by the times of the starts and ends of sequences of repeated symbols. The sequence can thus be represented as

$$A_s A_e B_s B_e C_s -------- C_e D_s ------ D_e E_s E_e F_s F_e --- < t >$$

While at first sight this representation seems qualitatively similar to the earlier one, it leads to a new way to represent the sequence. Considered at time t, a string of starts and stops can alternatively be represented by the set of analog *time intervals* $\tau[\mathrm{A}_s]$, $\tau[\mathrm{A}_e]$, $\tau[\mathrm{B}_s]$, $\tau[\mathrm{B}_e]$, $\tau[\mathrm{C}_s]$, ..., as illustrated schematically below.

$$< ----------\tau[A_s]---------------- >$$
$$< ----------\tau[A_e]---------------- >$$
$$< --------\tau[B_s]---------------- >$$
$$< ------\tau[B_e]---------------- >$$
$$< -------\tau[C_s]------------- >$$
$$< ---\tau[C_e]----------- >$$

etc.

$$< \tau[F_e] >$$

In this representation, the sequence which occurs in a finite time-window immediately prior to t is described by a vector τ(t) in a space whose coordinate axes represent the possible symbols $A_s, A_e, B_s, B_e, \ldots$. Most of the symbols will not occur at all, and their components are zero. The symbols which do occur have components $\tau[A_s], \tau[A_e], \tau[B_s], \ldots$. τ(t) implicitly depends on t, since as time passes new symbols are constantly occurring (making a previously zero component non-zero), non-zero components of τ are constantly lengthening, and components longer than a cutoff time τ_c are reset to zero.

A known sequence b can also be described by a vector of times $\tau_b = \{\tau_b[A_s], \tau_b[A_e], \tau_b[B_s], \tau_b[B_e], \tau_b[C_s], \ldots\}$. This vector does *not* change with time, since its components represent fixed time intervals. To determine whether a known sequence b is present in the data stream, the vectors τ_b and τ(t) must be compared. Recognition of the occurrence of sequence b within a uniform time-warp factor λ should take place at the times when

$$\tau(t) \cong \lambda\tau_b \text{ or } \tau[K] \cong \lambda\,\tau_b[K] \quad \text{for all symbols K}. \tag{3.8}$$

In this representation, the *mathematics* of recognizing a given known word has been made equivalent to the problem of finding a match (if any) between the present analog vector τ, representing what has been heard shortly before the present time t, and one of the known analog template vectors. *The word-recognition problem* in the presence of uniform time-warp (within a word) *is equivalent to the analog match problem of olfaction.*

If the brain solves the sequence recognition problem by this means, it must first produce a transformation of the sequence into the occurrence time representation. This requires a "neural" way to generate signals which represent the time-dependent analog variables $\tau[A_s], \tau[A_e], \tau[B_s]$. Neurons which respond strongly to a particular stimulus, but which do not have a sustained response to a sustained stimulus, are common in sensory areas of the brain. Such cells are said to adapt and to detect changing or moving stimuli. Under some circumstances, however, the response of these cells can be described as generating a *transformation of representation*. Consider a cell which responds to the acoustic feature "D" in a sequence. An "adapting" neuron would produce action potentials at a decreasing rate after the initiation of the sequence of D's. Suppose the beginning of a sequence of symbols D initiates a *stereotyped* response of increasing interspike intervals. Then the length of each interspike interval is determined by how much time has elapsed since the beginning of the string of D's (i.e., the time since D_s). The observation of a particular time separation between two successive action potentials uniquely describes the time since D_s. The output of such a cell *represents* the time-dependent quantity $\tau[D_s]$.

Thus we see that problem fundamental to olfaction and one fundamental to the recognition of sequences are related by a simple transformation which cells can carry out. Our brains seem to have huge numbers of different capabilities, but perhaps there are fewer independent and unrelated capacities than we might think.

Problems which seem different, can, in fact be related through transformations, and thus solved similarly.

Another computational problem which seems to be similarly solved in three sensory modalities concerns learning what an object is. We are superb at looking at a visual scene and parsing it into objects. The same thing is done when we hear a violin note as an entity, not as a chord consisting of its harmonics, or when we separate one voice from another in a crowded room. We may learn what objects are, but in order to do so, we need a referent for learning. This referent must come from our genes and development, not from our experience. Organisms need appropriate hard-wired processing on which to base our further initial learning. What elementary idea might be built in, to enable the system to bootstrap this object-separation capability?

The simplest unifying idea for object recognition is "what fluctuates [or moves] together is linked together" as an object. An elementary demonstration of the power of this algorithm can be seen in the following demonstration. Use a piece of randomly speckled paper as a background. Lay upon it a cut-out shape, such as a square or a silhouette of a face, made from the same kind of random speckling. The shape will be invisible, since there is no difference between it and the background. Now jiggle the shape slightly. While the jiggling is going on, the shape will be clearly visible. When the jiggling stops, the shape becomes invisible again. Co-motion==object is a very powerful algorithm in our visual processing. A similar effect occurs when three narrow frequency bands of sound noise are played, centered at f, 2f, and 3f. The listener hears a chord of three notes. When, however, the frequency center of each of the tones is jiggled together (as vibrato), the percept of a chord is lost, and instead what is heard is a single sound (a single sound object) having an acoustic texture. A similar idea may lie behind an ability to separate odors coming from spatially separate sources, but which are mixed by natural turbulent air flow.

There seems a likely unity in the fundamental algorithm used by these three sensory systems for the definition of objects. In spite of the fact that we do not know exactly how this algorithm is implemented on neural hardware, we might ask whether psychophysics experiments can be done to demonstrate a commonality of object recognition illusions in these three senses.

3.6 CONCLUDING REMARKS

Understanding how the brain works is a problem with strong connections to physics. Dynamical systems and ideas of collective properties are the two most physics-like general aspects. From the world of analog computing hardware we also know that for a computational system to get the most out of hardware, the hardware and algorithm must be closely related. Aspects of the "device physics" will then be essential to understanding biological computational function.

Evolutionary systems have no shame—they are able and willing to use anything that works, no matter how much it seems a kluge. At the same time, the larger nervous systems will require robustness, and thus no one cell or algorithm seems to tell a whole story. In physics, we can often say things like "electronic properties of matter are responsible for high temperature superconductivity," knowing that nuclear physics, elementary particle physics, and gravitational physics will be essentially irrelevant. In human-engineered systems, we can often say that this part (or algorithm) has a particular function. This separation of scales and of functions, which is so much a part of much of our experience of physics and engineering, is less present in the neurobiology of higher organisms. With relatively few exceptions, there are multiple areas of brain, and multiple algorithms or approaches, involved in most natural calculations. A single base change in a DNA molecule can result in a cell type which is not present, and a fundamental change in brain structure and function. Thoughts produce changes, if not at the DNA level, then certainly at the RNA level. These facts make the problem of how the brain works intrinsically more complex, and much less analytically tractable, than the most complex of ordinary physical systems.

Important aspects of dynamical systems, collective properties, and device physics seem to make understanding brain computation a problem for physics. At the same time, the amount of broken symmetry, the number of facts which must intrinsically be dealt with, is immensely larger than in conventional physics. In the past, physics as a discipline has tended to shed areas which require too many details, such as climatology or materials science. There will be relatively few important contributions to neurobiology which are not intimately related to at least a subset of details. It remains to be seen whether physics as a discipline will take up understanding the brain, or whether the field will merely continue to be populated by physics expatriates.

3.7 DISCUSSION

Session Chair: Albert Libchaber
Rapporteur: Konstantin Savvidi

BERG: I work on single cell nervous systems, bacteria that made their way in the world in interesting ways, though they never heard about the action potential. Do you have a feeling for what is the simplest neurological system that experimentalist might want to study? You don't have to have 10^{11} neurons to make headway in understanding something about specific circuits.

Do you need an alert monkey or a fruit fly, or can you get something from *C. elegans* or work with something even simpler?

HOPFIELD: The simpler the system which still has the computational circuit you are interested in the better off you are. From the point of view of how neural networks compute, I think *C. elegans* or *Drosophila* are obvious things to use, and on the mammalian side the mouse. Given the power of molecular biology, whenever possible one should choose organisms whose genetics has been thoroughly analyzed and is under good control. This lies behind my choice of organisms which also represent a span from simplest computational network that it makes sense to study to a large system that will be in great part collective in its operation.

BLOCK: You called yourself an expatriate physicist. I recall the unflattering definition of the biophysicist as one who speaks physics to biologists, and biology to physicists. Another, more flattering definition is that a biophysicist is a biologist who builds his own apparatus. Now physics departments are getting serious about biophysics, inspired by you and others. How do the physics departments need to change to support biology, or how do biology or other departments need to change, or will biophysics perhaps always be for crazy people who find their own way?

HOPFIELD: I think every institution is going to feel its own way in that regard. Attitude is a deep problem. When you hear a physicist say, "my problem is more important because it is fundamental," or "this is what physicists ought to do," you know you're in trouble. When I was at Princeton the general attitude of the physicists was: "We're sure you doing interesting things, but please don't tell us too much about them."

I think biophysics is a problem which has much in common with the kind of physics which Jim Langer was talking about. Physics includes a study of the very small, the very large, and the very complex. This third leg of physics has just started to emerge. There is a great tendency for people in physics departments to say that this subject is for others. In my view, each physics department has the option to say it's not part of us—and contracting—or including it, broadening its intellectual scope, and expanding.

LEIBLER: Let me just make a couple of comments. The first is from an experimentalist's point of view. Physicists like to study crystal growth because it allows them to do what physicists like the most, which is first to isolate, then to simplify, and then to measure. The problem with the brain is that, first of all, isolation is very hard, because at some level of complexity and detail you stop seeing the very phenomena you want to study. This is the trouble with *C. elegans*: studying what is he thinking when we do not know what he is thinking, or if he is thinking at all. So in a sense the usual methodology of physics is a little bit difficult in biology, which is a frustration to the experimentalist.

The second comment is more conceptual. I enjoyed very much the part of your talk when you talked about evolution. We must not forget that biological systems evolve, and evolution is, as Jacob said, tinkering; this is your story of the kettle.

In biology, unlike what we find in, say, condensed matter physics, there is a history part. There are two possible scenarios. One is very pessimistic, and states that because of the history component all details would be important. The other is the more optimistic, saying that exactly because of the evolution there are some rules or properties of the system which we may eventually find, some properties like robustness which are necessary to the existence of things as we see them now. And maybe this is really what we have to discover. I think, as is usual in biology, both scenarios may be equally true.

SMITH: If I were a painter I would say you were doing some kind of pantheism. You were talking about psychophysics, and about wet lab physiology, and about molecular biology, and about computing aspects... Is that the message you want to send, that you must look at all those aspects together?

HOPFIELD: If you want to address the overarching problem of how we think, you have to link these various levels. On the other hand, that's a problem which is not going to be answered at the next conference or probably in my lifetime. So what do we do? Fortunately, the the sub-areas themselves have sufficient technical problems and unexplored science that working in one of these sub-areas without worrying too much about interface to others is probably tenable as a scientific enterprise, and helps make progress toward the possibility of broader understanding. The neurochemist who isolates a neurotransmitter by grinding up a brain, throwing away all computational circuitry as part of his experimental paradigm, is still helping make progress on larger issues.

ANDERSON: You bring out a point of importance both to the theory of mind and to the philosophy of science: that it is a dead end to separate the perceiving mind from the object it perceives. The hardware of the mind is wired by interaction with the environment, and the mind actually interacts with the environment it perceives. The same is true of the way science proceeds, in contrast to the philosophers' idea of science as separate from the things it is studying.

HOPFIELD: The point that the brain keeps operating on the environment is quite central. The visual system must in some sense understand a picture. But it is not a fixed picture. Even when the scene is still, your eyes flick around from point to point several times a second; you are actively exploring. You are not just passively thinking "Here are all the pixels in an image, now interpret the picture."

CHAKRAVARTY: You remarked that there is a lack of scales in biology. Is there a way to understand why this is so? Actually, I do not even know why there are scales in physics.

HOPFIELD: Well, there *are* certainly scales in physics. I do not know "why" either, but, for example, the fine structure constant is so wonderfully small that quantum electrodynamics is essentially unimportant for condensed matter physics. Similarly, nuclear physics can be ignored.

The problem in biology is illustrated by the fact that things which occur in a single molecule can have a dramatic effect at the macroscopic level. That is something we are not accustomed to dealing with in ordinary physical systems.

JAFFE: What is your reaction to the assertion by modern linguists that certain mental states and abilities are innate and heritable? I'm thinking especially of the concept of "universal grammar." You seemed to describe the brain as a dynamical system, programmed by early experience.

HOPFIELD: While early experience is very important, I certainly said nothing about the brain being a *tabula rasa*. The situation is quite the contrary. We have many particular structures that are intrinsically built into us. On top of that, we then interact with the environment, and that interaction results in tuning and selecting or modifying circuitry in various ways. Stereoscopic depth perception is an interesting example. We are born with a potential capability for stereoscopic vision because there are neurons early in our visual system which are driven by both eyes. However, if you don't learn to see stereoscopic images correctly within your first two years (because, for example, of problems with your eye muscles), you will never gain stereoscopic vision even if the eye muscle problem is then later repaired by surgery.

One could similarly believe that you have a built-in capacity for languages of some particular structures, and within that broad capability your environment then determines your selection of a set of phonemes, vocabulary, and so on. The mind is in this view not a *tabula rasa* but rather a delimited arena on which a great deal of selection and shaping will be done by the environment.

JAFFE: I was struck by the number of bits available in the DNA code, versus the complexity of the object you're trying to program.

HOPFIELD: It's only the apparent complexity of the object that you're trying to program. One can imagine developing a complex structure by a program which tells a neuron to grow 100 connections to cells of class A lying within 10 microns, 10 connections to cells of class B within 30 microns, etc. Such a program can itself contain a lot less bits than what it produces, just as a program which computes the digits of π can be quite short compared to the generated string.

FISCH: My question has to do with where this field is headed. I remember thirteen years ago you gave a very stimulating Hamilton lecture, in which you showed how memories can be formed as fixed points of dynamical systems. You were looking for some unifying principle. Now in this lecture you are pointing to nature as a bag of tricks: edge detection, speech recognition, and so on. Where do you think this synthesis of physics and biology is heading in the next 20 years?

HOPFIELD: I believe that for any particular computation that you want to do there are a couple particular tricks at the cellular, synapse, or circuit level which will be very significant. For the rest of the details you can get by with rather gross approximations and still be able to come up with the essence of how computation is being done. Thirteen years ago I described a particular bag of tricks involving attractors which got you so far and then didn't seem to be capable of getting much further. The reason that it couldn't get much further is that it did not contain the realities of neuroanatomy relevant to systems other than associative memory. The attractor concept is broadly useful, but one must add to it classes of neuroanatomy capable of carrying out classes of computation. I think the progress is going to be in adding some details of particular systems.

For collective phenomena, hydrodynamics seems such a beautiful subject because all the details at the molecular level seem to disappear into constructs like viscosity. You don't see any trace of the microscopic level; all you have is this wonderful Navier-Stokes equation. Wouldn't be wonderful if neurobiology were like this!

But alas, the collective phenomena of neurobiology are more like those of chicken soup. To clarify the point, let me finish with a story. I was having a lunch with Feynman, and at one point he rotated his bowl of soup (his invariant lunch) and asked: "Have you ever rotated a cup of soup, Hopfield?" I rotated my cup of tea and nothing of any interest seemed to happen, it just followed. "Look what happens in my chicken soup," he said. "I rotate the soup, the soup follows, I stop the bowl and the soup rebounds in the opposite direction! Why is that?" Lunch with Feynman was often an exam.

Chicken soup is a system that contains very long molecules and some more macroscopic clumps. Ordinary hydrodynamics doesn't work. Navier-Stokes fluids will not rebound. So you must add one or two more details describing more of the essential physics. You can then write down equations much more complex than ordinary hydrodynamics that can describe chicken soup. In that same fashion I expect we will have to add a few more details on how particular parts of the brain are built, how particular neurons behave, in order to define and understand the equations on which some particular neural computations are based.

CHAPTER 4

EMERGENCE AND EVOLUTION OF PATTERNS

HARRY L. SWINNEY

Center for Nonlinear Dynamics and Department of Physics
University of Texas, Austin, TX

4.1 NEAR EQUILIBRIUM: THE BASE STATE

We consider macroscopic systems such as fluids, liquid crystals, reacting chemical liquids or gases, solids, and biological systems, which are driven away from thermodynamic equilibrium. The systems are *dissipative*—energy must be supplied through the imposition of a gradient (e.g., in temperature, velocity, or concentration) to maintain the system away from equilibrium. The distance away from equilibrium can often be characterized by one or more dimensionless control parameters, e.g., the Reynolds number for a fluid, $R = VL/\nu$ for a fluid flow with a characteristic velocity V, length scale L, and kinematic viscosity ν.

The dynamical behavior of macroscopic systems driven away from thermodynamic equilibrium is usually governed by partial differential equations. Again taking a fluid as an example, the equation of motion for the velocity field $\mathbf{u}(\mathbf{r}, t)$ and the pressure field $p(\mathbf{r}, t)$ is the Navier-Stokes equation,

$$\frac{\partial \mathbf{u}}{\partial t} + \mathbf{u} \cdot \nabla \mathbf{u} = -\nabla p + \frac{1}{R} \nabla^2 \mathbf{u}, \tag{4.1}$$

which is Newton's second law for a continuum fluid. For an incompressible fluid, we also have

$$\nabla \cdot \mathbf{u} = 0. \tag{4.2}$$

Equations (4.1) and (4.2) are dimensionless—$\mathbf{u}$, $\mathbf{r}$, t, and p are expressed in units of V, L, L/V, and ρV^2. Thus two systems with the same geometry but with a different size, velocity scale, and viscosity will behave in the same way if the Reynolds numbers are the same.

Even when a system is driven so far from equilibrium that the behavior becomes chaotic or turbulent, varying erratically in space and time, the system is still usually completely described by deterministic equations of motion such as (4.1), i.e., stochastic effects such as thermal fluctuations are usually negligible.

The solution of the nonlinear equation of motion near equilibrium is called the *base state*. For simple geometries the base state can usually be determined analytically, while for complicated geometries the base state must be determined numerically. The base state has the symmetry of the boundary conditions; hence for time-independent boundary conditions, the base state is time independent. Sufficiently close to equilibrium, the base state is stable—any perturbation, no matter how large, will decay and the system will asymptotically approach the base state. For sufficiently small R, *any* initial condition will evolve to the base state; in contrast, for large R, the uniqueness property is lost, and there may be several or even many different stable solutions for a given R.

4.2 INSTABILITY OF THE BASE STATE

We now consider the linear stability of the base state solution $\mathbf{u}(\mathbf{r}, t)$. If an infinitesimal perturbation of the base state, $\delta\mathbf{u}(\mathbf{r}, t) = \mathbf{A}e^{\sigma t}e^{i\mathbf{k}\cdot r}$, decays, then the base state is stable with respect to small perturbations. As the distance away from equilibrium increases, a critical value of R, R_c, is reached where the growth rate of the perturbation is zero at a wavenumber k_c, and for $R > R_c$, the perturbation grows, as Fig. 4.1(a) illustrates. R_c can be found by solving the equation of motion, linearized in $\delta\mathbf{u}$. The base state loses stability at R_c and the solution spontaneously loses the symmetry of the boundary conditions—perturbations at wavenumber k_c grow, developing into a spatial pattern with wavelength $\lambda = 2\pi/k_c$.

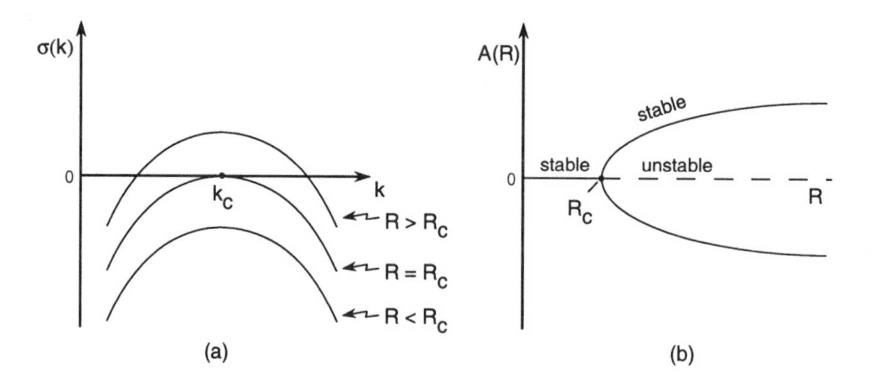

Figure 4.1: (a) Growth rate of a perturbation, $\sigma(k)$, as a function of wavenumber for Reynolds numbers below, at, and above the onset of instability. (b) The amplitude of the mode that becomes unstable at $R = R_c$. The transition at R_c is called a pitchfork bifurcation. The base state remains a solution of the equation of motion for $R > R_c$, but it is no longer stable.

In 1923, G.I. Taylor conducted a linear stability analysis of flow between concentric rotating cylinders [1]. The base state for this system is an axisymmetric velocity field that is invariant under translation in the axial direction. Only the azimuthal component of the velocity is nonzero, $u_\theta = Ar + B/r$, where the constants A and B are determined by the no-slip boundary conditions at the inner and outer cylinders, $u_\theta(r_1) = \Omega_1 r_1$ and $u_\theta(r_2) = \Omega_2 r_2$, where the subscripts 1 and 2 correspond to the inner and outer cylinders, respectively. Taylor calculated the critical values of the cylinder rotation rates (Ω_1, Ω_2) where the base state becomes unstable. Taylor also built a concentric cylinder apparatus and used dye to visualize the flow patterns that form at the critical rotation rates. He observed that beyond the onset of the instability, axisymmetric toroidal vortices form, encircling the inner cylinder. The vortices are stacked like a pile of doughnuts in the axial direction; at a boundary between one vortex pair, fluid flows inward towards the inner cylinder, while at the next boundary in either axial direction, fluid flows outward towards the outer cylinder. The observed wavelength of a pair of vortices agreed well with the value calculated from the linear stability analysis. Moreover, Taylor found remarkable agreement between the predictions of the stability analysis and the laboratory observations, as Fig. 4.2 shows.

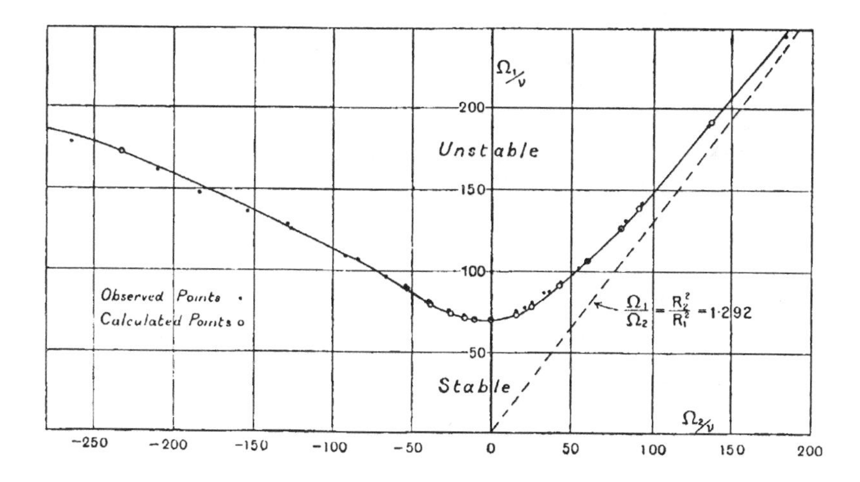

Figure 4.2: G.I. Taylor's stability diagram for flow between concentric cylinders with the inner cylinder rotating with angular velocity Ω_1 and the outer cylinder rotating with angular velocity Ω_2 [1]. The solid dots are Taylor's measurements and the open circles are from Taylor's linear stability analysis. The cylinders are co-rotating in the right-hand quadrant and counter-rotating in the left-hand quadrant; 0 corresponds to thermodynamic equilibrium.

This was the first experimental test of a linear stability analysis for any system driven away from equilibrium. As Taylor said, "All [previous] attempts to calculate the speed at which any type of flow would become unstable have failed"[1].

The procedure for finding the instability of the base state is straightforward, but in practice there can be subtleties in both experiments and in analyses. For example, one of the earliest instabilities studied was a thin layer of fluid heated from below. Nearly a century ago Henri Bénard [2] discovered that hexagonal convection cells form at a well-defined threshold; one of Bénard's photographs is shown in Fig. 4.3(a). In 1916, Rayleigh conducted a linear stability analysis for the problem, assuming that buoyancy effects caused the convection, but the threshold that Rayleigh predicted did not agree with Bénard's observations [3] (note that Rayleigh's analysis of convection preceded Taylor's analysis of flow between cylinders). Forty years elapsed before it was recognized that the instability observed by Bénard was not caused by buoyancy but by surface tension gradients [4]. Even then, experiments remained in disagreement with theory until recently, when the experimental difficulties were finally resolved; the observations shown in Fig. 4.3(b) are in good accord with theory [5] [6].

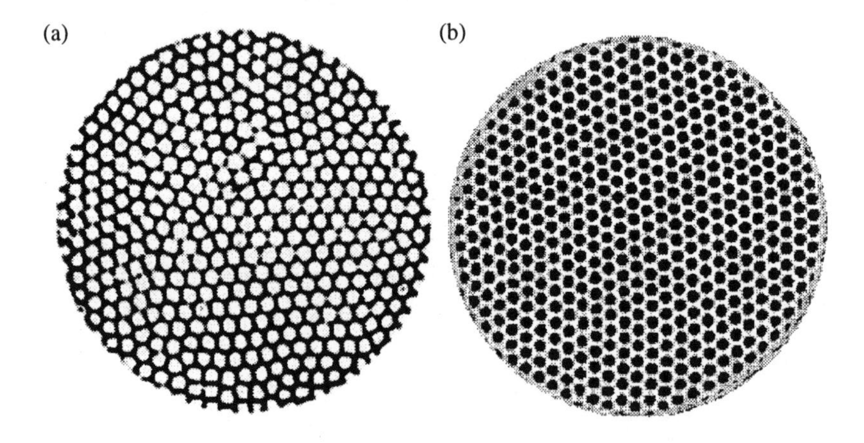

Figure 4.3: Hexagonal convection patterns observed in thin liquid layers heated from below: (a) Bénard (1900) [2] and (b) Schatz et al. (1995) [5].

When the growth rate of a perturbation, σ (Eq. (4.3)), is complex rather than real, instability of the uniform state leads to an oscillatory rather than stationary secondary state. The pattern that forms at the instability often consists of traveling waves in the form of spirals. Rotating spirals were observed in a chemical reaction-diffusion system more than two decades ago. The chemical reagents were poured into a petri dish, and a pattern with many rotating spirals formed and evolved as the chemicals were consumed [7]. Spiral wave patterns have also been observed in convecting fluids [8], Xenopus oocytes (the spirals in these frog eggs are waves of calcium concentration) [9], heterogeneous catalysis [10], and even heart muscle [11] (where the interest is motivated by a possible connection with sudden cardiac death [12, 13]).

A linear stability analysis yields the critical parameter value at which a base state becomes unstable, but perturbation theory must be used to determine how the amplitude A of the the most unstable mode (the mode with wavenumber k_c) varies above the instability threshold. In the simplest cases it is sufficient to truncate the amplitude expansion at cubic order,

$$\frac{dA}{dt} = \sigma A + bA^3, \tag{4.3}$$

where the quadratic term is absent because the cubic term is the lowest order term that feeds back on the unstable mode. Expanding the coefficients about R_c, we have

$$\sigma(R) = \sigma_0 + \sigma_1(R - R_c) + \cdots, \tag{4.4}$$

$$b(R) = b_0 + b_1(R - R_c) + \cdots. \tag{4.5}$$

Since $\sigma(R_c) = \sigma_0 = 0$,

$$\frac{dA}{dt} = \sigma_1(R - R_c)A + b_0 A^3, \tag{4.6}$$

where $b_0 < 0$ for stability (for $b_0 > 0$, higher order terms would be needed), and $\sigma_1 > 0$ since the system is unstable for $R > R_c$. The steady state amplitude, $dA/dt = 0$, is given by

$$A \propto \sqrt{R - R_c}. \tag{4.7}$$

The square-root growth in the amplitude of the secondary state is illustrated in Fig. 4.1(b). This square-root growth of the amplitude has been confirmed in many experiments, e.g., for Taylor vortex flow, see [14].

Beyond the threshold of instability there is a band of unstable wavenumbers (cf. Fig. 4.1(a)) and the amplitude $A(\mathbf{r}, t)$ of the pattern can vary slowly in space and time. For a given type of instability, there is universal form for the equation describing the amplitude near threshold. For example, a pattern of two-dimensional stationary rolls (also called stripes) that emerges at an instability from a uniform state is governed by

$$\tau_0 \frac{\partial A}{\partial t} = \frac{R - R_c}{R_c} A + \xi_0^2 \left(\frac{\partial}{\partial x} - \frac{i}{2k_c} \frac{\partial^2}{\partial y^2} \right) A - g_0 |A|^2 A, \tag{4.8}$$

where the constants τ_0, ξ_0, k_c, and g_0 specify the detailed properties of individual systems [15]; in principle these constants can be derived from the equations of motion, but this has been done for only a few cases. Much has been learned from comparison of predictions of amplitude equations with experiments (see especially the work on Rayleigh-Bénard convection, that is, fluid convection in a box heated from below [16]).

The possibility that instability could lead to the spontaneous formation of spatial patterns in chemical systems was first proposed in 1952 by Alan Turing[1] in a paper entitled "The Chemical Basis of Morphogenesis" [17]. Turing considered systems of reaction and diffusing chemical species, governed by

$$\frac{\partial \mathbf{c}}{\partial t} = D\nabla^2 \mathbf{c} + \mathbf{f}_\mu(\mathbf{c}), \tag{4.9}$$

where $\mathbf{c}(\mathbf{r}, t)$ is a vector of concentrations of different species, $\mathbf{D}$ is the diffusion coefficient matrix, $\mathbf{f}$ is a function describing the chemical kinetics (nonlinear in the cases of interest), and μ represents the control parameters (e.g., reagent concentrations, temperature) [18]. Turing said, "Such a system, although it may originally be quite homogeneous, may later develop a pattern or structure due to an instability of the homogeneous equilibrium." Turing's analysis stimulated considerable theoretical research on mathematical models of pattern formation in chemical and biological systems (e.g., see [19]), but Turing-type patterns were not observed in controlled laboratory experiments until 1990 [20]. The absence of experiments was due to a lack of a device in which a reaction-diffusion system could be maintained in well-defined nonequilibrium conditions. Such a device was developed in the late 1980s [21, 22, 23, 24]. It is simply a thin gel layer with its surfaces in contact with well-stirred reservoirs that are continuously refreshed. This type of reactor was used to observe the Turing patterns shown in Fig. 4.4 [23]. As a control parameter (reagent concentration or temperature) was varied, a critical value was reached where there was a spontaneous transition from a uniform (nonpatterned state) to the hexagonal pattern shown in Fig. 4.4(a). At other reagent concentrations, patterns of stripes rather than hexagons form; see Fig. 4.4(b).

What sets the length scale in the patterns in nonequilibrium systems? The size of the lattice constant in a crystalline lattice is determined by the atomic potential, but the length scales in patterns such as in Figs. 4.3 and 4.4 are many times larger. In fluid patterns the length scale is determined by the geometry. For example, a toroidal Taylor vortex near the onset of instability in flow between concentric rotating cylinders has an axial length very nearly equal to the gap between the cylinders. For chemical patterns the wavelength λ is determined not by the geometry but by the intrinsic properties of the reaction-diffusion system: $\lambda \propto \sqrt{D\tau}$, where D and τ are a characteristic diffusion coefficient and time scale (e.g., a period of an oscillation of the homogeneous reaction) [24].

Bifurcations from a spatially uniform state to stationary arrays of hexagons, stripes, and squares have been observed in many systems, e.g., hexagonal arrays like those in Figs. 4.3 and 4.4 have recently been observed in an optical beam in

[1] Editor's note: Turing received his Ph.D. in mathematics from Princeton in 1938.

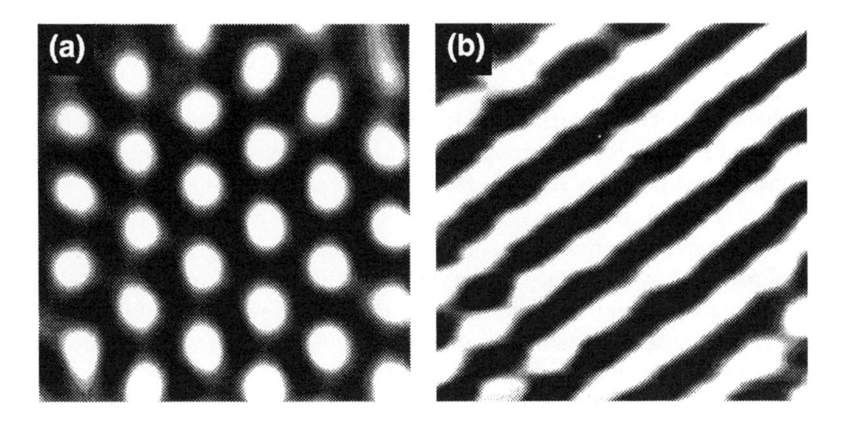

Figure 4.4: (a) Hexagonal and (b) stripe patterns of chemical concentration that form spontaneously in a continuously fed laboratory thin gel reactor [23]. The darker regions represent higher concentrations of triiodide, which is one of the species in the reaction. Each image is 1 mm $\times$ 1 mm.

a nonlinear medium [25] and in a gas of electrons in a strong magnetic field [26]. Another example is shown in Fig. 4.5: a vertically oscillated layer of liquid spontaneously forms stationary surface wave patterns when the acceleration amplitude exceeds a critical value; the patterns at the onset of instability can be squares or stripes, as Fig. 4.5 illustrates.

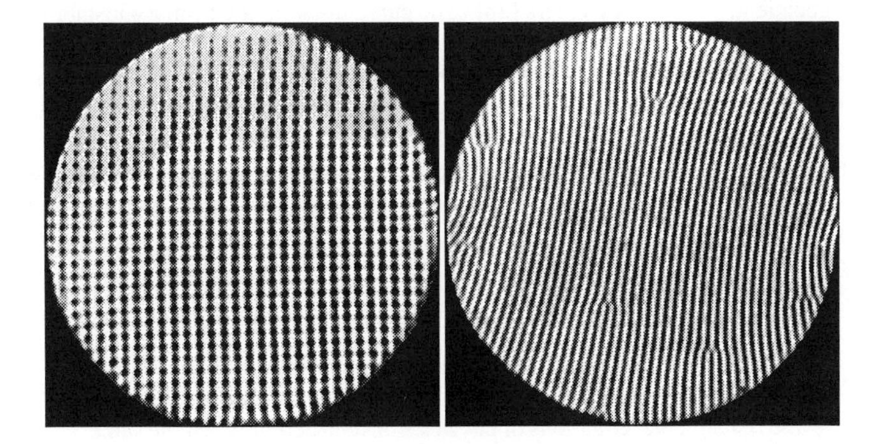

Figure 4.5: (a) Square and (b) stripe stationary wave patterns in a vertically oscillated layer of liquid just below the liquid-vapor critical point; (a) is at 0.08 K below the critical temperature and (b) is at 0.02 K below the critical temperature [27]. The cell diameter is 10 mm and the container oscillation frequency is 60 Hz.

4.3 FAR BEYOND THE ONSET OF INSTABILITY: EXPERIMENTS, SIMULATIONS, AND SYMMETRY

Perturbation theory provides a satisfactory description of the transition from a uniform state to spatial patterns in systems driven away from equilibrium, as described in the previous section. However, at control parameter values far beyond that at which the base state becomes unstable, perturbation theory fails. Experiments and numerical simulations reveal, for increasing R or other relevant control parameters, secondary and higher instabilities, each of which leads to a spatiotemporal pattern that breaks a space and/or time symmetry of the previous state. This cascade of instabilities often leads ultimately to states that are disordered in both space and time, even turbulent.

There is no general theory to describe the behavior of systems far beyond primary instability. In contrast to equilibrium systems, there is no function like the free energy that can be determined and minimized to find the state of a nonequilibrium system for a given set of control parameters. The many attempts to find an extremum principle for systems far from equilibrium have all failed. However, it is just this regime far from equilibrium that is often of interest in nature and technology, where one would like to be able to predict the behavior of, e.g., the atmosphere and oceans (weather and climate), combustion, mixing and separation processes, and biological systems.

In the absence of a general theory, experiments and simulations provide a guide to the kinds of phenomena that can occur. Figure 4.6 is an experimentally determined phase diagram showing the instabilities observed in flow between independently rotating cylinders [28]. The curve marking the primary instability (the lowest curve) is the one determined by Taylor (cf. Fig. 4.2).

Very little is understood about the different regimes shown in Fig. 4.6, especially those well beyond the primary instability. But one might argue that since we know the equation of motion, the problem is really solved. This is not true. The Navier-Stokes equation has been known for more than a century, but, despite a century of theoretical interest in flow between concentric cylinders, there was no hint of the complex phase diagram shown in Fig. 4.6 until it was determined in experiments. Thus knowledge of the equations of motion is not enough. Freeman Dyson recently made this point [29]:

> Oppenheimer in his later years believed that the only problem worthy of the attention of a serious theoretical physicist was the discovery of the fundamental equations of physics. Einstein certainly felt the same way. To discover the right equations was all that mattered. Once you had discovered the right equations, then the study of particular solutions of the equations would be a routine exercise for second-rate physicists or graduate students.... It often happens that the understanding of the mathematical nature of an equation is impossible

> without a detailed understanding of its solutions. The black hole is a case in point. One could say without exaggeration that Einstein's equations of general relativity were understood only at a very superficial level before the discovery of the black hole.

One could even go further—knowledge of the solutions of the equations of motion may not be enough. Philip Holmes found this was the case in his analysis of the constrained Euler buckling problem [30]. The solution he obtained, involving pages of elliptic integrals, did not provide much insight. The bifurcation structure for this problem was not understood until the bifurcations were identified in numerical simulations; then the mathematical solution could be examined to determine the behavior in the neighborhood of each bifurcation. Holmes' point is that experiments or simulations are often still needed, even when a closed form solution is known.

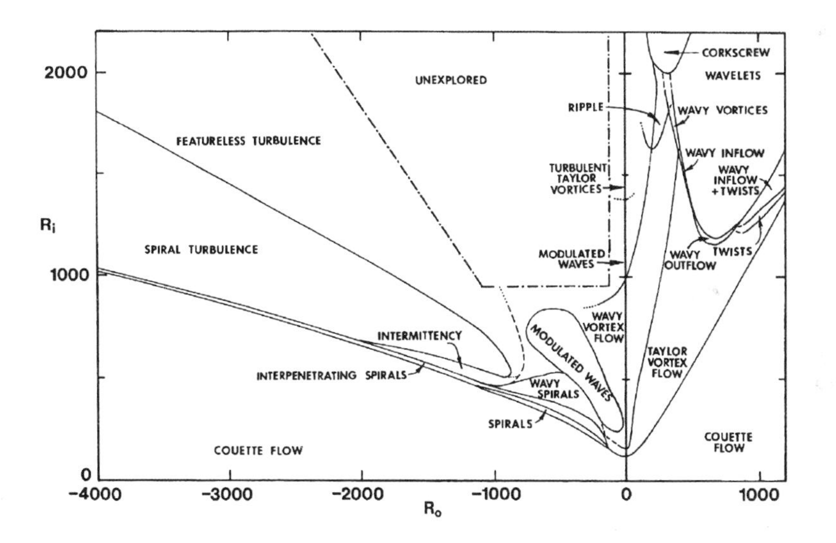

Figure 4.6: Regimes observed in flow between concentric independently rotating cylinders. R_i and R_o are Reynolds numbers proportional to the rotation rates of the inner and outer cylinders respectively [28]. The phase diagram is different for cylinders of different radius ratios; this diagram is for cylinders with radius ratio 7/8.

Experiments and numerical simulations of models help in understanding the kinds of bifurcations and patterns that can occur far from equilibrium. By "models" we mean systems of nonlinear partial differential equations that are simpler and hence easier to explore numerically and analytically than the full equations of motion (e.g., the Navier-Stokes equation). Amplitude equations (see 4.3) become models (often called complex Ginzburg-Landau models) when studied beyond the range where perturbation theory is applicable. Two other models that have been widely studied are the Swift-Hohenberg equation and the Kuramoto-Sivashinsky equation [15].

Figure 4.7 shows a phenomenon that, after it was observed in a laboratory experiment, led to the kind of interplay between model studies and laboratory experiments that is typical in investigations of nonequilibrium systems. The figure illustrates the development of a transverse instability of a chemical front. The instability was discovered in a search for Turing patterns in a reaction-diffusion system that had not been previously studied. Several reaction-diffusion models with two species were then examined to search for a similar front instability. The laboratory reaction is much more complex than the two species models; five chemical species are fed to the laboratory reactor, and the reaction produces many intermediate species and reaction products. However, each of the studies of models with only two species yielded a front instability similar to that illustrated by Fig. 4.7 [32, 33, 34, 35].

One of the simulations that yielded a front instability revealed another phenomenon that had not been seen in the experiments: patterns of spots (domains with a chemical concentration different from the background) that undergo a continuous process of "birth" through replication and "death" through overcrowding [34]. A search for this phenomenon in laboratory experiments was successful, as Fig. 4.8 shows. The experiments show that if a spot is isolated, it grows until the center drops back into the low pH state, thus forming a doughnut-like pattern. A subsequent study revealed this behavior in the numerical simulations as well. Front instabilities are common in systems far from equilibrium, but further experiments, simulations, and analyses are needed to determine what general lessons can be gleaned from the studies of the chemical fronts shown in Figs. 4.7 and 4.8.

An analysis of the consequence of symmetries can lead to insights into particular instabilities and to general insights on pattern formation [37]. An example of use of symmetry to analyze the stability of a particular flow is a study by Iooss [38], who considered what symmetries would be possible for secondary flows that bifurcate from Taylor vortex flow, given the possible axial reflection and translation symmetries. The selection rules derived from symmetry yielded only four possible states, which were just the four states found in an experiment [28]: vortices with waves on both the inflow and outflow boundaries (Fig. 4.9(a)), vortices with waves on only the inflow boundaries (Fig. 4.9(b)), vortices with waves on only the outflow boundaries (not shown), and vortices with flat inflow and outflow boundaries with waves in the vortex core (Fig. 4.9(c)).

An example of a general use of symmetry is illustrated in Fig. 4.10, which shows a sequence of images of disordered surface wave patterns, where the spatial correlations extend only about one wavelength [39, 40]. However, an analysis by Dellnitz et al. [41] predicted that, even though the instantaneous patterns may have no apparent order, the time-averaged patterns should exhibit the symmetry

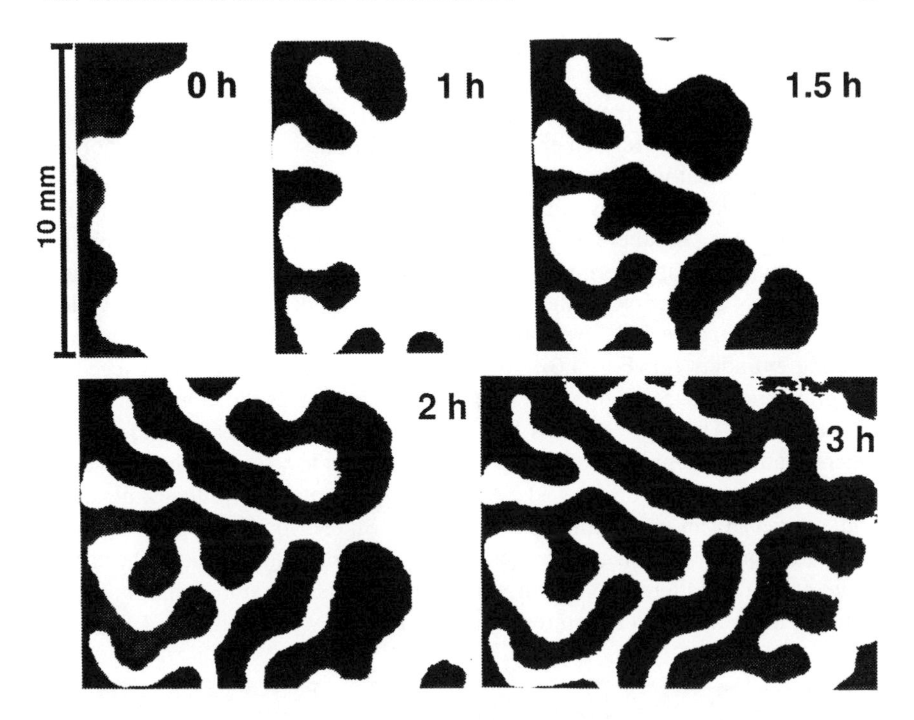

Figure 4.7: Time evolution of pattern observed in a bistable reaction-diffusion system (a ferrocyanide-iodate-sulfite reaction) [31]. The white state has low pH (about 4) and the black state has high pH (about 7). There is a transverse instability of the front that separates the black and white states, and the front evolves until at 3 hours (last picture) a stable, stationary labyrinthine pattern is achieved.

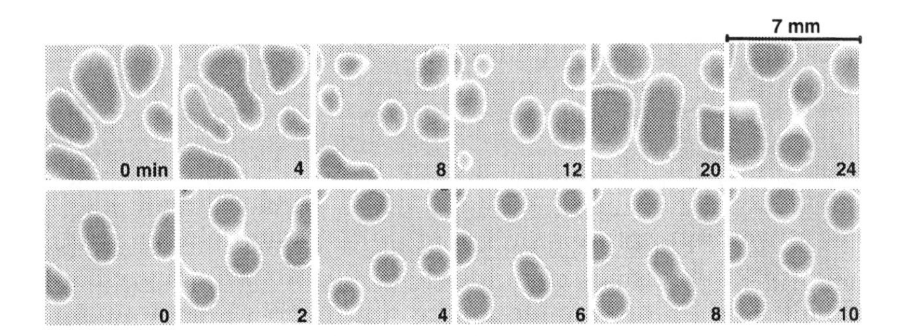

Figure 4.8: Replicating spot patterns in a laboratory experiment (upper row) and in a numerical simulation of a two-species model (lower row) [36]. In the experiment, the darker regions have higher pH (about 7) than the lighter regions (about pH 3).

of the boundaries; the time-averaged pattern in Fig. 4.10 is in accord with this prediction.

For some systems there are no equations of motion or even mathematical models, and one must rely on experiment to determine what kinds of phenomena can occur. Vertically oscillated containers of small particles are a good example. The models available apply only for a limited set of conditions and have no real predictive value, and there are no equations of motion analogous to the Navier-Stokes equation. But experiments on thin granular layers in evacuated containers, conducted as a function of the acceleration amplitude and frequency, yield a bifurcation structure [42] every bit as rich as that in flow between concentric rotating cylinders (cf. Fig. 4.6). When the acceleration magnitude exceeds g, the gravitational acceleration, the layer of particles leaves the container for part of the cycle. The layer remains flat until the acceleration reaches about $2.5g$, where spatial patterns spontaneously form, squares at low frequencies and stripes at high frequencies; see Fig. 4.11. At higher acceleration amplitudes, hexagons and a variety of more complex patterns form. Stable, localized, standing wave structures form when the acceleration is decreased just slightly below the critical value at which patterns first form with increasing acceleration amplitude [43]. Some of these localized structures are shown in the lower row in Fig. 4.11.

Figure 4.9: Secondary flows arising from instability of Taylor vortex flow: wavy vortex flow with waves on all vortex boundaries, (b) wavy inflow boundaries, and (c) "twisted" vortices (waves in the vortex core but not on the vortex boundaries) [28]. The flow is visualized using a dilute suspension of small flat flakes (Kalliroscope).

We have not discussed chaos or routes to chaos. Systems described by a few coupled *ordinary* differential equations are described by low dimensional phase space attractors, and these systems often exhibit bifurcation sequences that have universal properties, e.g., the period doubling route to chaos, a periodic to quasiperiodic to chaotic sequence, and intermittency [44]. However, in the spatially extended systems of interest here, low dimensional attractors and the routes to chaos are not usually observed, except for small systems (only a few characteristic wavelengths in size) not too far beyond the primary instability. Nevertheless, dynamical systems concepts such as strange attractors, fractal basin boundaries, and multiplicity (multiple stable states for the same control parameters), are often helpful in interpreting the behavior of spatially extended systems,

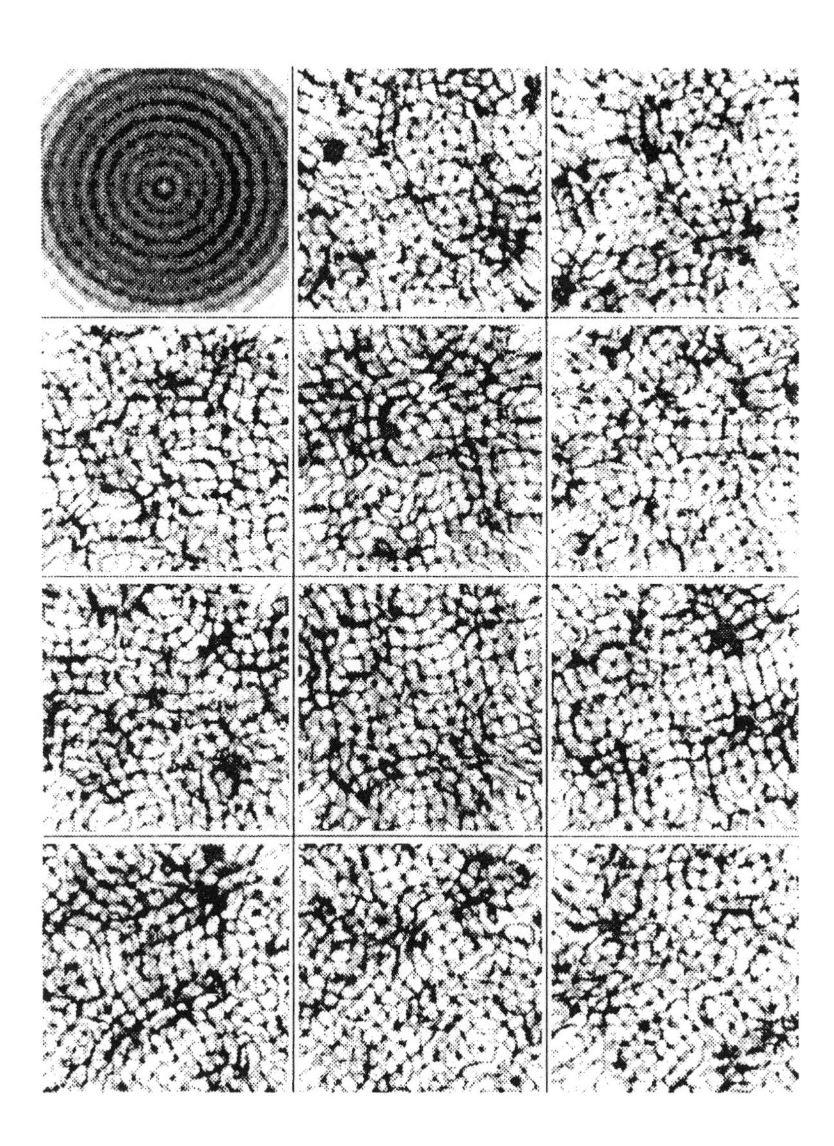

Figure 4.10: A sequence of disordered surface wave patterns observed in a vertically oscillated liquid layer [40]. The individual pictures are snapshots, illustrating the lack of order, but the time-averaged pattern at the top left has the symmetry of the container.

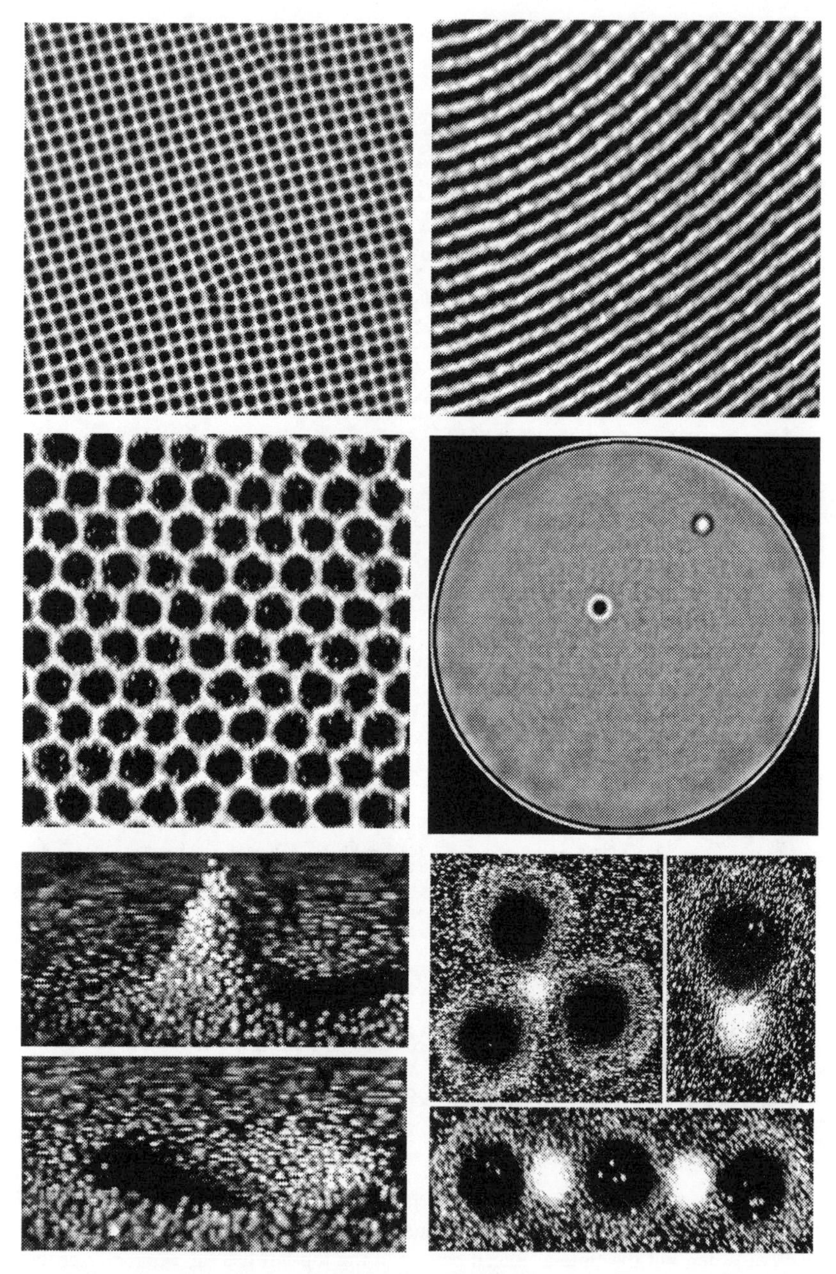

Figure 4.11: Patterns formed in a vertically oscillated thin layer of bronze spheres (0.16 mm in diameter) in a 127 mm diameter container. The images of squares (acceleration $3.2g$, frequency 20 Hz), stripes ($3.3g$, 67 Hz), and hexagons ($4.0g$, 67 Hz) are each 50 mm $\times$ 50 mm [42]. The image on the right in the middle row shows a layer with two localized standing wave structures (*oscillons*) of opposite phase; the white dot is seen as a peak when viewed from the side (upper image at the bottom left), while the black dot is a crater (lower image at the bottom left). The pictures on the bottom right show bound states (oscillon "molecules")—a tetramer, a dimer, and a polymer chain [43].

especially in the parameter range where disorder first develops as the system is driven beyond the primary instability.

4.4 PERSISTENCE OF ORDER

Very far from equilibrium ($R \gg 1$), a flow becomes turbulent—disordered in space and time and with a wide range of spatial scales. Yet a surprising degree of order often persists. Figure 4.12 shows turbulent Taylor vortices at a value of R nearly two orders of magnitude larger than that at which Taylor vortices first form [45]. The vortices are clearly discernible, and they are still apparent when R is increased by another order of magnitude.

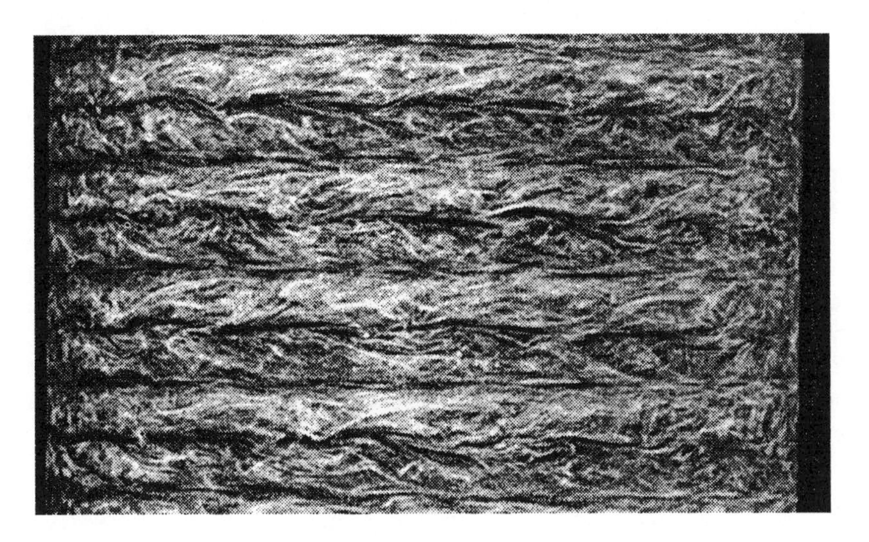

Figure 4.12: Photograph of turbulent flow between concentric cylinders with the inner cylinder rotating and the outer cylinder at rest [45]. Four vortex pairs are discernible. $R = 7600 = 92R_c$.

The Great Red Spot of Jupiter is another example of the persistence of order in a turbulent flow. This large vortex was first observed by Robert Hooke in 1664 [46]. The width of the vortex is about twice the diameter of the earth, and typical velocities are of order 50 m/s; hence the Reynolds number is enormous. The pictures obtained by space craft (Voyager 2 in 1979 and Galileo in 1996) show that the Jovian atmosphere is strongly turbulent. There are many such vortices in the eastward and westward jets in the Jovian atmosphere; the Great Red Spot is just the largest of the vortices. The earth's atmosphere is also at high Reynolds number; however, coherent structures are readily apparent on a weather map, which shows large cyclonic low pressure vortices and anti-cyclonic high pressure vortices, and a wavy jet stream which encircles the globe in the stratosphere at middle latitudes in the northern hemisphere.

The persistence of coherent vortices and jets in strongly turbulent flows can profoundly affect the transport properties of the flow. A recent experiment on a rapidly rotating flow containing jets and vortices showed that, for the case examined, the mean square displacement of tracer particles, $\langle(\delta r)^2\rangle$, did not vary linearly with time as in normal diffusive processes, rather, $\langle(\delta r)^2\rangle \propto t^\gamma$ with $\gamma = 1.6$ [47, 48]. The persistent coherent structures were responsible for this "anomalous diffusion": particles would become trapped for a while in a vortex and then escape and travel very long distances in a jet. By tracking large numbers of individual particles it was possible to determine the probability distribution function for a single step of length L. The result was $P(L) \propto L^{-\mu}$ with $\mu = 2.2$ for the case examined. Hence the mean square step size, $< L^2 >= \int L^2 P(L) dL$, diverged and the central limit theorem could not be used to analyze the transport.

Probability distribution functions with divergent second moments were studied by Paul Lévy, starting in the 1930s, and applied to physical systems by Shlesinger and others a half century later [49]. The experiment described in the previous paragraph was the first to directly observe "Lévy flights," which lead to a probability distribution function with a divergent second moment. The experiment was conducted on a simple isothermal flow with circularly symmetric forcing and circularly symmetric boundaries. This is a far cry from a real ocean or atmosphere, but the persistence of order (i.e., jets and vortices) in planetary flows may lead to mixing properties rather different from those described by normal diffusive processes. In some cases the jets and vortices could lead to greatly enhanced mixing. In other cases, the persistent structures could lead to barriers to transport, analogous to the invariant curves (KAM tori) in Hamiltonian dynamical systems [50].

4.5 PROSPECTS

As we have described, the nonlinear partial differential equations governing a macroscopic system driven away from thermodynamic equilibrium can be solved for a base state, which is always stable sufficiently close to equilibrium. The stability of the base state can be determined by solving a linear equation involving an infinitesimal perturbation of the base state. At a critical control parameter value, the base state becomes unstable and spatial structure emerges with a wavelength $2\pi/k_c$, which is determined from the linear stability analysis. The form of the pattern that emerges and the development of this pattern with increasing control parameter can be determined from a weakly nonlinear perturbation analysis ("amplitude equations").

Far beyond the primary instability, each system behaves differently. Details matter, as Langer emphasizes in his chapter in this book. There is no universality. However, the situation is not bleak. While there are no universal sequences leading to complex spatiotemporal behavior, similar bifurcations and patterns are frequently found in numerical simulations of simple models and in experiments on diverse systems. Simulations, experiments, and analyses are providing concepts and tools that are broadly applicable to systems driven far from equilibrium.

The study of nonequilibrium systems is inherently interdisciplinary. The problems often cross the boundaries of the traditional disciplines of mathematics, physics, chemistry, biology, and engineering. Important mathematical questions remain open, for example, what is the fundamental difference between patterns formed in a bounded (or periodic) domain, where the differential equations have compact groups of symmetry, and patterns formed in systems where boundaries are unimportant, where the equations have noncompact symmetry groups? One general area of applications concerns the control of industrial processes, which usually operate far from thermodynamic equilibrium. To produce more product, larger gradients are imposed, thus driving a system further from equilibrium and leading to instabilities and loss of control. Methods are now being developed to control processes in far from equilibrium conditions, even on unstable states remote in phase space from the attractor for a given set of control parameters; this would be desirable in situations where the unstable state is more efficient than the stable state [51].

In closing, we raise the question of whether or not the study of the emergence of patterns in systems driven away from equilibrium is a "fundamental" problem. It does not involve questions regarding the fundamental interactions between particles, but it does seem fundamental in the sense stated by Philip Anderson in his essay, "More is Different" [52]:

> The ability to reduce everything to simple fundamental laws does not imply the ability to start from those laws and reconstruct the universe. ... at each level of complexity entire new properties appear, and the understanding of the new behaviors requires research which I think is as fundamental in its nature as any other.

ACKNOWLEDGMENTS

I gratefully acknowledge the collaboration and support of the members of the Center for Nonlinear Dynamics of the University of Texas. I also thank Jerry Gollub (Haverford College) and Marty Golubitsky (University of Houston) for many helpful discussions over the past two decades.

My research on pattern formation has been made possible by the support of DOE, ONR, NASA, the Welch Foundation, and the University of Texas.

REFERENCES

[1] G.I. Taylor, Phil. Trans. Roy. Soc. (London) **A223** 289 (1923).

[2] H. Bénard, Rev. Gen. Sci. Pure Appl. **11** 1261 (1900).

[3] Lord Rayleigh, Philos. Mag. **32** 529 (1916).

[4] M.J. Block, Nature **178** 650 (1956); J.R.A. Pearson, J. Fluid Mech. **4** 489 (1958).

[5] M.F. Schatz, S.J. VanHook, W.D. McCormick, J.B. Swift, and H.L. Swinney, Phys. Rev. Lett. **75** 1938 (1995).

[6] The experiments reveal a small hysteresis: when the temperature drop across the layer is decreased after it was first increased until the system made a transition from the conducting state to the convecting state, the transition from the convecting to the conducting state occurs for a 2% smaller temperature drop across the layer than for the up transition. Nonlinear stability analyses yield a similar hysteresis.

[7] A.N. Zaikin and A.M. Zhabotinksy, Nature **225** 535 (1970).

[8] E. Bodenschatz, J. de Bruyn, G. Ahlers, and D.S. Cannell, Phys. Rev. Lett. **67** 3078 (1991).

[9] J. Lechleiter, S. Girard, E. Peralta, and D. Clapham, Science **252** 123 (1991).

[10] S. Jakubith, H.H. Rotermund, W. Engel, A. von Oertzen, and G. Ertl, Phys. Rev. Lett. **65** 3013 (1990).

[11] L. Glass, Physics Today **49**(8) 40 (1996).

[12] A. Garfinkel, M.L. Spano, W.L. Ditto, J.N. Weiss, Science **257** 1230 (1992); A. Garfinkel et al., J. Clinical Investigation **99** 305 (1997).

[13] R.A. Gray, J. Jalife, A. Panifilov, W.T. Baxter, C. Cabo, J.M. Davidenko, and A.M. Pertsov, Circulation **91** 2454 (1995).

[14] J.P. Gollub and M.H. Frielich, Phys. Fluids **19** 618 (1976).

[15] M.C. Cross and P.C. Hohenberg, Revs. Mod. Phys. **65** 851 (1993).

[16] G. Ahlers, "Over Two Decades of Pattern Formation, a Personal Perspective," in *25 Years of Nonequilibrium Statistical Mechanics*, edited by M. Rubí (Springer, 1995).

[17] A.M. Turing, Phil. Trans. Roy. Soc. (London) **237** 37 (1952).

[18] Q. Ouyang and H.L. Swinney, in *Chemical Waves and Patterns*, edited by R. Kapral and K. Showalter (Klewer, Dordrecht, 1995), p. 269.

[19] J.D. Murray, *Mathematical Biology*, (Springer-Verlag, New York, 1989).

[20] V. Castets, E. Dulos, J. Boissonade, and P. de Kepper, Phys. Rev. Lett. **64** 2953 (1990).

[21] Z. Noszticzius, W. Horsthemke, W.D. McCormick, H.L. Swinney, and W.Y. Tam, Nature **329** 619 (1987).

[22] W.Y. Tam, W. Horsthemke, Z. Nosticzius, and H.L. Swinney, J. Chem. Phys. **88** 3395 (1988).

[23] Q. Ouyang and H.L. Swinney, Nature **352** 610 (1991).

[24] Q. Ouyang, R. Li, G. Li, and H.L. Swinney, J. Chem. Phys. **102** 2551 (1995).

[25] S. Residori, P.L. Ramazza, E. Pampaloni, S. Boccaletti, and F.T. Arecchi, Phys. Rev. Lett. **76** 1063 (1996).

[26] K.S. Fine, A.C. Cass, W.G. Flynn, and C.F. Driscoll, Phys. Rev. Lett. **75** 3277 (1995).

[27] S. Fauve, K. Kumar, C. Laroche, D. Beysens, and Y. Garrabos, Phys. Rev. Lett. **68** 3160 (1992).

[28] C.D. Andereck, S.S. Liu, and H.L. Swinney, J. Fluid Mech. **164** 155 (1986).

[29] F. Dyson, New York Review of Books (May 25, 1995), p. 31.

[30] P. Holmes, talk at the University of Texas, October 28, 1996.

[31] K.J. Lee, W.D. McCormick, Q. Ouyang, and H.L. Swinney, Science **261** 192 (1993).

[32] D.M. Petrich and R.E. Goldstein, Phys. Rev. Lett. **72** 1120 (1994).

[33] A. Hagberg and E. Meron, Phys. Rev. Lett. **72** 2492 (1994).

[34] J.E. Pearson, Science **261** 289 (1993).

[35] G. Li, Q. Ouyang, and H.L. Swinney, J. Chem. Phys. **105** 10830 (1996).

[36] K.J. Lee, W.D. McCormick, H.L. Swinney, and J.E. Pearson, Nature **369** 215 (1994).

[37] M. Golubitsky, I.N. Stewart, and D. G. Schaeffer, *Singularities and Groups in Bifurcation Theory: Vol. II*, Applied Mathematical Sciences **69** (Springer-Verlag, New York, 1988); I.N. Stewart and M. Golubitsky, *Fearful Symmetry: Is God a Geometer?* (Blackwell Publishers, Oxford, 1992).

[38] G. Iooss, J. Fluid Mech. **173** 273 (1986).

[39] B.J. Gluckman, P. Marcq, J. Bridger, and J.P. Gollub, Phys. Rev. Lett. **71** 2034 (1993).

[40] J.P. Gollub, Proc. Natl. Acad. Sci. USA **92** 6705 (1995).

[41] M. Dellnitz, M. Golubitsky, and I. Melbourne, in *Bifurcation and Symmetry*, edited by E. Allgower et al., ISNM **104** 99 (Birkhauser, 1992).

[42] F. Melo, P.B. Umbanhowar, and H.L. Swinney, Phys. Rev. Lett. **75** 3838 (1995).

[43] P.B. Umbanhowar, F. Melo, and H.L. Swinney, Nature **382** 793 (1996).

[44] E. Ott, *Chaos in Dynamical Systems* (Cambridge University Press, 1993).

[45] D.P. Lathrop, J. Fineberg, and H.L. Swinney, Phys. Rev. A **46** 6390 (1992).

[46] R. Hooke, Phil. Trans. Roy. Soc. (London) **1** 2 (1666).

[47] E.R. Weeks, J.S. Urbach, and H.L. Swinney, Physica D **97** 291 (1996).

[48] T.H. Solomon, E.R. Weeks, and H.L. Swinney, Physica D **76** 70 (1994).

[49] For a review, see J. Klafter, M.F. Shlesinger, and G. Zumofen, Physics Today **49**(2) 33 (1996).

[50] J. Sommeria, S.D. Meyers, and H.L. Swinney, Nature **337** 58 (1989); R.P. Behringer, S.D. Meyers, and H.L. Swinney, Phys. Fluids A **3** 1243 (1991).

[51] Control theory is a well-developed subject with an extensive literature. An attempt to use a dynamical systems method to control a remote unstable state is described in V. Petrov, M.F. Schatz, K.A. Muehlner, S.J. VanHook, W.D. McCormick, J.B. Swift, and H.L. Swinney, Phys. Rev. Lett. **77** 3779 (1996); see references therein for other work on control.

[52] P.W. Anderson, Science **177** 393 (1972).

4.6 DISCUSSION

Session Chair: Philip Anderson
Rapporteur: Anastasia Ruzmaikina

ANDERSON: Thank you very much, I hoped that a number of people might have some comments to offer to Harry's talk. Let me start with Jerry Gollub; his name was mentioned again and again.

GOLLUB: It is difficult to add much after such a beautiful presentation which spans such a wide range of phenomena including many disciplines. I would like to emphasize how exciting and stimulating it has been to be in contact so frequently with people working in engineering, or applied math, or biology. It has been a truly exhilarating aspect of the field, which, I think, is likely to continue. One of the most difficult aspects of the subject is that one is often faced with new phenomena that have no obvious explanation, at least immediately. Often the experiments precede the attempt to understand. That is exciting, but it is also uncomfortable. One of my colleagues is fond of saying, "I can never believe any experiment I cannot understand." We are often anxious about that. I don't have much more to say, thanks for a beautiful talk!

ANDERSON: In recent years Sasha Polyakov has been playing with some of these systems, and he starts from the other half of physics. Perhaps he could tell a little bit about his experience of being confronted with this kind of problem.

POLYAKOV: Thank you. One really can't help feeling childish fascination looking at this picture of different beautiful systems. But switching to my adult mode, I start thinking about what I can really do as a theorist apart from going to my kitchen and trying to repeat these experiments. And, as a matter of fact, from a theorist's point of view, I also think that onset of chaos is beautiful and interesting. Most fascinating for me is the developed chaos, the limit as Reynold's number tends to infinity. This is the theory of developed turbulence—not necessarily of fluids—however, turbulence in a more general sense. I think that it is now an exciting time in this field, or it will be soon because there are two ingredients present which have not been there before. One is that we have reached considerable art in handling Quantum Field Theory, and we have learned a lot of lessons on field theories; there is an enormous stock of hidden tricks which we have now. On the other hand, we have now an exactly solvable model on which we can test this field theoretic approach to turbulence. By that exactly solvable model, I mean turbulence without pressure. You simply drop the pressure term from the Navier-Stokes equation, and then you try to study the inertial range. And in some sense this model is solvable. What I want to communicate are the results of these studies. Surprisingly, I thought I would create some helpful controversy, but I have to say that it is remarkably in accord with what professor Swinney showed us. Namely, we do get a steady state in this model which describes statistics; it is possible to compute probability distribution functions very explicitly and to check those field theoretic methods which I described, but what is most remarkable is that there is still order in the limit. When the Reynolds number is infinite this model predicts some periodic structure in the fluid. It is very easy to imagine that when you modify the model those structures will become more complicated than a simple crystal, but essentially you have the steady state, you have the structures which maximize probability and which are periodic. It is very similar to statistical mechanics, except there is one considerable difference and that is universality. Actually, that was a surprise to me. I thought it would be very similar, but it turned out to be quite different. Namely, the system is extremely sensitive to all control parameters. There one finds a family of steady states depending on parameters, and when you change external conditions you can continually switch from one steady state to another. That is very much like the normal critical phenomena. To conclude, that also seems in accord with what professor Langer told us yesterday. Universality in nonequilibrium systems, at least at infinite Reynolds number, seems to be very low. On the other hand, there is certainly some universality in the sense that we have a finite number of parameters which control the system. But since I want to be controversial anyway, I think that you need a state of the art field theory to do something in the field of developed turbulence; plus, it is impossible to do something without getting some input from real and numerical experiments. I think it is a safe prediction to say that in the close future we will see a very substantial progress in this field. Thank you.

ANDERSON: The next person I want to call is Per Bak.

BAK: I would like to challenge your philosophy somewhat. It seems to me that you are viewing the patterns in non-equilibrium systems like a zoo, where we view one animal at a time, admire it and describe it, and then go on to the next animal. Shouldn't we as physicists be more ambitious and look for general features in the experiments, and in the world surrounding us. Certainly there is no lack of important non-equilibrium phenomena, biology among them, that we do not understand. Wouldn't it be a shame if a whole were simply a sum of all the details?

SWINNEY: One point of my talk was to show the general features of the patterns for different systems, but I don't think we have an understanding why such similar patterns appear in so many diverse systems. There are lots of general questions. For example, how can we understand the similarities and differences between fluid and chemical turbulence? In the Navier-Stokes equation the nonlinearity is in the term $\mathbf{u} \cdot \nabla \mathbf{u}$, which leads to cascades of spatial scales in turbulent flow. In contrast, for reaction-diffusion systems driven far from equilibrium, where the nonlinearity comes from the chemical kinetics, the turbulent patterns have complex spatial and temporal structure but not a wide range of spatial scales. An understanding of the differences between turbulence in different types of systems is a subject for future work. One reason for my discussing the two-species Gray-Scott model was to show that an abstract model, which does not accurately model any of the laboratory chemical systems I described, displays many of the general features observed in the experiments on different reaction-diffusion systems. I certainly agree that as physicists we should search for general unifying principles. By examining different animals in the zoo in experiments and models, we can gradually learn what features are common and understand the conditions necessary for those features to occur. This is a problem for the next 50 years, to be reported at the conference in 2046.

ANDERSON: Great. There is one general aspect in the systems that Per is looking at. That is what you call a Reynolds parameter, and in many cases it is, in some sense, a ratio of a macroscopic scale and a microscopic scale. Certainly it is true in turbulence and in many other systems, but some of your other systems do not have this difference in scales. This kind of Reynolds number system in fully-developed turbulence is where the scales are completely distinct and it has some hope of being scale invariant. There are other cases where we do not have such obvious separation of scales.

LANGER: I obviously had more than my share of time yesterday, so I am not going to say very much, but what I had been thinking of asking, I realized, was Per Bak's question with exactly reverse spin. Harry, you have shown us this lovely world of different phenomena: they are complex, they are organizing, they are disorganizing in some sense. You, in your laboratory (with a few good competitors, of course), may be the most successful in going out and deciding to look at a number of different things. And almost every single time you do this you find something that is remarkably interesting, and often each thing is different from the other. My question is: "What is your guiding principle, how do you and the rest of the world organize their research in topics like these? If you could tell us just a little bit of your secret of success, maybe in 50 years from now we will know where to be going.

SWINNEY: I can answer that easily: I look at what people like Jim Langer are looking at—results from theoretical and numerical studies guide my selection of laboratory problems. I like problems that are simple and well-defined, where one student or postdoc can conduct a table top experiment with precision control of all the parameters. Problems with high symmetry are nice because the symmetries simplify both experiments and theory. Frequently I choose problems that have a long history, and much is already known, but we approach them with an emphasis on instabilities and scaling, varying parameters in small steps to characterize the transitions and to look for unifying features. As I mentioned, mathematical models provide a useful guide to experiments, but one has to be selective—with computing being so cheap there is a tendency to use computers to generate many pretty pictures that don't mean beans, just as one can generate many pictures from experiments that don't aid much in understanding. As Jerry Gollub mentioned, an exciting aspect of this research area is that we can work with mathematicians who try to prove things (say, extend KAM theory or prove certain consequences of symmetry) and at the same time we interact with engineers in industry, so that you get stimulation both from mathematicians who say "symmetry implies this" and from engineers who may be concerned with very specific applications.

MORRISON: I would like to stand up for exploring the "zoo" of phenomena: we need always to explore as far as we can. I would like to extend the beautiful examples presented here by mentioning artists in this same field: I suggest the work done in the last few years by artists making exhibits in Exploratorium, Frank Oppenheimer's Science Museum in San Francisco.

ANDERSON: Well, that is, perhaps, a very good comment to close on. Thank you.

CHAPTER 5

HIGH TEMPERATURE SUPERCONDUCTORS: FACTS AND THEORIES

T.V. RAMAKRISHNAN

Department of Physics
Indian Institute of Science, Bangalore, India

I would like to convey in this talk a sense of what makes the cuprate superconductors uniquely interesting, and will describe briefly some ideas that have been put forward to understand what is happening. My talk is likely to be different from those of some others in this meeting, for several reasons.

Firstly, I will talk about *one* family of systems exhibiting a wealth of surprising but presumably related phenomena, and will not give an overview of a large area.

Secondly, I hope that this problem does not spill over into the next century. It is almost certain that a remarkable variety of experimental results overconstrains candidate theories. It is possible that many of the basic ideas and methods necessary for a theoretical description are already present. However, history is not a very optimistic guide here. Superconductivity was discovered in 1911, but a fully microscopic theory took forty-six years. For the first fifteen years, quantum mechanics was nonexistent, but for nearly thirty years almost all the leading physicists of the time had a try at understanding it.

Even if the behavior of high temperature superconductors is understood soon, there is a serious sense in which this episode is only a beginning. The cuprates are one of the many families of systems in which, because of strong local interactions (correlations) or low dimensionality, or both, electrons behave in quite new ways, very unlike the metals and alloys we know. The evolution of our understanding could well take several decades.

Thirdly, I am not an expert in this field, only an interested observer and an occasional participant.

I will focus mainly on experimental results. By now we have a remarkably detailed picture of the properties of cuprate metals, and of the many ways in which they are unlike any metals that we have known. I will then briefly touch on some theoretical proposals, and end with a mention of other unusual electronic systems.

5.1 THE SETTING

About ten years ago, in a dull black ceramic oxide which was barely metallic, superconductivity was discovered by Bednorz and Müller [1]. This subject has since launched a hundred thousand papers [2]. We are yet to fully understand why these metals are unlike others, and what makes them go superconducting at unusually high temperatures.

5.1.1 THE ATOMIC ARRANGEMENT

The physical setting in which these phenomena take place is shown in Fig. 5.1. Structurally, the systems are distorted perovskites with octahedral coordination. The electronically active charge carriers are in the central square planar sections of the octahedra. A copper ion is at the center of each square, with oxygens at the corners. The Cu-O distance in most of the cuprates is nearly the same, about 1.9 Å. The square is only slightly distorted. The Cu-O planes ($a - b$ planes) are arranged in basically two ways along the c-axis. In single-layer compounds, e.g., $La_{2-x}Sr_xCuO_4$, (often abbreviated as 214) the layers are about 7 Å apart, with intervening (La,Sr) ions. In bilayer compounds, e.g., $YBa_2Cu_3O_7$ (abbreviated 123), there are two Cu-O layers close together (about 3.4 Å apart, with a Y ion in between). Two successive bilayers are separated by about 11.7 Å, with an intervening chain of Cu ions.

5.1.2 ELECTRONIC STRUCTURE

The active electronic states, i.e., those with energy close to the Fermi energy, have an $(x^2 - y^2)$ local symmetry around copper, being made up of $3d$ electrons of Cu. Copper is divalent here (Cu^{++}) and the electronic configuration is d^9. Out of these nine d electrons, only one need be considered active in the energy regime of interest to us, because of crystal-field splitting and strong local correlations. Oxygen-p orbitals covalently admix with Cu d-orbitals to produce states with local $d_{x^2-y^2}$ symmetry around the Cu site. On replacing the trivalent La with divalent Sr in La_2CuO_4, a combination of chemistry and electrostatics transfers this hole to the filled oxygen p_x, p_y orbitals. Since this hole has local Cu $d_{x^2-y^2}$ site symmetry, it is sufficient to consider only one band of states, namely those formed out of the $d_{x^2-y^2}$ orbitals of Cu. (This one-band Hubbard model is not universally accepted as adequate, even for low-energy behavior. Many authors believe that oxygen p bands need to be explicitly included).

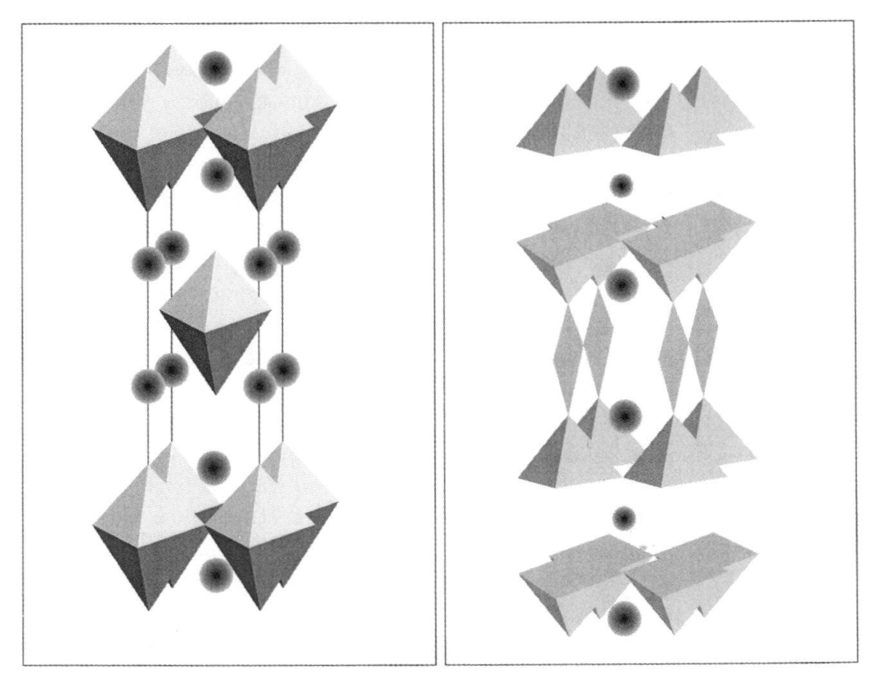

Figure 5.1: Structure of cuprate superconductors: (a) $La_{2-x}Sr_xCuO_4$:- The corner sharing octahedra (with Cu at the centre, and O at the corners) and the La, Sr atoms are shown. The central square plane of the octahedron (with Cu at the centre) is electronically active. (b) $YBa_2Cu_3O_7$:- Two close Cu-O bilayers (with Y atoms in between) are separated by Ba atoms, and a chain like arrangement of Cu-O planes.

5.1.3 MOTT INSULATOR

The cuprate metals can be regarded in general as 'doped' derivatives of insulating parent compounds. For example, the first family, namely $La_{2-x}Sr_xCuO_4$, is obtained by substituting Sr for La, and has a superconducting ground state for $0.08 \leq x \leq 0.22$. The parent compound (La_2CuO_4) is an antiferromagnetic insulator. It is an insulator even above the Ne′el temperature T_N, i.e., in the absence of sublattice magnetic order. It ought to be a metal, because there is one d electron per unit cell, and according to the band theory of solids any crystal with an odd number of electrons per unit cell has an incompletely filled band of electronic states. Such Mott insulators signal the dominance of interactions or local correlations and the inadequacy of approaches where interactions are treated as a perturbation on electronic kinetic energy. Since the fact that superconducting cuprates are doped Mott insulators may be central to their unusual behavior, and since the Mott or strong correlation phenomenon requires new ideas and methods basic to a large class of poorly understood systems, I digress a little to describe it.

A lattice of orbitally nondegenerate electrons is characterized by the amplitude t_{ij} for an electron to hop from an orbital at one site i to another at site j (usually the amplitude t for hopping to one of the z nearest neighbours is the largest, and is the only one considered). There are additional energy terms having to do with electron interactions. One which is qualitatively crucial is the extra energy cost U of putting two electrons (with up and down spins) on the same site. If this is larger than the kinetic energy zt gained by electron motion, and there is exactly one electron per site, electrons do not move but stay put at their home sites because any real hopping motion necessarily produces doubly occupied sites with extra energy $(U - zt)$ per site. The system is an insulator not because of lattice periodicity, but due to interaction effects.

In cuprates, U [or the energy difference $(E_{d^8} + E_{d^{10}} - 2E_{d^9})$] is large, between 6 to 10 eV. The electron kinetic energy zt is much smaller, being 2 eV or so. Thus both undoped and doped cuprates are strongly correlated; doubly occupied site states are energetically unfavourable, and need to be projected out of the Hilbert space. Such local projective constraints on many-body dynamics can lead to novel physical behavior. They are also difficult to implement. The large U limit is not accessible perturbatively: for example, the effective nearest neighbour (antiferromagnetic) spin-spin interaction has a size (t^2/U), i.e., it depends inversely on U!

In the last ten years or so, sustained efforts have been made to understand strong correlation phenomena, e.g., the Mott (metal to insulator) transition with increasing correlation for half filling (i.e., one electron per site), the nature of the metallic state away from half filling and for large U, the effect of various relevant perturbations such as disorder, etc.. In the limit of one spatial dimension [3], as well very large dimensions [3], very good understanding of several regimes exists. Obviously, one would like to know the extent to which these results can be applied to the quasi-two dimensional cuprates.

5.1.4 DOPED MOTT INSULATOR

On doping (namely substituting Sr for La), La_2CuO_4 exhibits a variety of phases (Fig. 5.2). The antiferromagnetic phase disappears at an astonishingly low 2 to 3% of Sr. This is incomprehensible on the assumption that antiferromagnetic Cu-Cu bonds are removed by dilution. For $x \geq 0.02$ to 0.03, there is a glassy insulator or spin glass phase, which gives way to a metallic phase for $x \geq 0.07$ or 0.08. This metal which persists till $x \sim 0.22$ is anomalous, as we shall see. It has a superconducting ground state. The superconducting transition temperature T_c is maximum at $x \sim 0.15$ or so; this is called optimum doping x_0. For $x < x_0$ the system is described as underdoped, and for $x > x_0$ as overdoped. For x beyond a certain doping level (~ 0.22 for $La_{2-x}Sr_xCuO_4$) superconductivity disappears; one then has a metallic phase that is quite different in character from the metal in the regime around optimal doping. We will concentrate on the optimum doping regime for most of the talk. One major unsolved problem in the field is a basic understanding of the global (x, T) phase diagram exhibited in Fig. 5.2. This is common to many cuprates, though not so well studied for all of them. The optimally doped system may be nominally stochiometric, e.g., in $YBa_2Cu_3O_7$; and oxygen deficiency, for example, could take one to the underdoped regime and even to the antiferromagnet. These are hole-doped systems; there is another class, namely the electron doped cuprate superconductors, e.g., $Nd_{2-x}Ce_xCuO_4$ which are different in many significant ways and which we shall not discuss. The apparent lack of symmetry between hole and electron doping is an unresolved question. However, the latter systems are much less extensively studied, materials problems being more serious.

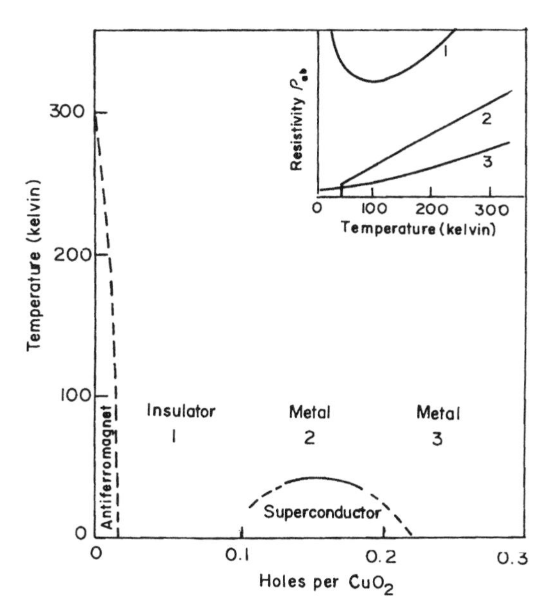

Figure 5.2: The electronic phase diagram of $La_{2-x}Sr_xCuO_4$ in the $(x - T)$ plane, where x is the number of holes per CuO_2 unit. The antiferromagnetic phase, the glassy insulator, the unusual metal (metal 2), the superconductor and the overdoped metal (metal 3) are shown.

5.2 THE INSULAR METAL

The layered cuprates are unique in being peculiar metals in-plane, and unusual *insulators* perpendicular to it. The transition on cooling is to a *three dimensional* (albeit anisotropic) superconductor. I shall present three experimental indications of this behavior, and will then go into some detail about the characteristics of the in-plane metallic state.

5.2.1 ELECTRICAL RESISTIVITY

To appreciate the unusual electrical character of the cuprates, we recall that electrical resistance or resistivity has a natural quantum scale, corresponding to the mean free path ℓ for current relaxing collisions becoming short enough to equal the electron wavelength λ_F which is of the order of the lattice distance d_i. This geometrical, quantum upper limit to metallic resistivity was first explored in detail by Mott. It is about 25,000 Ω for a metallic sheet (or for each layer of the weakly coupled layered cuprate) and about 5 $\mu\Omega \cdot$ cm for resistivity along the c-axis at or near optimal doping. Across this value, the mode of electrical conduction changes from random scattering of propagating electron waves to hopping of localized electrons.

The Mott maximum resistivity $\rho_{\rm max}$ (or minimum metallic conductivity) does not depend on the bandwidth or on the density of states at the Fermi energy. For example, in a tight binding band of width zt in a lattice of spacing d_i, it depends only on d_i (and somewhat on band filling) but not on zt! It can also be shown that the many body renormalization of electron wave function (described by the quasiparticle residue z_k) does not affect $\rho_{\rm max}$. Thus, polaronic or other band narrowing effects do not influence $\rho_{\rm max}$. However, if the scattering process is inelastic (energy changing), it randomly mixes electrons with different energies, and the Mott limit is not applicable. At low temperatures, inelastic processes are unlikely, so that the 'residual' or low temperature resistivity is a real separator between metals and insulators. (There are localization effects which at rather low temperatures lead to a characteristic increase of resistivity with decreasing T, near critical disorder.) Further, it is generally the case that, in a metal, the resistivity decreases with decreasing T while in an insulator it increases, either because thermally-assisted hopping becomes less probable or because the effective mobile carrier density decreases.

Let us look at the resistivity of the cuprates in the light of these ideas. Figure 5.3 shows ρ_{ab} and ρ_c for single crystal $Bi_2Sr_2CaCu_2O_{8+\delta}$ (abbreviated Bi 2212) [4, 5]. In the ab plane, the resistivity is relatively small and decreases with decreasing temperature. The resistivity near T_c ($\simeq$ 200 $\mu\Omega \cdot$ cm) corresponds to a sheet resistance of 1700 Ω, and a mean free path $\simeq$ 10 to 13 lattice spacings. This is metallic behavior. The resistivity ρ_c along the c-axis does have a 'metallic' T dependence, i.e., it decreases with decreasing temperature. However, the size is enormous, being several hundred times larger than the maximum metallic resistivity $\rho_c^{\rm max} \simeq$ 5 m$\Omega \cdot$ cm, and corresponding to a mean free path of (1/300) times the lattice spacing! This is incomprehensible in terms of a picture of electronic states propagating in the c direction and scattered by disorder. It requires one to describe electrons as extended in the ab plane, but localized in the c direction,

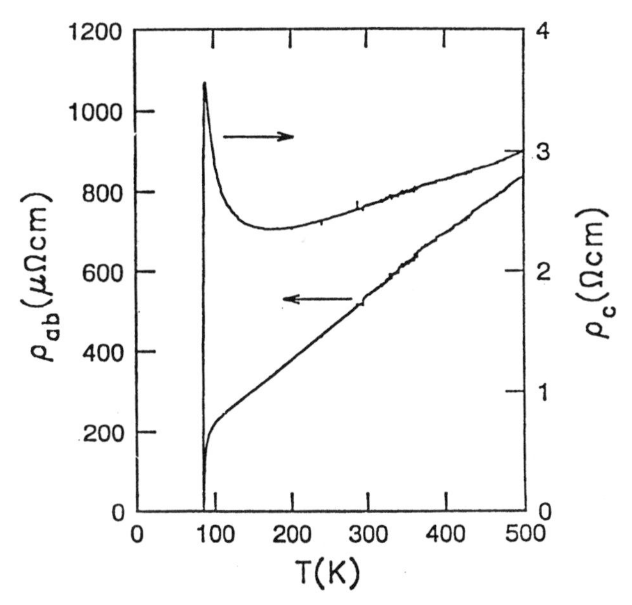

Figure 5.3: The resistivity of single crystal $Bi_2Sr_2Ca\ Cu_2\ O_{8+\delta}$ as a function of temperature, for current and voltage along the same direction in the ab plane (ρ_{ab}), as well as along the c-axis (ρ_c). From Ref. [5].

with occasional random hops from layer to layer. As the temperature decreases below about 130 K or so, the resistivity begins to rise rapidly as if the ground state is insulating along the c-axis, but this rise is cut off by the transition to superconductivity. A clearer indication of the low temperature state is provided by the high magnetic field experiments of Ando, et al. [6] on single layer, lanthanum-doped $Bi_2Sr_2CuO_y$ (La doped 2201) compounds with a T_c of about 12 K or so. Here, a very large pulsed magnetic field of about 50 T along the c-axis suppresses the superconducting transition, and enables one to probe the normal state down to low temperatures. We notice (Fig. 5.4) that while ρ_{ab} flattens out to a small residual value of 70-80 $\mu\Omega \cdot$ cm, ρ_c increases with decreasing temperatures down to about 0.2 K. The resistivity values ρ_c are in the range 2 $\Omega \cdot$ cm to 5 $\Omega \cdot$ cm! All this strongly suggests that the system is not really a metal along the c-axis. It is not a conventional insulator either, eg., with an exponentially increasing resistivity as temperature decreases. The c-axis resistivity varies over a wide range among the cuprates, from about 3 to 4 m$\Omega \cdot$ cm in optimally doped YBCO ($YBa_2Cu_3O_7$ or 123) to several $\Omega \cdot$ cm or larger in Bi 2212 compounds. The smallest values are already close to the maximum metallic resistivity of about 5 m$\Omega \cdot$ cm. The temperature dependence can be metallic or insulating. The superconducting transition temperature T_c seems to bear no simple relation to

ρ_c. (Optimally doped 123, with a ρ_c of about 3 m$\Omega \cdot$ cm and 2212 with a ρ_c of about 3 $\Omega \cdot$ cm, both have a T_c of 90 K or so.)

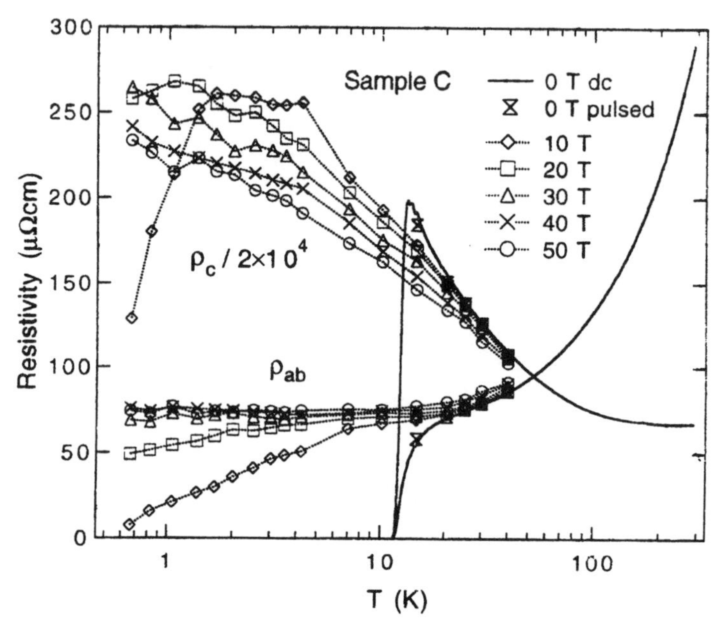

Figure 5.4: The resistivity of single crystal La doped $Bi_2Sr_2CuO_y$ (2201), in the ab plane as well as along the c-axis, for magnetic fields along the c-axis as shown. From Ref. [6].

5.2.2 OPTICAL CONDUCTIVITY

The ac optical conductivity $\sigma(\omega)$ measures the absorption of light of different frequencies ω by electrons excited from occupied to unoccupied states. For a free electron metal, it has the Drude (Lorentzian) form $Re\sigma(\omega) = \sigma(0)[1/(1+\omega^2\tau^2)]$ where τ is the relaxation time. Figure 5.5 shows the real part of the ac conductivity of single crystal $La_{2-x}Sr_xCuO_4$ for $x = 0.15$, as measured by Uchida et al., [7]. For electric field vector in the ab plane, the conductivity is large, and decreases with increasing frequency, crudely consistent with the Drude form but with a frequency dependent τ. For electric field along the c-axis, the conductivity is more than a few hundred times smaller and is independent of frequency, over a very broad energy range of 0.2 eV or more. The smallness and frequency independence over a wide energy scale are not appropriate to a metal, but to a lossy dielectric. A surprising feature of the c-axis conductivity for bilayer systems is the absence of absorption at energies of the order of the intra-bilayer coupling. This coupling

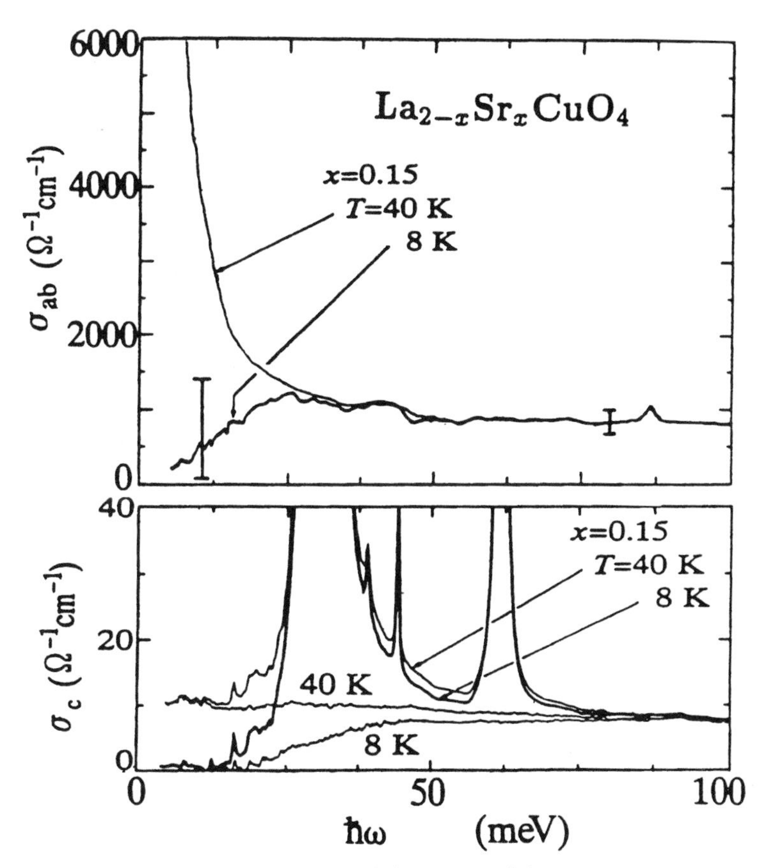

Figure 5.5: Optical conductivity Re $\sigma_{ab}(\omega)$ and Re $\sigma_c(\omega)$ (for electric field vector in plane and along c-axis, respectively) as a function of energy $\hbar\,\omega$, for single crystal $La_{2-x}Sr_xCuO_4$ with $x = 0.15$. From Uchida et al., Ref. [7].

energy is about 0.1 to 0.2 eV for YBCO. One can imagine eigenstates that are even or odd combinations of states on the two layers, and dipole-allowed transitions between them. These are expected to have a large oscillator strength and an energy $\sim$ 0.1 to 0.2 eV. They are never seen, so that something must be wrong with the picture of such a coherent or quantum superposition of layer states.

5.2.3 Angle-Resolved Photoemission Spectroscopy (ARPES)

The absence of bilayer splitting, or splitting in energy of states coupled by bilayer tunnelling, is even more directly seen in angle-resolved photoemission. This is a direct spectroscopic probe of electronic states, in which an incident photon

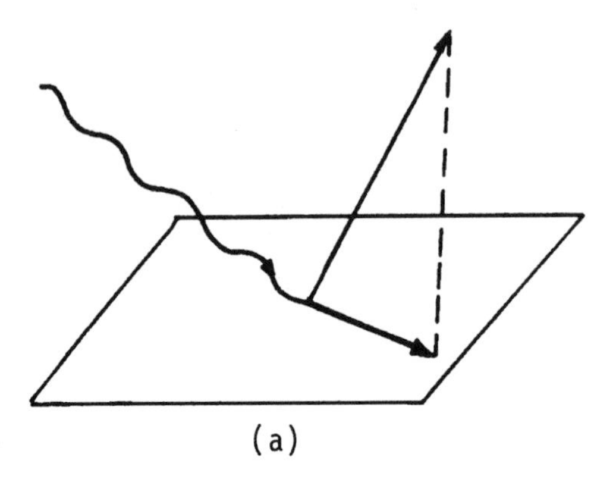

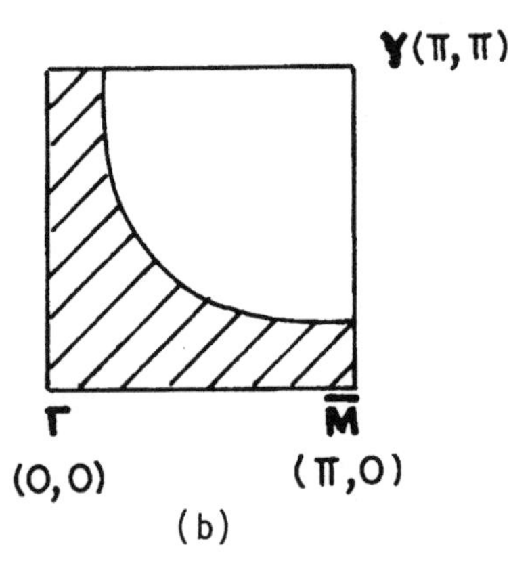

Figure 5.6: (a) Schematic description of angle resolved photoemission (ARPES). A photon of energy $h\nu$ is incident on the ab plane of the cuprate. An electron of energy E is emitted in a particular direction which fixes the parallel wavevector $\vec{k}_{||}$.
(b) The Fermi surface of 2212, as deduced from photoemission. One quadrant of the square Brillouin zone is shown, with the symmetry points $\Gamma\,(0,0)$, $\bar{M}\,(\pi,0)$, and $Y\,(\pi,\pi)$ indicated. The occupied states are shaded.

knocks out an electron from one of the occupied states (Fig. 5.6a). The energy of this electron and its direction of emission (and thus the electron momentum) are measured. Since momentum $\vec{k}_{||}$ parallel to the crystal face is expected to be conserved, one thus probes $\rho\,(E, \vec{k}_{||})$ or the spectral density of occupied electronic states for a well-defined momentum $\vec{k}_{||}$ in the ab plane (for a properly oriented single crystal). The observed Fermi surface is shown in Fig. 5.6b. Figure 5.7 shows the electronic spectral density for $\vec{k}_{\|}$ in the direction $\bar{M}Y$, and near the Fermi surface, for single crystal Bi-2212. For a perfectly defined quasiparticle, the spectral density has a δ-function part, $\rho(E, \vec{k}_{\|}) = z_k \, \delta\,(E - E_{\vec{k}_{\|}})$ where $E_{\vec{k}_{\|}}$ is the energy of the quasiparticle with momentum $\vec{k}_{\|}$. The spectral density in the normal state (even for an electron with momentum very close to the Fermi momentum) is seen to be very broad, though peaked at an excitation energy $E - E_F \simeq 0$. The spectral density sharpens to a resolution limited width below T_c! Now if there is bilayer splitting, one expects two peaks (if both the states are occupied). From the shape (Fig. 5.6b) of the observed Fermi surface (locus of zero binding energy crossings in the ARPES spectrum as a function of $\vec{k}_{\|}$) and from the symmetry of the interlayer coupling matrix element $t_{12}(\vec{k}_{\|})$ it can be shown that this should happen near the $\bar{M}$ point. Careful measurements of the superconducting state ARPES spectrum here for various incident photon angles and polarizations [9] show exactly the same spectrum within experimental resolution. Since for varying incident photon conditions the matrix element coupling the two layer states is different, there ought to be a variation in the relative intensities of two peaks if there is bilayer splitting. The fact that only one intensity curve is observed (with a sharp peak and a higher binding energy broad peak) argues against bilayer splitting of electronic energy levels.

The three kinds of experimental results discussed above are sensitive to interlayer coupling in different ways. The c-axis resistivity is determined by the weakest interlayer coupling, e.g., by the weak coupling between bilayers in bilayer systems. The ac conductivity is affected by both the weak and the strong interlayer coupling. In ARPES, bilayer splitting depends mainly on the strongest interlayer coupling. In each case, there is no evidence for states mixing coherently or quantum mechanically along the c-axis due to interlayer coupling, and propagating. The observations are difficult to reconcile with the often-assumed picture of an anisotropic metal (e.g., one derived from band theory), in which residual interactions can be treated perturbatively. A meaningful zero-order theory should have the effective 'confinement' of electrons to planes and incoherent interplane hopping as part of it. We now try to look at the behavior of electrons in the plane.

5.3 THE METAL IN-PLANE

I describe now the properties of the metallic state in the ab plane, as seen experimentally. There is an embarrassment of riches here; a bewildering variety of often novel behavior is revealed by a number of different probes. Many of these properties are clearly connected. What follows is a sampler. For the normal state some transport properties, optical conductivity, single particle excitations, and magnetic fluctuations are described. I summarize some of the evidence for the existence of a second, lower energy scale in cuprates; this is the so called

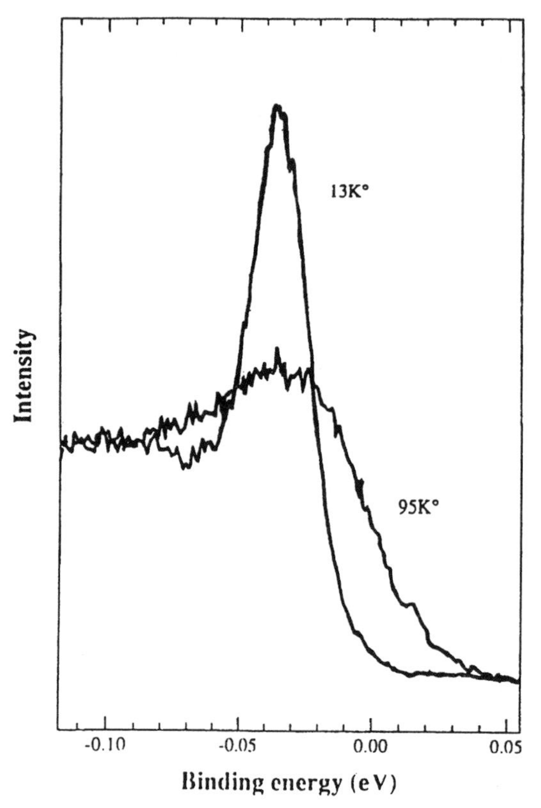

Figure 5.7: The photoemission spectrum of Bi 2212 with $T_c = 87$ K. The intensity of the outcoming electron beam as a function of binding energy (in eV) is shown, for angle such that one is at the Fermi surface crossing along the $\bar{M}Y$ direction (see Fig. 6b). From Randeria et al., Ref. [8].

pseudogap (Section 5.4). I then describe the superconducting state (Section 5.5). Finally, some theoretical ideas developed to make sense of these properties are briefly mentioned (Section 5.6).

5.3.1 RESISTIVITY

The in-plane resistivity of $La_{2-x}Sr_xCuO_4$ is shown in Fig. 5.8 as a function of composition and temperature. The data are for both single crystalline and polycrystalline specimens. The resistivity evolves continuously with composition. In the underdoped cuprate (x=0.04, 0.05) the resistivity is large and decreases with temperature but rises rapidly at lower temperatures; the ground state is an insulating spin glass, i.e., a frozen random collection of spins. At slightly higher doping, there is a transition to superconductivity at low temperatures. For larger x, e.g., around optimal doping $x \simeq 0.15$, the resistivity is linear in temperature, with a nearly zero intercept. The behavior is observed over a very wide temperature range $T_c < T < 800$ K. This linear resistivity seems to be a defining characteristic of the near optimally doped cuprate metal. In all families of hole-doped cuprates near optimal doping, the resistivity is proportional to temperature; the slope is nearly independent of T_c which varies over more than a decade, from about 5-10 K to 120 K [6]. The intercept seems to be zero.

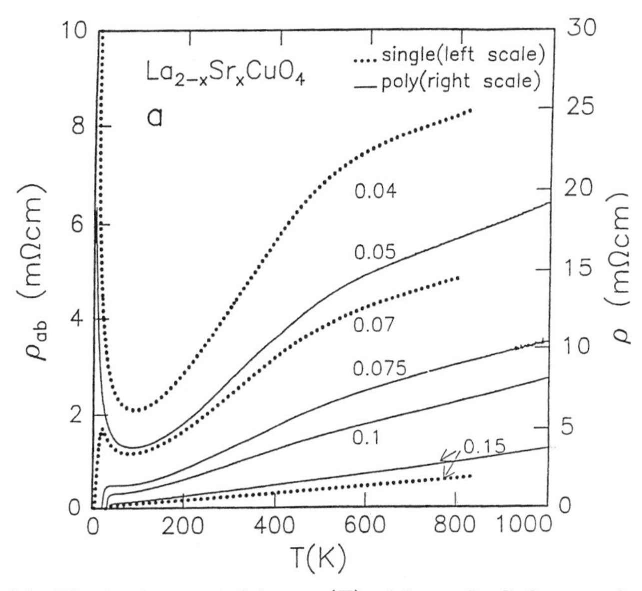

Figure 5.8: The in-plane resistivity $\rho_{ab}(T)$ of $La_{2-x}Sr_xCuO_4$ as a function of temperature T for different Sr concentrations x. From Ref. [10].

The linear resistivity implies several things. One can view the vanishing of resistivity with $T \to 0$ as indicating a superconducting quantum critical point, with a very large quantum critical regime. A conventional way of describing temperature dependent resistivity in a metal is in terms of scattering of electrons by excitations, e.g., phonons or spin waves. If the excitations have a characteristic energy ω_o, the resistivity depends as some higher power of T for $k_BT \ll \hbar\omega_o$, and linearly with T for $k_BT \gtrsim \hbar\omega_o$. The absence of such a crossover temperature implies that the excitations have a zero energy scale. (It could of course be that the coupled electron-excitation picture is not appropriate at all.) The constancy of the slope $(d\rho_{ab}/dT) \simeq 1.2\,\mu\Omega\cdot\text{cm/K}$ [5], independent of T_c, implies that whatever process causes the resistivity in plane, does not determine the superconducting transition temperature. This is a serious difficulty for purely two dimensional models of the superconducting transition and the transition temperature T_c. The linear resistivity and its slope are intrinsic to the plane; T_c seems extrinsic, as emphasized by Anderson [11].

Finally, extrapolating to higher temperatures, we find that the resistivity equals the Mott maximum at a temperature of order 1500 K. This is close to a well-known energy scale of the cuprate system, namely the antiferromagnetic exchange coupling J.

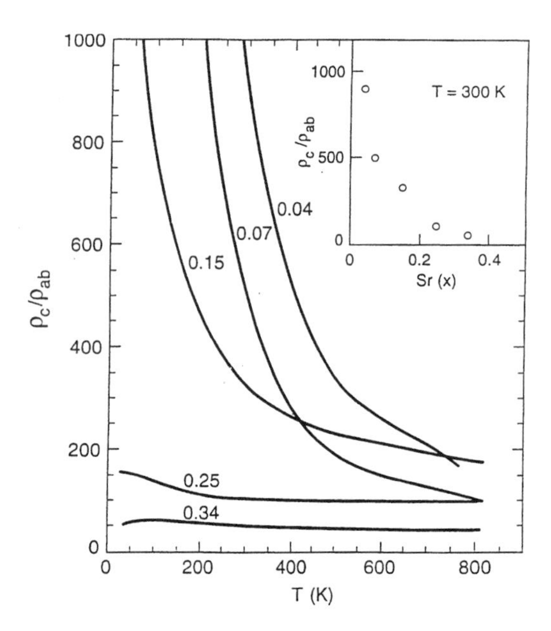

Figure 5.9: The anisotropy of resistivity, as measured by the ratio (ρ_c/ρ_{ab}) for $La_{2-x}Sr_xCuO_4$. From Ref. [12].

The c-axis resistivity is large and varies over a wide range, as mentioned earlier. It is quite instructive to exhibit the anisotropy (ρ_c/ρ_{ab}). This is shown in Fig. 5.9 for $La_{2-x}Sr_xCuO_4$. The ratio is about 50 and is temperature-independent for the overdoped cuprate, while ρ_{ab} does depend on temperature. This kind of resistivity behavior is well described by a model due to Kumar and Jayannavar [13], who regard electrons as inelastically diffusing in-plane at a rate Γ_i, and occasionally hopping to the next layer, with an amplitude t . The c-axis resistivity ρ_c, in the limit $t \ll \Gamma_i$, is given by $\rho_c \simeq \rho_{ab}\{tn(\epsilon_F)\}^2$ where $n(\epsilon)$ is the density of electronic states in the ab plane. In this model ρ_c and ρ_{ab} have the same temperature dependence, while ρ_c can be very small; its size is a direct measure of the electronic anisotropy. From the observed anisotropy of 50, one infers a ratio of 7 for the in-plane to out-of-plane hopping matrix elements. This is in the same range as that obtained from electronic structure calculations. This can be regarded as the bare anisotropy; for lower values of doping the anisotropy is seen to tend to this value at high temperatures, in some kind of incoherent limit. As temperature decreases, the anisotropy increases, e.g., for $x \simeq 0.15$, to values as high as 1,000! There seems to be some kind of dynamic blocking of interlayer movement, which becomes more effective at lower temperatures.

Similar but less spectacular effects are seen in bilayer cuprates. At low enough temperatures, if superconductivity is turned off, the c-axis resistivity always increases, as seen in Fig. 5.4. Is this an Anderson localization effect? The actual c-axis transport behavior is one of the more difficult mysteries connected with cuprate superconductors.

5.3.2 THE HALL EFFECT

The Hall resistivity of cuprates is unique among metals. The Hall resistivity in general has the value $\rho_H = (B/ne)$, where n is the carrier density (electron or hole). In a metal, the carrier density is expected to be constant, independent of temperature. One finds that in the cuprate metals, ρ_H varies inversely with temperature and is hole-like in sign [Harris et al., Ref. [14]] for $\vec{E}$ and the Hall current in the ab plane, and $\vec{B}$ along the c-axis. This is very difficult to understand; the naive conclusion that the carrier density has a component that is linear in T is hard to sustain.

A very suggestive interpretation, due to Anderson [15], is that longitudinal and transverse electrical currents in-plane relax via different processes. (Here longitudinal and transverse mean the extra, field-induced, currents along and perpendicular to the electron momentum; the former is due to the electric field and the latter to the magnetic field.) The longitudinal relaxation rate $(\hbar/\tau_\ell)$ is proportional to $k_B T$ (from the linearly T-dependent resistivity).

The ratio of the longitudinal to the transverse currents, symbolized as the cotangent of a postulated Hall angle, depends on the transverse relaxation rate $(\hbar/\tau_{tr})$:

$$(j_\ell/j_{tr}) = \cot\theta_H = (1/\omega_c^*\tau_{tr}) \,.$$

Thus this dimensionless quantity is a direct measure of the transverse relaxation rate. Experimentally [14], one finds that $\cot\theta_H = aT^2 + b$, i.e., the rate is proportional to T^2 (Fig. 5.10). There is a constant term b, absent in clean systems but induced by impurities such as Zn that substitute for Cu and disturb the plane magnetically. The fit is found to be very good, over several hundred degrees, for a wide variety of cuprates in which both σ_ℓ and σ_{tr} vary in different ways.

A crucial challenge for any fundamental or effective low energy theory of the cuprates is an explanation of the two temperature dependences, T and T^2, for longitudinal and transverse relaxation rates respectively (or for the resistivity and the inverse Hall angle), as well as the sign of the Hall resistivity. The size of the T^2 term is another puzzle; on expressing $(\hbar/\tau_{tr})$ as $k_B(T^2/T_F^*)$, we find that the characteristic temperature T_F^* is rather small, of the order of a hundred degrees or

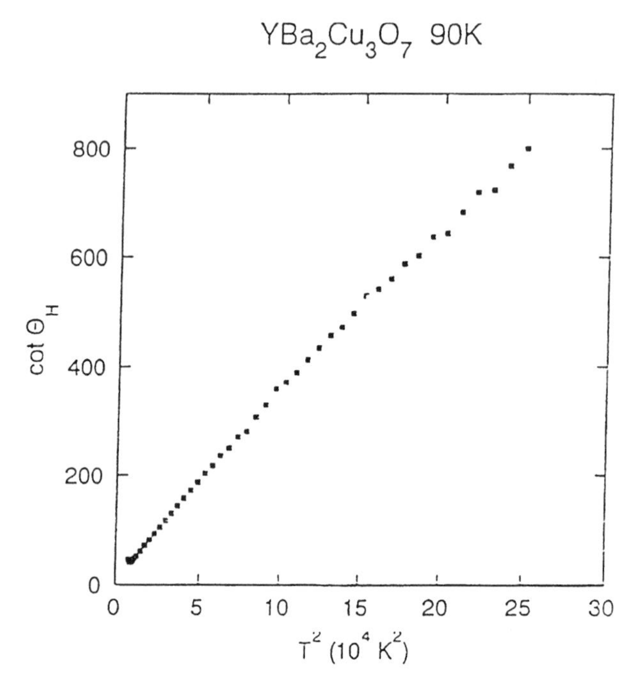

Figure 5.10: The cotangent of the Hall angle in the ab plane vs. (temperature)^2 for single crystal $YBa_2Cu_3O_7$. From Ref. [14].

so for $m^* \simeq m$. The Hall resistivity along the c-axis (i.e., for mutually perpendicular electric and magnetic fields in the plane, and the current along the c-axis) is electron-like, temperature-independent, and has a magnitude corresponding to $n \simeq (1 - x)$ per unit cell, for doping level x [5].

In the electron-doped superconductors, the ab plane Hall resistivity is electron-like.

5.3.3 OPTICAL CONDUCTIVITY

Optical conductivity explores electronic states and transitions on a much larger energy scale than dc transport. The real part of the conductivity can be written generally as

$$Re\,\sigma(\omega) = \frac{ne^2(\omega)}{m(\omega)} \left[\frac{1}{1 + \omega^2 \tau(\omega)^2} \right],$$

where the relaxation time $\tau(\omega)$ and the mass $m(\omega)$ are phenomenological fitting parameters. In a detailed study of a number of cuprate oxides, El Azrak and Bontemps [16] find that $1/\tau(\omega)$ depends nearly linearly on frequency over a range $0.03 < \hbar\omega < 1$ eV, instead of being nearly independent of it. This is consistent with the T-linear behavior of dc resistivity, but the surprising fact is that this behavior extends over a very wide range of frequencies and for many systems. The data are for thin films.

Experimental results [17] on untwinned single crystals of $YBa_2Cu_3O_7$, with the electric field vector along the a axis, confirm the linear frequency dependence of $1/\tau(\omega)$ with a coefficient that agrees with linear resistivity, but over a range of frequencies $\hbar\omega < 0.2$ eV.

5.3.4 SINGLE PARTICLE SPECTRUM

The spectrum of electronic excitations in cuprate superconductors as seen from photoemission is quite unusual in several ways.

The measured spectral density $\rho(E, \vec{k}_{\|})$ integrated over energy E has to equal the occupation number $n_{\vec{k}_{\|}}$. The Fermi function $n_{\vec{k}_{\|}}$ obtained this way, e.g., for single crystal Bi-2212 [8] does show approximately a step-function-like behavior. From this, it is possible to deduce an experimental Fermi surface, namely a surface in $\vec{k}$ space that contains occupied states. This surface is shown in Fig. 5.6b. The Fermi surface is large; the number of occupied electron states enclosed by it is $n_e = (1-x)$ per unit cell from the area of the enclosed Fermi surface. This is just the electron count, namely the number of $d_{x^2-y^2}$ electrons in the highest, partially occupied, band. It is what one expects from Luttinger's theorem which states that interactions between fermions do not affect the Fermi surface volume, or the total number of occupied states. The theorem was originally proved perturbatively; it seems to be valid even in the limit of strong correlations, probably because it is a state-counting result. However, some approaches to the Hubbard model near half filling would lead to a hole Fermi surface with area x per unit cell, since one starts from a Mott insulating ground state or vacuum in which there is one localized electron per lattice site. The doped holes move around, and the resulting metal could then have x holes per unit cell. This is not what is seen. The possibility of a small-area hole-like Fermi surface for small x evolving to a large electron like Fermi surface at larger x [18] is an important issue in the electronic structure of cuprates. We notice from Fig. 5.6b that the closed (in the extended zone) $\vec{k}$-space orbits near the observed Fermi surface are hole-like.

Another feature of cuprate electronic structure revealed by ARPES experiments on a wide variety of hole-doped systems is a nearly common electron dispersion E *vs.* $\vec{k}_{\|}$ in the high symmetry directions. A striking feature of the dispersion is the occurrence of relatively flat bands, near the Fermi energy, for $\vec{k}_{\|}$ in the neighbourhood of the $[\frac{\pi}{a}, 0]$ (and symmetrical) points (Fig. 5.11). These saddle points (minima in the (π,π) to $(\pi,0)$ direction and maxima in the $(\pi,0)$ to (0,0) direction) extend a considerable distance in $\vec{k}$ space; the electron energy is a few tens of meV below the Fermi energy. The origin of this flat band region is a major question. In principle, one can always tune nearest-neighbour and next-nearest-neighbour hopping integrals so as to engineer such a saddle-point and an associated van Hove singularity in the density of states. Such an approach has been advocated and developed. However, in reality this is very unlikely to happen over such a wide range of systems and carrier concentrations, and always just below the Fermi energy. It seems much more likely to be a many body-effect in which, due to interactions, the effective single-particle density of states develops a peak near the Fermi energy. In this connection, it is interesting that the superconducting gap (below T_c) is largest in this region of k space. Quite intriguing is the ARPES spectrum of the electron doped superconductor Nd Ce CO. The flat band in this case is in the same region of k space, but lies $\sim$ 0.4 eV below the Fermi energy.

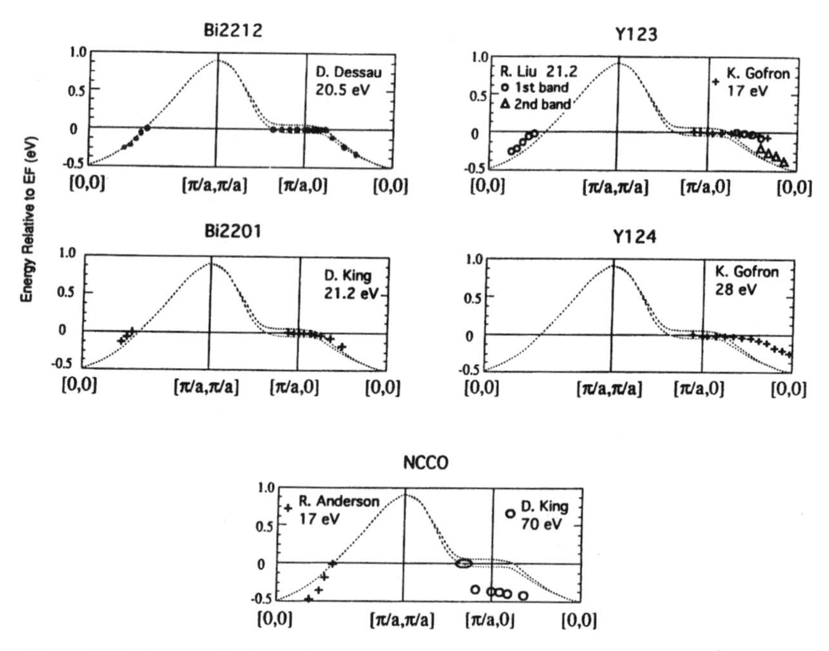

Figure 5.11: A compilation of electron energy E vs in-plane momentum $\vec{k}_{\|}$ in some symmetry directions, from Ref. [19]. The dotted curves are analytical fits.

In good metals, for occupied states near the Fermi energy, one finds a relatively sharp ARPES peak, and a weak, smooth background. In the cuprates, the observed spectral density in the normal state is a very broad ($\Delta E \geq 0.1$ eV) and asymmetric peak (Fig. 5.7), suggesting that quasiparticles are not well defined in the normal state. This is consistent with the linear resistivity, which could imply that the quasiparticle decay rate is linear in frequency, or proportional to quasiparticle excitation energy. The spectral density peaks are broader than this expectation, however, and are significantly asymmetric, with a long, low-energy tail. Interestingly, the quasiparticle peak sharpens to an instrument resolution-limited width in the superconducting state (Fig. 5.7).

5.3.5 MAGNETIC EXCITATIONS

The metallic cuprate is a doped antiferromagnetic (Mott) insulator, so that one expects significant magnetic correlations involving the Cu^{++} (spin 1/2) moments to persist in the metallic phase, the actual extent depending on the doping level or metallicity, and temperature. A variety of probes, e.g., susceptibility, NMR and neutron scattering, looking at a wide range of spatial and temporal scales, suggests that the metal is unusual magnetically as well.

The parent compound, La_2CuO_4, is a Heisenberg antiferromagnet with an in-plane nearest neighbour coupling energy $J \simeq 1500$ K. With hole doping, the Ne′el temperature drops precipitously to zero; presumably this is because hole motion leads to 'melting' of the antiferromagnetic long-range order. The details of what happens magnetically and electronically in the low hole-doping regime are not well understood. In the metallic regime, the system is a Pauli paramagnet, implying that local magnetic moments are shortlived. The Knight shift which measures conduction electron spin polarization in a magnetic field, is temperature-dependent in many systems [20] especially underdoped ones, and decreases significantly with temperature. However, the temperature dependence is the same at different nuclei, e.g., ^{17}O and ^{63}Cu. This implies that, statically at least, there is only one spin fluid. However, the spin relaxation rates $(1/T_1)$ at ^{63}Cu and ^{17}O nuclear sites have unusual and different temperature dependences. The nuclear spin relaxes its orientation by coupling to the electronic spin on that site. The number of electronic spin-flip excitations in a degenerate Fermi system is proportional to $k_B T$, for obvious reasons. The consequent Korringa relation, namely $(1/T_1) \propto T$ is seen in many metals, but is not observed for ^{63}Cu. Its relaxation rate rises with temperature up to about 400-600 K, and then saturates [21]. This saturation value seems to be about the same [21] for doped and undoped La_2CuO_4! This implies that, approximately at least, the low-energy, spatially-local, spin-excitation spectrum is the same at a temperature ≥ 600 K, irrespective of whether the spins are fixed in space but flip each other through exchange coupling, or whether they are itinerant. Another major surprise is that the temperature dependence of the ^{17}O relaxation rate is nearly linear, and is thus qualitatively different from that of copper. There is no satisfactory resolution of this anomaly [22, 23].

Inelastic neutron scattering measures the spectrum of spin excitations, $S(q, \omega)$, for a given wavevector transfer q and frequency transfer ω. Difficulties connected with counting rates, background subtraction, limited energy $(\hbar\omega)$ range, crystal size, etc. have been overcome in several systems, e.g., $La_{2-x}Sr_xCuO_4$ [24] and $YBa_2Cu_3O_7$ [25]. In the former, χ_q peaks not at the antiferromagnetic lattice vectors $\left[\frac{\pi}{a}, \frac{\pi}{a}\right]$ in plane, but at incommensurate vectors $\frac{\pi}{a}\left[1, 1-\delta\right]$ where δ is nearly proportional to doping. This does not seem to happen for $YBa_2Cu_3O_7$, where χ_q is maximum for commensurate reciprocal lattice vectors. The antiferromagnetic correlation length ξ_{AF} estimated from integrating $S(q,\omega)$ over a wide frequency range is about one or two lattice spacings for both 123 and $La_{1.85}Sr_{0.15}CuO_4$ [24]. The characteristic energy of spin fluctuations around $(\frac{\pi}{a}, \frac{\pi}{a})$ is about 20 meV in both systems (14 meV and 22 meV, respectively, for optimally doped 123 and 214 compounds). The small correlation length and its weak dependence on temperature argue against the system being near an antiferromagnetic instability. The small characteristic spin fluctuation energy is then somewhat mysterious.

5.4 LOWER ENERGY SCALE

An unexpected development in the last few years has been the experimental observation that there is a characteristic *lower* energy scale in cuprates. There is a diminution in the density of low-energy spin and charge excitations at energies of the order of a few hundred Kelvin. The reduction is also often described as a pseudogap. The pseudogap temperature scale T^* increases with underdoping, and is smallest at optimal doping, where it is nearly equal to T_c. The scale T^* appears to be a characteristic, smooth, crossover temperature, not a sharp phase transition temperature. I describe some of the data here.

Pseudogap-related effects were first noticed in inelastic neutron scattering [26]. They have since been seen in static magnetic susceptibility, Knight shift [20], resistivity, Hall effect, *ac* conductivity [17], specific heat [27] and, most recently, in ARPES [28]. I shall describe the last three here. Rotter, et al., [17] measured the optical conductivity $\sigma_{1a}(\omega)$ of untwinned single crystals of $YBa_2Cu_3O_{7-\delta}$, for the electric field vector polarized along the a axis. This eliminates possible absorption from electrons in the chains (which are oriented in the b direction). The measurements were made at different temperatures and oxygen concentrations δ (T_c decreases as δ or underdoping increases). The results are shown in Fig. 5.12. Above T_c, the *ac* conductivity increases with decreasing frequency ω. Below T_c, there is a decrease in absorption or in $\sigma_{1a}(\omega)$, starting at around $\omega \simeq 1000\ \text{cm}^{-1}$ in the 93 K ($\delta = 0$) crystal because of the superconducting gap in the single-particle excitation spectrum. (The missing spectral weight appears as a zero-frequency δ function, whose strength is proportional to the superfluid density). Comparing Fig. 5.12a with Figs. 12b,c, we see that in the underdoped compounds (T_c=82 K and T_c=56 K), there is a gradual relative suppression of $\sigma_{1a}(\omega)$, starting at frequencies ω higher than 1000 cm^{-1}, and at temperatures above T_c. This suppression is seen to become more pronounced with increasing δ. One way of quantifying the effect is to plot $\sigma_{1a}(\omega)$ at a fixed frequency for different temperatures. This is done in the lowest panel of Fig. 5.12, where $\sigma_{1a}(\omega)$ is plotted for the $T_c = 56$ K sample, as a function of (T/T_c). For comparison, appropriately scaled $(1/T_1T)$ data for O and Cu in samples with nearly the same

T_c are shown. It is clear that the characteristic decrease in all these starts well above T_c, and is nearly identical!

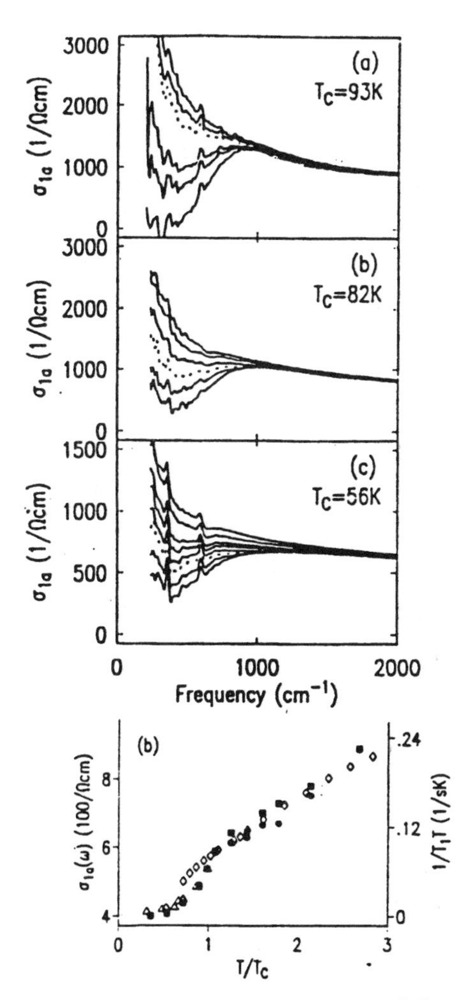

Figure 5.12: The real part of the optical conductivity $\sigma_{1a}(\omega)$ for $\vec{E} \parallel$ to the a axis, in single crystal $YBa_2CuO_{7-\delta}$. The three panels are for increasing underdoping δ, as reflected in the decreasing T_c (=93 K, 82 K, 56 K). The curves in each panel are downward in order of decreasing temperatures. The dotted curve is $\sigma_{1a}(\omega)$ at T_c. ω is measured in wavenumbers. The lowest panel compares $\sigma_{1a}(\omega)$ at 250 cm^{-1} (filled squares) and 500 cm^{-1} (filled circles) with $(1/T_1T)$ for ^{63}Cu (open diamonds) and ^{17}O (open triangles) appropriately scaled. Data are for different samples all with $T_c \simeq 60$ K. From Ref. [17].

Careful measurements of the electronic specific heat of $YBa_2Cu_3O_{7-\delta}$ as a function of δ and of temperature clearly show that there is a decrease in the number of thermally accessible excitations at temperatures of order a few hundred degrees [27]. These measurements are difficult because the electronic heat capacity is a small percentage of the total. By subtracting the lattice heat capacity against a standard, and making corrections for variation in the former with δ, T, etc., it is possible to obtain the electronic specific heat, and thus the entropy, to an accuracy of a few per cent. The electronic entropy is shown in Fig. 5.13 for different compositions δ as a function of temperature T. We note that the entropy for optimal doping is linear in T. This is as expected for a degenerate Fermi system. The entropy decreases rapidly below T_c, due to the formation of superconducting gap. We notice that the entropy of the underdoped 123 systems

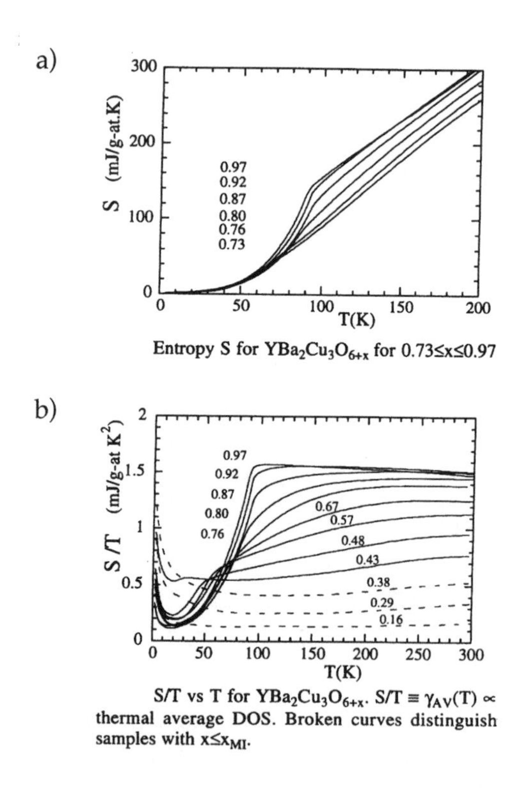

Figure 5.13: (a) Electronic entropy ΔS vs. temperature T for $YBa_2Cu_3O_{7-\delta}$ samples, with δ increasing as indicated. From Ref. [27]. (b) The spin susceptibility vs. temperature, for the same range of compositions.

is less than that of the optimally-doped one at temperatures well above T_c, and that the difference increases with increasing δ. Also, for the underdoped ones, the decrease in entropy below T_c is not particularly pronounced. All this implies that there is a lower density of electronic excitations in the underdoped compounds well above T_c; this could be due to a 'pseudogap' which develops into a real gap below T_c. Indeed, Loram et al., [27] could fit their electronic entropy data to a model of electrons with a d-wave gap of magnitude $\Delta = 1100\ (\delta + 0.03)$ K, where δ is the oxygen concentration or degree of underdoping. An analysis of the temperature-dependent spin susceptibility shows that there is a closely comparable pseudogap in the spin excitation spectrum as well.

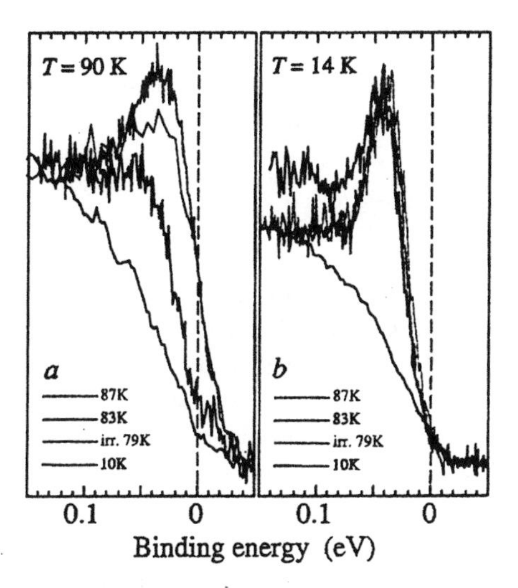

Figure 5.14: ARPES spectra of two single crystals of Bi-2212 superconductor, (a) $T_c \simeq 83$ K, (b) $T_c \simeq 10$ K. From Ref. [28].

A most interesting sidelight on the phenomenon is thrown by direct ARPES measurements of the single particle excitation spectrum. Ding et al., [28] and Loesser et al., [29] have seen clear evidence for a pseudogap in the spectral density of electronic states, above the superconducting transition temperature. Since the spectral density even for states close to the Fermi energy is broad, one has to focus on some feature, and compare it with a standard, e.g., photoemission from Pt. The results are shown in Fig. 5.14 for two samples of underdoped Bi-2212; one with $T_c = 83$ K and another with $T_c = 10$ K. There is a sensible shift of the feature (initial rise) away from zero binding energy even above T_c, suggestive of a pseudogap. This shift decreases with increasing temperature, and becomes unnoticeably small at a temperature of about 170 K for the sample with $T_c = 83$ K and at about 300 K for the $T_c = 10$ K sample. Another interesting feature is that the pseudogap magnitude depends on the direction along the planar Fermi surface, being a maximum at the Fermi surface intersection along the direction $\bar{M}Y$, and a minimum along the ΓY direction. As we shall see in the next section, the superconducting gap has the same angular dependence. This suggests strongly that the pseudogap is the precursor of the well-developed superconducting gap, and gives some support to a preformed pair picture. Why the pseudogap temperature T^* has the observed rapid increase with underdoping is not clear.

It is often stated or implied that pseudogap behavior is specific to underdoped *bilayer* cuprates. It is true that the temperature dependence of spin susceptibility, Knight shift, etc. in these systems show the pseudogap behavior most clearly. However, Batlogg et al., [47] have presented transport and other evidence for such a lower energy scale in single layer cuprates $La_{2-x}Sr_xCuO_4$ as well, and show that T^* rises rapidly from about 300 K to 600 K as x decreases from the optimum value of 0.15 to 0.10. It is thus possible that the lower energy scale is a generic feature of all metallic cuprates.

5.5 THE SUPERCONDUCTING STATE

The strange cuprate metal becomes a superconductor on cooling. I now describe some of its features. The superconducting state is a coherent superposition of electron pairs (Cooper pairs). This is known, say, from flux quantization or Shapiro step experiments. In the former, the magnetic flux entering a superconductor is seen to be quantized in units of $(hc/2e)$. The Cooper pair size or superconducting coherence length in-plane, namely ξ_{ab} inferred from magnetization and H_{c_2} near T_c is small, being about 15 Å at zero temperature. This is to be compared with the coherence length of clean low T_c superconductors, which is a few thousand angstroms. The difference cannot be attributed entirely to that in T_c. For example, $\xi = (\hbar v_F/k_B T_c)$ for a clean, weak coupling BCS superconductor. Clearly, the observed ξ_{ab} is smaller by a factor of five to ten than the value expected from this equation. The small size of ξ implies that order parameter fluctuation effects are not negligible. The pair size of 15 Å is about four times the average interelectron spacing, so that a simple picture of close pairs (bosons) well separated from each other is not realistic. The coherence length ξ_c perpendicular to the planes (along the c-axis) is very small, of order 3 Å or so. The smallness of this length in relation to ξ_{ab} reflects the extreme anisotropy of the superconductor; the layers are weakly coupled. Indeed, since ξ_c is smaller than the smallest relevant atomic length along the c-axis, namely the interlayer spacing $d(\simeq 7$ Å for single layer and $\simeq 12$ Å for bilayer compounds), and since the mobile electrons are in the planes, not much meaning can be ascribed to such a number. A continuum picture is not appropriate, and a model of layers coupled by pair tunnelling (Josephson coupled layers, as in the Lawrence-Doniach model [49]) is more realistic.

The other fundamental length scale characterizing the charged superconductor is the magnetic or London penetration depth λ, which measures the distance over which magnetic perturbations are screened. In the *ab* plane, this is about 1500 Å at $T = 0$ for optimally-doped bilayer superconductors; it is much longer along the c-axis (depending on the anisotropy of the superconductor). λ is related to the superfluid density n_s; we have $\lambda = (m^* c^2 / 4\pi n_s e^2)^{1/2}$ where m^* is the effective mass of carriers in the plane.

An interesting empirical, approximately linear relation between n_s as inferred from the measured λ_{ab}, the hole density x, and T_c has been pointed out by Uemura [50], and has been interpreted to support a model of bosons condensing at T_c. The superfluid density, e.g., for optimal doping $x \simeq 0.15$ in $La_{2-x}Sr_xCuO_4$ is $n_s = 0.06$ (assuming $m^* = m$). This is noticeably smaller than the number of holes (0.15) and much smaller than the number n of electrons (0.85) per unit cell. In a clean BCS superconductor, $n_s = n$. We see that the cuprates are 'soft' superfluids. It is quite likely that there are large zero point fluctuations of the order parameter phase that reduce n_s. Phase fluctuations, which in isotropic superconductors have an eV energy scale, being plasma oscillations, are known [7] to have very low energies of order 50-100 K in the layered cuprates, and can suppress n_s [33]. The low collective-oscillation frequency is another unusual feature of the cuprate superconductors.

The cuprate superconductors have a highly anisotropic order parameter, the gap $\Delta_{\vec{k}}$ varying a great deal with direction of the in-plane vector $\vec{k}$ on the Fermi surface. The anisotropy in the gap magnitude $|\Delta_{\vec{k}_{\parallel}}|$ can be directly seen in angle-resolved photoemission [34] which measures the quasiparticle energy $E_k = \sqrt{\epsilon_k^2 + |\Delta_k|^2}$. The gap is a maximum along the $\bar{M}Y$ direction and a minimum along the ΓY direction. (See Fig. 5.15.) The gap seems to vanish within experimental accuracy at the minimum, and to have a $\vec{k}$ dependence of the form $|\cos k_x - \cos k_y|$ for k_x and k_y in-plane. This is the simplest interpretation of the ARPES spectrum for Bi-2212 after the superlattice bands are eliminated in the analysis. The $|\cos k_x - \cos k_y|$ amplitude is what one expects for nearest neighbour pairs, in which the pair amplitude along the x direction has a sign opposite to the pair amplitude along the y direction. This lattice-symmetrized version of d wave spin singlet pair order (i.e., the $d_{x^2-y^2}$ symmetry order parameter) has a vanishing amplitude along a direction 45° to the x or y axes. This kind of order parameter is expected since the basic interaction between electrons on the same site is not attractive, but strongly repulsive. Also, the nearest neighbour electron correlation is antiferromagnetic (spin singlet). Since there is no gap against single particle excitations in some directions, their density is not exponentially small for $T << T_c$ as in s-wave superconductors, but is much larger, being linear in T.

While ARPES is sensitive only to the magnitude of the order parameter, Josephson junction related effects depend on the phase as well. A number of ingenious

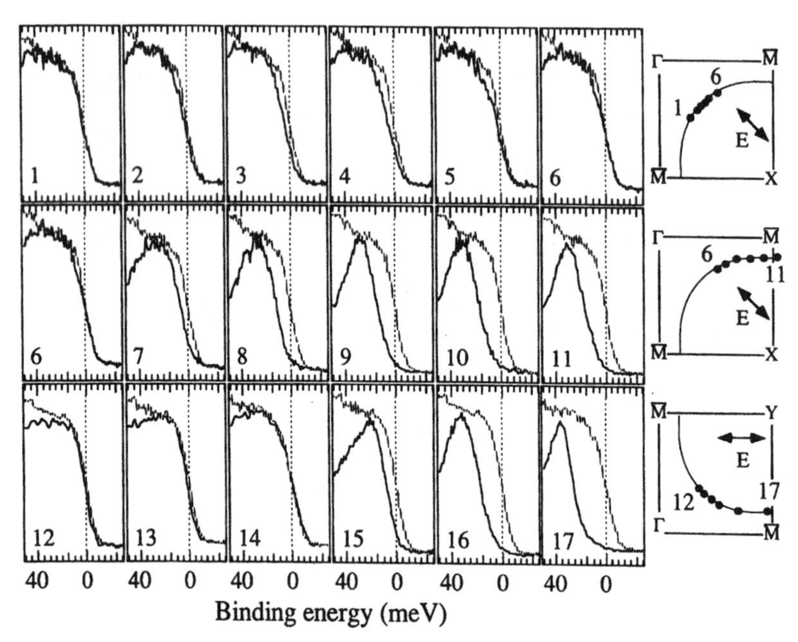

Figure 5.15: The magnitude of the energy gap for electronic states near E_F, at values of momenta correponding to the points (numbered 1 to 15) shown in the inset, which shows the Fermi surface.

and difficult atomic interface junction experiments [35] demonstrate the existence of an order parameter that changes sign. Again, the simplest interpretation is that the order parameter is $d_{x^2-y^2}$-like. In contrast, the tunnelling experiments of Dynes and coworkers [36], in which paired electrons in a 123 superconductor tunnel across the ab plane into a conventional nodeless superconductor (namely Pb) show a nonvanishing (if smaller than expected) tunneling current. For an ideal cuprate superconductor whose order parameter has equal positive and negative parts, this current ought to be zero. A reconciliation of this finding with other phase-sensitive measurements is an important issue.

A number of indirect measurements, e.g., the temperature dependence of the spin lattice relaxation rate [20], and the London penetration depth [37] are also consistent with a gap that has nodes.

An interesting feature of the superconducting state is that the quasiparticles are very well defined. This is clearly evident from the ARPES spectrum in the superconducting state [8] which is very sharp. A measurement of the Hall thermal conductivity in the mixed state [38] can be analyzed to argue that the quasiparticle mean-free-path in the superconducting state is very long.

The superconducting state has a magnetic collective mode [39] with energy $\simeq$ 41 meV, and wave vector $(\frac{\pi}{a}, \frac{\pi}{b}, \frac{\pi}{c'},)$, where c' is the distance between the adjacent layers in bilayer $YBa_2Cu_3O_7$. The mode is seen in inelastic neutron scattering.

5.6 THEORIES

Immediately after the discovery of superconductivity at liquid nitrogen temperatures, the central problem seemed to be the high transition temperatures. However, as the unusual properties of the metal have come to be established, an understanding of the normal state has emerged as the critical issue, with the expectation that a basis for high transition temperatures will emerge. The characteristically different properties for the metal suggest that it is not a Landau Fermi liquid, which is a generic description of a degenerate system of interacting fermions e.g., electrons. A number of non-Fermi liquid models have been proposed and explored. There are also a number of Fermi liquid-based approaches. At present, there is no agreement on a theoretical description. Perhaps none has the combination of conceptual depth, comprehensive scope, and calculational strength that is expected of a theory for such a strikingly unusual, experimentally overdetermined physical system. I shall therefore not describe theoretical approaches in any detail, but try to outline the ideas underlying some of these.

There are many reasons for believing that the metallic cuprates are not conventional or Landau Fermi liquids. A Fermi liquid (an interacting Fermi system) is qualitatively similar to a Fermi gas (free fermions). Quasiparticles with low excitation energy are long lived (decay rate is much less than excitation energy) and have the same quantum numbers as free fermions (e.g., momentum $\vec{k}$, spin σ). Physical properties have the same low temperature or energy dependence; e.g., the specific heat is linear in temperature, but the coefficient is different from that of a free Fermi gas. The observed metallic in-plane and insulating out-of-plane behavior of cuprates is difficult to reconcile with the existence of quasiparticles. The spectral density of a quasiparticle is roughly $\rho(\vec{k}, E) = z_k\,\delta(E - E_{\vec{k}}) + \phi_{\vec{k}}(E)$ where z_k is the coherent quasiparticle strength, and $\phi_{\vec{k}}(E)$ is a smooth function of energy. The coherent part ought to enable the electron to hop from layer to layer and form a band in the c direction. As pointed out earlier, this does not happen; there is no coherent propagation along the c-axis. Effectively, z_k seems to be very small or zero for E near the Fermi energy. Further, the spectral density as measured by ARPES is very broad; for E close to zero (Fermi energy) the width is 0.2 eV or more (Fig. 5.7). It is nowhere near a resolution-limited δ-function. The temperature dependence of, say, the spin-lattice relaxation rate is qualitatively unlike that of a Fermi gas (Section 5.3.5). Many other temperature and frequency dependences are also anomalous. Interestingly, the state-counting argument seems to be still valid; the area of the Fermi surface (number of occupied states) coincides with the number of electrons (Fig. 5.6b).

Perhaps the most radical non-Fermi liquid approach is due to Anderson [40]. It has been developed by him and his coworkers to describe the wide range of normal, as well as superconducting state, properties. A number of ideas proposed by him, e.g., the crucial importance of strong correlations, the sufficiency of a single band based on the Cu $d_{x^2-y^2}$ orbital, the broad (non-Fermi liquid state) implications of the observed metallic behavior, have become widely-shared presuppositions. One consequence of strong correlation is that, in general, spin- and charge-excitations have very different dynamics. In one dimension, it is known that the electron is a composite of spin- and charge-excitations which have different velocities, and is not a well-defined quasiparticle. The two dimensional normal metallic state is argued to be similar qualitatively to a one-dimensional interacting electron system called a Luttinger liquid. Quantitatively, it is suggested that the 2d system is a collection of marginally interacting Luttinger liquids.

(A Luttinger liquid in one dimension has only a cusp-like singularity in $n(k)$ at the free Fermi gas k_F, but has a power-law low-energy behavior that is qualitatively different from the free gas.)

In-plane confinement of electrons is due to the negligibly small amplitude for an electron to tunnel from one Luttinger liquid to another. Removing an electron produces phase shifts in all the other states, leading to an orthogonality of the excited state to the ground state. A phase-space argument for the T-linear electrical resistivity and for the T^2 Hall-angle can also be made. It has, however, proved difficult to develop these insights and arguments into straightforward calculations. For example, the low energy equivalence of a two-dimensional interacting Fermi system to a collection of one-dimensional Luttinger liquids is not established in detail.

A related, novel idea is that superconductivity is due to the kinetic energy gained by interlayer Josephson (or electron pair) tunneling [41]. The smallness of the normal-state c-axis optical conductivity $\sigma_{1c}(\omega)$ implies, via the relation between the frequency integral of $\sigma_{1c}(\omega)$ and average kinetic energy, that there is no energy gain due to the latter. This could be because single-particle tunneling is blocked (dynamically?). It is pointed out that coherent c-axis pair tunneling can restore some of the lost kinetic energy. Such a state is superconducting. Thus, interlayer pair tunneling is the cause of superconductivity, though intralayer correlations can help [41]. The mechanism is compatible with various kinds of order parameter symmetry including d-wave. The observed lack of relation between $(d\rho_{ab}/dT)$ and T_c suggests that T_c might be not intrinsic to a layer but might be an interlayer phenomenon. The Josephson coupling kinetic energy gain estimated from the known London penetration depth λ_c turns out to be much smaller (as also noted in Ref. [42]), for example, than the condensation energy. The latter has been measured thermodynamically [27] in some cuprates. Zero-point phase fluctuations, in-plane correlations and bilayer coupling effects could be contributory causes.

Another non-Fermi liquid model is the coupled fermion-boson gauge theory developed by several authors [43], especially Lee and coworkers. In a strongly correlated lattice system, there is either a spin-1/2 electron on a lattice site (spinon, a fermion) or no particle (holon, a boson). At any site, therefore, the number of spinons and holons adds up to one. When an electron hops with amplitude t_{ij} from one site i to another site j, a spinon is removed from i and a holon added there; the reverse happens at site j. In addition, there is an antiferromagnetic coupling J_{ij} between spins at neighboring sites. This $t-J$ model, with

$$H = \sum_{i,j,\sigma} t_{ij}\, f^+_{i\sigma} f_{j\sigma} b_i b^+_j \; + \; J_{ij} \sum_{i,j} f^+_{i\alpha} (\vec{\sigma}_i)_{\alpha\beta}\, f_{i\beta} \cdot f^+_{j\gamma} (\vec{\sigma}_j)_{\gamma\delta} f_{j\delta}$$

and the local constraint $\sum_\sigma f^+_{i\sigma} f_{i\sigma} + b^+_i b_i = 1$ has been extensively investigated, especially in two dimensions. The mean-field-theory phase diagram, as a function of hole density $< b^+_i b_i >= x$, has features qualitatively similar to that of the cuprates [44], with mean-fields that can be related to the anomalous metal, spin-gap, and superconducting phases in the appropriate regions of the (x, T) plane. The anomalous properties of the metal originate from chiral gauge fluctuations coupled to the spinons and holons. The amplitude $< f^+_{i\sigma} f_{i\sigma} >= \xi_{ij} = \xi_o$ $\exp(a_{ij})$ has a phase a_{ij}. The transverse part of this, the one that does not vanish on taking a lattice circuit, represents chiral, out-of-plane fluctuations of the spin orientations from perfect Ne′el order. These couple to holons and spinons like fluctuating magnetic fields. The resulting system has a large Fermi surface and low-energy behavior not unlike that of cuprates in some respects, including qualitatively different in-plane and out-of-plane behavior. However, it is not obvious that chiral fluctuations are the most significant gauge-like degrees of freedom at sizeable doping. The theory is uncontrolled, with no small parameter, and too many degrees of freedom. Further, the resulting local constraints are implemented approximately. The effect of fluctuations appears large enough to qualitatively change parts of the phase diagram and to greatly suppress mean-field transition temperatures. Many properties are difficult to calculate reliably. The superconducting transition is intrinsic to the two dimensional system, and is described as a Bose condensation of holons, in addition to spinon propagation.

A well-developed Fermi liquid approach is the semi-phenomenological spin fluctuation theory of Pines and coworkers [45]. See also Moriya [46]. Here, the spin fluctuation spectrum is parametrized so as to fit the spin lattice relaxation data; the exchange of these spin fluctuations leads to high temperature superconductivity in the d-wave channel. A number of properties, especially magnetic ones both in the normal and superconducting states can be rationalized. However, there are important limitations of both principle and practice. The theory is in the Fermi liquid family, and thus cannot in any obvious way lead to the observed in-plane metallic and out-of-plane insulating behavior. It is a weak coupling theory in which the system is near a spin-density-wave instability, whereas there is clear evidence for strong local correlation and local magnetic moments as origins of the magnetic correlations. Internally, there are questions such as inconsistency with electron-spin gradient coupling [47], reliability of the postulated spin fluctuation spectrum [48], and difficulty in describing transport behavior [49]. The difficulty in reconciling the constancy of $(d\rho_{ab}/dT)$ for systems with very different T_c within a spin fluctuation exchange model has been mentioned earlier.

The above summary of the theoretical situation is certainly sketchy and quite likely prejudiced. It is meant to indicate the variety of approaches, as well as the inability to meet expectations raised by experiment and by earlier successful theories in condensed matter physics.

I now turn very briefly to the much larger subject of oxides and correlations.

5.7 FUTURE

Perhaps the nature of cuprates will soon be understood in depth. Certainly, the attention given and the experimental knowledge available suggest such a possibility. However, this is probably one of our first glimpses into "the second book of solid state physics" (a phrase due to J.R. Schrieffer).

In this book, dynamical consequences of correlations and interactions have to be included from the very beginning. Often low-dimensional (fluctuation) effects are also crucial. There is a whole zoo of unusual systems and phenomena that probably do not fall within the gamut of ideas used so far for describing Fermi liquid systems. The cuprates are the latest and, perhaps, best-studied examples. Other oxides, especially transition metal oxides [50] show a variety of electronic phases and instabilities, e.g., metal insulator transitions [51, 3], charge ordering and mixed valence, colossal magnetoresistance involving strong coupling between spin order and electron motion [52]. Old systems are being dusted off the shelf and found to exhibit strange behavior; new ones are being synthesized, some to order (e.g., spin chains and ladders) and often to spectacular effect. Because of difficulties with the materials, our knowledge often grows slowly. Our theoretical understanding remains poor.

Another extensively investigated family is rare earth metals and alloys exhibiting heavy fermion behavior [53]. Here again there is an interplay between superconductivity and antiferromagnetism, a variety of phases including small gap semiconductors (Kondo insulator), non-Fermi liquid ground states and metamagnetism. Strong correlations are clearly crucial, and there is a sense of the growing inability of theory to cope with experiment.

Organic metals represent yet another domain [54] where novel phenomena abound, aided by low dimensionality, often weak intersite coupling, and correlations.

It is apparent that extant resources of many body physics are being stretched beyond limits. Perhaps the ideas and methods needed will first be conclusively established for the cuprate superconductors, and this will be the beginning of a new phase in condensed matter physics.

REFERENCES

[1] G. Bednorz and K.A.Müller, Z. Phys. B **64** 189 (1986).

[2] There are a number of review articles, books and conference proceedings marking the development of the field. A few recent ones are the following: *Proceedings of the International Conference on Materials and Mechanisms of Superconductivity: High Temperature Superconductors IV, Grenoble, 1994*, (North Holland, 1994 Vols.I-V.) (Physica C **235-240** 1994), *Physical Properties of High Temperature Superconductors, Vols. I to V*, edited by D.M. Ginsberg (World Scientific, 1991-96).

[3] For a review of interacting electron systems in one dimension, see H J Schulz, *Proceedings of Les Houches Summer School LXI*, edited by E. Akkermans, G. Montambaux, J. Richard, J. Zinn-Justin (Elsevier, Amsterdam, 1995), p.533. For a comprehensive review of the Hubbard model in large dimensions, see A. Georges, G. Kotliar, W. Krauth and M.J. Rozenberg, Rev. Mod. Phys. **68** 1 (1996).

[4] Y.F. Fan, O.K.C. Tsui and N.P. Ong (unpublished, see Ref. [5]).

[5] N.P. Ong, Y.F. Fan and J.M. Harris, in *CCAST Symposium on High Temperature Superconductivity and* C_{60}, Beijing 1994 (Gordon and Breach, New York, 1995).

[6] Y.S. Ando et al., Phys. Rev. Lett. **77** 2065 (1996).

[7] S. Uchida, K. Tamasaku and S. Tajima, Phys. Rev. B **53** 14558 (1996).

[8] M. Randeria et al., Phys. Rev. Lett. **74** 4951 (1995).

[9] H. Ding et al., Phys. Rev. B **54** R9678 (1996).

[10] H. Takagi et al., Phys. Rev. Lett. **69** 2975 (1992).

[11] P.W. Anderson, Science **256** 1526 (1992); Proc. Nat. Acad. Sci. **92** 6668 (1995).

[12] H.L. Kao et al., Phys. Rev. B **48** 9925 (1992).

[13] N. Kumar and A.M. Jayannavar, Phys. Rev. B **45** 5001 (1992).

[14] J.M. Harris, Y.F. Ren and N.P. Ong, Phys. Rev. B 46 14293 (1992).

[15] P.W. Anderson, Phys. Rev. Lett. **67** 2992 (1991).

[16] A. El Azrak and N. Bontemps, Phys. Rev. B **46** 9846 (1994).

[17] L.D. Rotter et al., Phys. Rev. Lett. **67** 274 (1991).

[18] X.-G. Wen and P.A. Lee, Phys. Rev. Lett. **76** 503 (1995).

[19] Z.X. Shen and D.S. Dessau, Phys. Rep. **253** 1 (1995).

[20] See for example the review by D. Brinkmann and M. Mali, in *NMR, Basic Problems and Progress 31*, (Springer Verlag, Berlin and Heidelberg, 1994), p. 172.

[21] T. Imai, C.P. Slichter, K. Yoshimura and K. Kosuge, Phys. Rev. Lett. **70** 1002 (1993); M. Matsumara et al., J. Phys. Soc. Japan, **65** (1996).

[22] B.S. Shastry, Phys. Rev. Lett. **63** 1288 (1989).

[23] R.E. Walstedt, B.S. Shastry and S.-W. Cheong, Phys. Rev. Lett. **72** 3610 (1994).

[24] S.M. Hayden et al., Phys. Rev. Lett. **76** 1344 (1996).

[25] J.M. Tranquada et al., Phys. Rev. B **46** 5561 (1992).

[26] J. Rossat-Mignod et al., Physica C **185-189** 86 (1991).

[27] J.W. Loram et al., Phys. Rev. Lett. **71** 1740 (1993); Physica C **235-240** 135 (1994).

[28] H. Ding et al., Nature **382** 53 (1996).

[29] A.G. Loesser et al., Science **273** 325 (1996).

[30] B. Batlogg et al., Physica C **235-240** 130 (1994).

[31] J.M. Lawrence and S. Doniach, *Proceedings of the 12th International Conference on Low Temperature Physics, Kyoto, 1970* edited by K. Kanda (Keigaku, Tokyo, 1970).

[32] Y.J. Uemura et al., Phys. Rev. Lett. **62** 2317 (1989).

[33] B.K. Chakraverty and T.V. Ramakrishnan (to be published).

[34] H. Ding et al., Phys. Rev. Lett. **74** 2784 (1995).

[35] D.A. Wollmann et al., Phys. Rev. Lett. **71** 2134 (1993); C.C. Tsuei et al., Phys. Rev. Lett. **73** 593 (1994); J.R. Kirtley et al., Phys. Rev. Lett. **76** 1336 (1996).

[36] A.G. Sun et al., Phys. Rev. Lett. **72** 2267 (1994); Phys. Rev. B **54** 6734 (1996).

[37] W. Hardy et al., Phys. Rev. Lett. **70** 399 (1993).

[38] K. Krishana, J.M. Harris and N.P. Ong, Phys. Rev. Lett. **75** 3529 (1995).

[39] H.F. Fong et al., Phys. Rev. Lett. **75** 321 (1995).

[40] P.W. Anderson, Science **235** 1196 (1987); G. Baskaran, Z. Zou and P.W. Anderson, Solid State Comm. **63** 973 (1987); D.G. Clarke, S.W. Strong and P.W. Anderson, Phys. Rev. Lett. **72** 3218 (1994); See also Ref. [11] and P.W. Anderson, forthcoming book.

[41] J.M. Wheatley, T.C. Hsu and P.W. Anderson, Phys. Rev. B **37** 5897 (1988); S. Chakravarty, S. Sudbo, P.W. Anderson and S. Strong, Science **261** 337 (1993).

[42] A.J. Leggett, Science **274** 587 (1996).

[43] See P.A. Lee and and N. Nagaosa, Phys. Rev. E **46** 5621 (1992); M.U. Ubbens and P.A. Lee, Phys. Rev. B **49** 6853 (1994); N. Nagaosa, Phys. Rev. B **52** 10561 (1995).

[44] See, for example. H.Fukuyama, Physica C **235-240** 63 (1994) and references therein.

[45] D. Pines, Physica C, **235-240** 113 (1994) and references therein.

[46] T. Moriya, Y. Takahashi and K. Ueda, J. Phys. Soc. Japan **59** 2905 (1990).

[47] J.R. Schrieffer, J. Low Temp. Phys. **99** 397 (1994).

[48] K. Levin, Q. Si and Y. Zha, Physica C **235-240** 61 (1994) and references therein.

[49] R. Hlubina and T.M. Rice, Phys. Rev. B **51** 9253 (1995); ibid **52** 13043 (1995).

[50] N. Tsuda, K. Nasu, A. Yanase, K. Siratori, *Electronic Conduction in Oxides*, (Springer Verlag, Berlin, 1991).

[51] See for example, N.F. Mott, *Metal Insulator Transitions*, (Taylor and Francis, London, 1974, 1990).

[52] See for example, Y. Tokura, J. Appl. Phys. **79** 1 (1996).

[53] See for example P. Coleman, Physica B **206** and **207** 872 (1995).

[54] Proceedings ICSM94, *Synth. Metals and Organic Conductors*, edited by J.P. Farges (M. Dekker, New York, 1994). P.M. Chaikin (to be published).

5.8 DISCUSSION

Session Chair: Philip Anderson
Rapporteur: Olexei Motrunich

ANDERSON Before we let the floor open to general discussion, I would like to ask several people to comment. Although we've seen a lot of experimental facts, perhaps we should start with Phuan Ong who is a real experimentalist.

ONG: The discovery and subsequent research in this area point to several things of relevance to physics and the education of graduate students. One thing that was very surprising was that in this country there was a neglect of the ability to synthesize crystals. This has been neglected for 20 years, and when superconductivity was discovered we were essentially 10 years behind the Japanese. This was brought out very clearly by the level of competition that evolved in the next 5 years. The other problem is the education of graduate students. The phenomena are becoming so complicated that it is not sufficient to just teach students how to measure things but to look at them and be able to understand what theorists say. I don't think we do this very well. Graduate students who go to experimental physics are usually not taught theoretical physics very much and this is a deficiency that, perhaps, we should look into.

RAMAKRISHNAN: I agree and I think that the gap hasn't narrowed as far as the crystal growth is concerned.

ANDERSON: I was going to say that Phuan is being a little bit excessively modest. As far as I remember, in spite of the Japanese lead, Phuan found somewhere a French-speaking Chinese postdoc who for a while had the best crystals of YBaCuO in the world. And that was one thing which let us have a certain amount of lead in the theoretical interpretation. But admittedly he was not American trained, he was French trained and Chinese. I couldn't agree more. Perhaps I should call on a theorist next.

BASKARAN: Ramakrishnan gave a beautiful summary of the experimental results and Phil made some important opening remarks. In two or three minutes I will summarise some of my impressions arising from the involvement in the high T_c game for the last 10 years.

Any good and fundamental development in condensed matter physics leads to rethinking of old problems, creates new notions and new materials. This has happened with the discovery of superconductivity in cuprates. Many laboratories, all over the world, started looking at oxides, sulphides, nitrides and others with the primary aim of discovering higher T_c, of course. It has paid off: doped $BaBiO_3$, an old low T_c oxide became a high T_c oxide with K doping. It is unlikely, in the absence of high T_c fever, that they would have doped the fullernes and gotten the high T_c in Rb_3C_{60}. The compound LuNiBC, which culminated from the discovery of a related compound by the TIFR group, clearly has its origin in the high T_c search. And so is the low T_c material $SrRuO_3$, with all its potential non-fermi liquid physics. The field of quantum magnetism was revived with the study of old nickelates and new compounds - thanks to Anderson's idea of RVB in this context together with Duncan Haldane's idea of topological terms, gaps etc. From giant we went on to colossal magneto resistance. The rich world of both inorganic and organic materials was exposed to physicists.

In theory, as Phil pointed out there is perhaps a mini revolution that is going on. The notion of spin-charge separation and quantum number fractionization which came along with the quantized Hall effect and polyacetylene got revived in a more fundamental and deeper way in the context of high T_c cuprates. The gauge theory ideas that came along with the RVB theory brought out the rich physics and mathematics content of the problem. It spelled out that we have to start thinking afresh. The old notion of anyons of Wilczek and others, Polyakov and collaborators' idea of Hopf term (which is related to the Chern Simons term) along with Laughlin's idea of anyon superconductivity initiated intense collaboration between the condensed matter and field theory community.

The idea of tomographic Luttinger liquid of Phil, that has a deeper connection to the idea of anti bound states of some of us, was a beautiful and natural generalisation of Haldane's notion of one dimensional Luttinger liquid. The Fermi surface phase shifts in the cuprates and the strange exclusion properties of the spinons and in general the novel quasi particles have a non trivial description in terms of Haldane's new notion of exclusion statistics.

All the new compounds mentioned earlier, in some form or other exhibit departure from Fermi liquid behaviour. It is likely that the non Fermi liquid states that exhibit chiral symmetry breaking and anyon superconductivity are already there among these families in some corner of the phase diagram.

Another lesson that I have learned, which I will put in the form of a conjecture is: 'Non Fermi liquid states are seats of high temperature superconductivity'.

Let me put it on record and say, as Phil has pointed out elsewhere, that the basic brick work for the high T_c theory, both for the anomalous normal state and the superconducting state has been done. One need not wait for the beginning of the next century. As always, the future will bring quantitative development and of course new features and surprises and new problems will emerge.

ANDERSON: There are a couple of people, who haven't been closely associated with the group at Princeton, or not completely associated with the group. One of them is Elihu Abrahams, and then I would like to call on Frank Wilczek.

ABRAHAMS: I want to emphasize a point made in a rather restrained way by our chairman (not known for restraint) just before Ramakrishnan's excellent talk. The danger of having a talk on high-temperature superconductivity is that one may give the impression that once it is "solved" then there are no further major problems in the quantum theory of condensed matter.

Of course it is obvious why at this Princeton symposium it is that high-temperature superconductivity is singled out: There is always the hope that NJ Transit will replace the Dinky with the modern maglev train.

But one should not be misled. There is a wide variety of physical phenomena which are determined by correlated behavior driven in one way or another by strong electron-electron interactions - and we have no systematic way of treating them. These problems range from the (apparently mundane) one of the properties of doped semiconductors, through a variety of metal-insulator transitions, properties of metal oxides, ... , all the way to the physics of magnetism.

To quote an earlier remark by Sasha Polyakov: "You can not use obsolete methods to treat this new physics."

ANDERSON: Thank you. I couldn't agree more with everything you've said.

WILCZEK: I want to congratulate the speaker on his very clear presentation, which emphasized what to me is the central point, that there is almost certainly some qualitatively new physics—a new universality class—at work in the cuprate "high T_c" materials. It is quite impressive the normal state exhibits anomalous behaviors that are characteristic of a whole class of cuprate-based materials, but which appear to contradict the Landau Fermi liquid theory. Since the Landau theory has profound roots, and in conjunction with renormalization group ideas provides the conceptual foundation for much of solid-state theory, there is a big challenge here, and an opportunity to learn something fundamentally new.

There is a glorious history of concrete connections between between these areas and condensed matter physics, which I think goes well beyond what might be anticipated from generic considerations of the 'unity of knowledge', and which I believe is far from over.

A prominent example of these connections, in which I have been especially involved, and which continues to be the focus of much activity, involves the flow of gauge theory ideas back and forth between the different fields. A simple but profound circle of ideas around Aharonov-Bohm type effects and Berry phases has been very fruitful. These ideas are so simple that one is almost embarrassed to call them 'gauge theory', but their core concept is certainly the significance of non-integrable phases, which is the heart of gauge kinematics.

More substantial, from a theoretical point of view, is the use of ideas around anyons and Chern-Simons theory in the theoretical description of the quantum Hall complex of states. This has been a tremendous success story, and I think it is fair to say that there are now quite a few striking phenomena known, that would have been very difficult to discover without these insights. It's remarkable—astonishing, really—to see the sorts of symmetry we think of as fundamental in particle physics arising here in quite a concrete, tangible way as an emergent property of physical systems following microscopic laws of a less symmetrical character.

Will the cuprates add another chapter to the gauge story? There have been (at least) two rather different attempts to use gauge theory ideas in the context of cuprate superconductors. One involves a logical extrapolation of ideas from the quantum Hall effect. Its centerpiece is a brilliant idea from Laughlin, exploiting flux-trading ideas some of us had toyed with earlier, which defines a qualitatively new road to superconductivity for 2-space dimensional systems. This idea defines a valid theoretical universality class, which will—if there is any justice in the world—be realized in some class of materials. It leads to a specific prediction of T-violating effects, which can be estimated semi-quantitatively. Unfortunately experiments to search for such effects in the cuprates have yielded negative results, so it now appears almost certain that these ideas, at least in their simplest form, do not apply to the cuprates.

The other major attempt to apply gauge theories to the cuprates involves regarding constraints—that is, the Lagrange multipliers which implement constraints—as generating gauge fields. Specifically, one can model a no-double occupancy constraint, an idealization of strong short-range repulsion, in this way. Unfortunately the resulting gauge theories are of an unconventional form and very strongly coupled, so it appears difficult to make progress with them. I have not found anything in the literature on this subject convincing—there seem to be uncontrolled approximations and arbitrary assumptions galore—but although I have studied several of the most cited papers perhaps I have missed something here.

Finally I would like to mention a set of phenomena that the speaker did not mention, although it is quite striking and could turn out to be a real 'sleeper'. That is, the appearance of clear quasi-one dimensional features in the electronic properties of these materials. Anderson, if I understand him correctly, has been advocating for some time, at least partly on phenomenological grounds, that the anomalous behavior in the normal state must reflect effectively one-dimensional behavior, with enhanced infrared singularities in momentum space. At the same time, neutron diffraction measurements in the copper oxide planes of some of the materials indicate the existence of static or quasi-static narrow one-dimensional stripes in real space. More generally, I think we must be alive to the possibility of phases where inhomogeneous electronic structure, including effective reduction of dimensionality, occurs spontaneously.

RAMAKRISHNAN: There are stripe structures and there are systems without stripe structures.

CHAKRAVARTY: I very much hope that our speaker is right; that this subject doesn't continue beyond the year 2000. It is indeed true that more than 100,000 papers have been written on this subject. But I do want to make a comment. The problem is that I don't think that we have even developed new mathematics that we need to treat some of these systems in a really precise way. Things that I learned in graduate school are totally obsolete. When I teach solid state physics I don't know what to teach anymore because what I learnt doesn't apply, and I don't know what I should be teaching so that they can apply to understand this class of materials. And it is not just high-T_c cuprates, but a whole class of superconductors which have this unusual behavior.

ANDERSON: I agree with that last point very much. I think that there is a tendency to go on teaching graduate students one thing while we are all working using entirely different methods. Everyone who is actually working in the field is doing something else, something that is not taught to graduate students at all. That's a sign of a very healthy field but it is not a healthy thing towards our graduate students.

CHAPTER 6

THE ONGOING REVOLUTION IN MEDICAL IMAGING: GLORIOUS PAST, PRECARIOUS FUTURE

BRUCE J. HILLMAN, M.D.

Department of Radiology
University of Virginia School of Medicine and Health Sciences Center
Charlottesville, VA

The past hundred years has witnessed a development of a new medical science—diagnostic imaging. Diagnostic imaging has grown in its importance because of the collaboration of physicists and physicians in the development of new, clinically applicable technologies. The pace of important, clinically-relevant, technologic innovation has accelerated during the past two decades as innovations in unrelated fields have been brought to bear. The result has been the application of x-radiation, radionuclide emission, ultrasound, and nuclear magnetic resonance to medical imaging, as well as their digitization and computerization to bring out better diagnostic detail and expand their clinical applicability.

Few sciences can attribute their beginnings to a single event. However, medical imaging is one such science. During the evening of November 8, 1895, working after hours in his darkened home laboratory because his administrative responsibilities at the Physical Institute of the University of Wurzburg had reduced his laboratory time at work, Wilhelm Conrad Roentgen noticed that activation of his Hittorf tube caused fluorescence of a barium platinocyanide screen. The screen was too far away for this to be the result of cathode rays, which already had been well-investigated. Roentgen correctly assumed that the phenomenon was caused by a new kind of ray—the x-ray. In just 6 weeks, in a simple but brilliant set of experiments, Roentgen discovered and reported on most of what we know today about x-rays. In January 10, 1896, Roentgen reported on 17 properties of x-rays, among them such clinically important attributes as: the relationship between the density of the substance being traversed by the x-ray beam and the amount of blackening of the receptor; the ability of x-rays to induce fluorescence in a range of substances that might serve as receptors; the absence of refraction and reflection; and the imperviousness of x-rays to a magnetic field.

The discovery was transmitted worldwide. Both medical and popular dissemination followed quickly. Entrepreneurs, such as Thomas Edison, sold home x-ray units for entertainment purposes. The first medical applications were reported in 1896 and more than 1000 articles were published about the x-ray in that year. In one of those, Princeton physicist William Magie reported the x-ray filtering properties of aluminum that have been applied to reducing patients' medical dose absorption to this day. The first textbook of medical imaging, *The Roentgen Rays in Medicine and Surgery*, was published by Dr. Francis Williams in 1901. Even allowing for the important advances that have been facilitated by the introduction of imaging modalities based on other physical principles—such as ultrasound and nuclear, magnetic resonance—the x-ray remains the critical physical foundation for medical imaging.

Shortly after Roentgen's discovery, Henri Becquerel discovered the phenomenon of natural radiation emanating from potassium uranyl sulfate. Pierre and Marie Curie discovered more powerful radiation emitters, as well as describing both induced radioactivity and the basic principles of radioactive decay. A critical discovery to the eventual medical application of radionuclides was the discovery of the Joliot-Curies that the bombardment of some elements by atomic particles could produce new or "synthetic" radioactive elements. Collaborative efforts by physicists and physicians led to the application of this principle to design efficient medical radionuclides that had half-lives appropriate to studying physiologic processes and energies that were appropriate for clinically useful detectors.

Over the last hundred years, remarkable developments by physicists and physicians have led to better means of x-ray generation, clinically applicable radionuclides, more efficient receptors, higher resolution, higher contrast images, and better means of image display that allow physicians to interact with the images to bring out previously hidden diagnostic details. New imaging techniques allow for the new medical art of image-guided interventional procedures. Examples include stereotactic needle breast biopsy, balloon catheter angioplasty of narrowed blood vessels, image-guided drainage of abscesses, and a host of other procedures that have reduced patient morbidity, complications, time lost from work and social activities, and health care costs. Recently, applications of radionuclide imaging such as PET (positron emission tomography) scanning, as well as high-speed functional magnetic resonance imaging, are beginning to allow medical imaging to proceed to the investigation of not only gross anatomy and pathology, but also metabolism and microscopic processes.

A major watershed event in this progression of capabilities was a discovery by Godfrey Houndsfield in 1972. Making applicable the principle of backprojection, first detailed by Cormack in 1964, Houndsfield connected together the output of x-ray transmission to a primitive computer. This allowed the reconstruction of many "images" (really portions of images), made by a pencil-thin x-ray beam, into a coherent cross-section of the body. The result was images with much higher contrast sensitivity than had been achieved previously, allowing for better discrimination of abnormalities hidden in normal tissues. The clinical application, computed axial tomography, or CAT scan (now referred to as CT), has revolutionized medical imaging, allowing for much better diagnosis, the development of new image-guided interventional procedures, and setting the physical groundwork for the development of other new cross-sectional imaging technologies.

Two of these new technologies are ultrasound and magnetic resonance imaging. The origins of medical ultrasound can be traced to the 18th century discovery by Spellanzani that bats navigate by emitting and receiving reflected sound waves. The discovery of the piezoelectric effect by the Curie brothers in the late 19th century allowed for the eventual production of emitter-detector transducers that could be applied to medical use. The need to find sunken ships and eventual undersea combat applications renewed interest in ultrasound and led to the post-World War II application of these now more advanced technologic principles to medical imaging. Developments over the past twenty years have led to enormous diffusion of medical ultrasound into hospitals, clinics, and physicians' offices. Ultrasound is now used for both diagnostic and image-guided therapy; Doppler ultrasound now allows for the interrogation of blood vessels and other flowing bodily fluids, permitting less invasive diagnosis of disease than with such invasive technologies as angiography, as well as the quantification of flow.

Magnetic resonance imaging derives from the findings of Bloch, Purcell, and others that certain nuclei with unpaired nuclear elements respond to strong magnetic fields (nuclear magnetic resonance). A majority of the atoms align with the field and emit signals when their position and spin are impinged upon by radiofrequency emissions. These signals can be received and oriented into a geometry reflecting their location and molecular environment in the body through applying a fast Fourier transform, a mathematical concept first detailed by Princeton mathematics professors, Cooley and Tukey in 1965. Paul Lauterbur first detailed, in 1973, how nuclear magnetic resonance might be applied to medical imaging. Raymond Damadian built the first scanner, "Indomitable," in 1977 and, after hours of acquisition, produced a fuzzy image of the human brain. Commercialization and medical and physical developments since that time have exploited magnetic resonance imaging. Again, magnetic resonance imaging represents an advance in contrast resolution that surpasses even CT scanning in its ability to discriminate subtle abnormalities. It employs no ionizing radiation and often can be employed without the injection of contrast media (x-ray dye), usually required for CT scans. As noted above, new applications of magnetic resonance promise exciting capabilities in the investigation of metabolism and biologic microstructures.

The development of medical imaging has represented an exciting opportunity for physicists, physicians, and a host of related scientists to work together to bring to society important practical applications. The scientific foundations for new advances are more plentiful than ever. Yet there are important reasons for concern that the collaboration of physicists and physicians may be less effective, through no fault of its own, than it has been in the past. These reasons relate to important changes in the organization and financing of health care. The cost of health care has become, for many, intolerable. Health care costs have been increasing faster than other elements of the economy. Much of the higher costs have been laid at the doorstep of new technology. Health care is more often being delivered by large corporations, for which cost and quality tradeoffs are becoming the norm. In this climate, health care providers will have a harder time gaining access to new, potentially valuable technology unless it can first be rigorously shown that advances in medical imaging will also bring lower costs, significantly improve health, or attract populations of patients to the provider. The technologic, regulatory, and economic challenges facing a potentially valuable technology are exemplified by a current collaboration of the Princeton University Physics Department and the Department of Radiology at the University of Virginia. Princeton researchers have developed a practical means of hyperpolarizing noble gases for use as magnetic resonance contrast agents. They have developed a company to commercialize their technology and are working with the radiologists at UVA to investigate clinical applications and perform the clinical trials required by the Food and Drug Administration. These trials will be coordinated by a company owned by the UVA Department of Radiology. The process is an expensive one; given the greater regulatory constraints and the new demands of the marketplace, the challenge to become clinically successful is far greater than it was for the entry of CAT scanning or even magnetic resonance imaging. However, as Horace noted, "Adversity reveals genius." While future medical imaging innovations will face greater scrutiny, valuable new technologies produced by the proven collaboration of physicists and physicians will continue to progress to clinical applicability and thereby improve health.

GENERAL REFERENCES

Eisenberg, R.L. *Radiology—An Illustrated History*. Mosby, 1992.

Gagliardi, R.A. and McClennan, B.L. *A History of the Radiological Sciences*. Vol. I. *Diagnosis*; Vol. II. *Radiation Physics*. American College of Radiology, 1996.

6.1 DISCUSSION

Session Chair: William Happer
Rapporteur: Chris Erickson

CHARPAK: Following Dr. Bruce Hillman's talk on medical imaging, I wish to add a few comments on the state of the art in medical imaging.

In high energy physics laboratories an active research on detectors is driven by the needs appearing with every new generation of accelerators. The present trend is towards higher rate capability, better accuracy of the particle localisation, and higher capacity for dealing with very high multiplicity in the collisions. The new detectors, developed at high cost, have applications in all fields where the imaging of ionising radiations plays a role, like x-ray radiology. They can make a decisive impact by permitting a sizeable reduction of the radiation dose inflicted on patients in diagnostic radiology. They should permit high-accuracy localisation of each absorbed x-ray, to the order of 100 microns, with sometimes a measurement of the energy of the quanta.

The results which have been obtained in the pioneering work of a group from the Novosibirsk Budker Institute illustrate the potential of a method where the photons are detected one by one [1]. By exploiting the properties of multiwire chambers and drift chambers, a digital radiography system has been built which has demonstrated the possibility of drastic dose reduction.

Figure 1 shows a radiography of a thorax. The considerable dynamic range is shown by the images of the soft tissues and the rachis, stored in the computer at the same time. This eliminates the frequent need for repetition of pictures when highly contrasted tissues are being examined.

An English group working in collaboration with the Novosibirsk physicists has evaluated the dose reduction for large volume subjects, with low contrast, to factors of 5 to 300.

A firm in Paris has undertaken a clinical evaluation of the instrument, in a pediatric hospital [2]. About 200 patients have been examined, comparing the data from the electronic data with the observation obtained from a commercial radiography.

The analysis shows a strong reduction of the radiation doses inflicted on the patients for a similar amount of information:

- a factor 10 for front and 16 for profile images of the rachis.
- a factor 16 for front and 20 for profile images of the "basins."
- a factor 4 for front and 16 for profile images of the thorax.

The detector, built in Novosibirsk, is made of a pressurised xenon absorption space from which the electrons are transferred to a chamber where all the detecting anode wires are pointing to the x-ray source. Every wire is connected to a scaler. This permits high counting rates without any parallax error. Two slits in a lead foil define a fan beam exploring the patient with the measurement

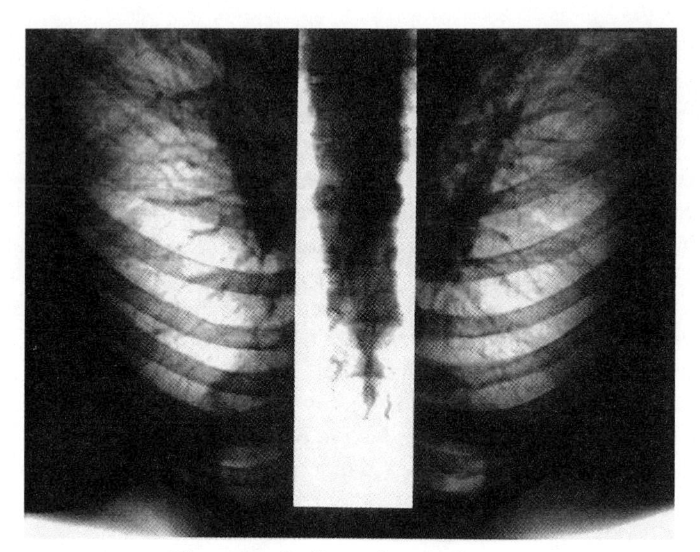

Figure 6.1: Radiography of a thorax.

at one position lasting 15 msec. The system scans the patient in 8 sec. The advantage of the collimation is that it eliminates much of the scattered radiations. The disadvantage of the scanning is that this type of radiography is incompatible with fast moving organs. The accuracy is better than a millimeter but worse than with film. It is very advantageous for the examination of pregnant women, of regions close to the gonads, and the study of scoliosis with prepubescent girls. The benefits are considerable for cases requiring frequent examinations like the survey of osteochondritis, epiphysiologisis, sequels of luxations, control under plaster . . . where the decrease in accuracy is irrelevant.

It is instructive to see in a pediatric hospital some examples of strong irradiations:

1. Jonathan M. born 1991, 199 radiologies (130 lungs + 50 abdomens + . . .)
2. Xavier N. born 1990, 117 radiologies (30 lungs + 14 abdomens, severe digestive pathology)
3. Alexander P. born 1991, 200 radiologies (140 lungs + 50 abdomens + . . .)
4. Raphael B. born 1990, 239 radiologies (190 lungs + . . .)
5. Mourtada B. born 1994, 194 radiologies (159 lungs + 31 abdomens + . . .)

The children have been subjected to irradiations 30 to 50 times larger than those given to the average population in the hospital, at a critical age.

This is why I think that even if we have not reached, at present, characteristics that could compete in all fields with the conventional radiology, it is necessary to have such instruments as additional tools in all departments of radiology where vulnerable patients have to be examined. If we could reduce, by a factor of two, the average dose of radiation inflicted by radiology in industrial countries, which is 1mSv, we would save, for 1.2 billion people, an irradiation of 600,000 person-Sieverts. This is exactly the total dose delivered to the northern hemisphere by the fall-out of radioactive elements from the catastrophe of Chernobyl!

I believe that with the progresses which are in view with new detectors for high energy physics we will make further steps, and that we will reach the requirements necessary to meet all the needs of radiography even in fields where an high accuracy is required, like in mammography. It is a matter of a few years and some reasonable investments.

GOLDHABER: Could the use of ultra-sound lead to cell or organ injury?

HILLMAN: At the frequencies and power levels that are used for diagnostic sonography nobody has shown any significant effects. If you have a situation where lower frequencies and higher powers are used, then there are clearly some dangers.

HAPPER: I have another question for Dr. Hillman. I am rather surprised that 14% of the GDP is going into medicine. I don't see that myself, and I am interested in how that is possible. Where does it go, and how much goes into high technology?

HILLMAN: An economist at Harvard named Joseph Newhouse, probably the best known health economist at this time, calculated what could be attributed to the increasing use of new technology, that is, to technology developed within the last ten years. He came up with something between 50 and 80% of the increase in cost. I may dispute that, because 80% surely sounds too high. But working with it every day one sees an enormous shift in complexity. We do about 215,000 exams a year at UVA, and the only thing for which the cost is going down is plain films; everything else—the more complicated imaging—is going upward at a rapid rate.

MAX: To what extent do HMOs do longitudinal studies of the long-term costs and benefits of using a new diagnostic procedure, one whose benefits may not show up for several years?

HILLMAN: At this point they do not seem to be interested in investing in that capability. Obviously if you control lives it is a wonderful resource to follow health care closely and see what impact it has, much as the Mayo Clinic has done for the past 100 years. There is that potential in HMOs. However, right now, they are very fixated on proxies of quality, such as what fraction of women over age 40 or 50 get mammograms and what fraction of children get vaccinations, which really are not quality at all. They do not address measuring the health care outcomes that we really are interested in. The methods are available, but they tend to be arcane and expensive, and so they are used only for the most interesting and important situations.

SMITH: A question to both speakers. It seems that in order to get close collaboration and evaluation of the potential worth of a line of research we may need to focus on the areas that have the greatest potential or promise.

HILLMAN: Absolutely, there is a whole new science—actually it's an old one, with a new name. It's called "pharmacoeconomics." The notion is that before one spends hundreds of millions of dollars to develop a product, one ought to model the costs and benefits as early as possible. This is not just a matter of getting a path to the market, one has to have a market niche as well. There are pharmacoeconomists who are currently doing that sort of thing.

CHARPAK: I found that the first reaction of a corporation like General Electric was that they were polite, but rather reluctant to enter into action because the impact on the business was not obviously immediate. I believe that if we can make this progress in accuracy at a low cost then the argument of low dose becomes a major selling point in the modern world. People are so concerned about radiation that perhaps a practitioner who could deliver the same result with five times less radiation would simply have more customers.

HILLMAN: I would like to respond to that. There are two populations of patients in which the physician should be conscious of the radiation dose, and those are in children and in people with chronic diseases that require frequent reexamination. That is not to say that all of us could not benefit from a lower dose per exam, but in general, the doses are sufficiently low that unless one falls into one of these two populations there is not that much anxiety. I suspect that is General Electric's take on this, that those two populations are sufficiently small that they are not going to be able to make a profit by developing the technology.

TAYLOR: You remarked that back in the 1980s the rules of the game began to change a little bit, particularly in this country because of the alarm over rapidly increasing medical costs. Could you comment on the degree to which things are different in other parts of the world?

HILLMAN: The symptom is the same worldwide, it is just less severe elsewhere. It is true that we do have the largest attribution of share of GDP of any of the developed countries, but in fact, all of the systems are now experiencing a crisis in health care expenditures, because the rate of rise in those countries is exactly like ours, and it seems to be related to the same kinds of influences.

REYNOLDS: My concern is with sono-imaging. It is known that microscopic bubbles form in tissues. Laboratory experiments show that very high temperatures exist, perhaps as high as 100,000 C. When power and temperatures are discussed in the medical applications spatial averages are usually quoted, rather than cellular dimensions. I hope that careful follow-up will be made to see if deleterious consequences follow.

HILLMAN: I was one of the earlier people into sonography, and I had the same concern myself. I had seen the laboratory data on microbubble formation, and I was very reluctant, given the experience with x-rays in the early part of this century when nobody understood the devastation it could reap. On the other hand, there is now about a thirty-year experience with ultrasound, and by all appearances, at least at the macroscopic population level, it appears that the frequencies and the amplitudes used in diagnostic ultrasound do not cause any kind of organism damage.

REFERENCES

[1] Multiwire proportional chamber for digital radiographic installation, S.E. Baru, A.G. Khabakhpashev and L.I. Shekhtman, Novosibirsk, Inst. Nucl. Phys. 22 March 1989.

[2] Biospace Radiologie, 75 rue du Faubourg Saint Antoine, 75005 Paris, France.

Chapter 7

Cosmological Challenges for the 21st Century

Paul J. Steinhardt
Department of Physics and Astronomy
University of Pennsylvania, Philadelphia, PA

7.1 Introduction

The 21st century shows every promise of being an age of historic discovery in cosmology due to an extraordinary influx of new technologies and new ideas. As the century approaches, the field is replete with controversial issues great and small, as demonstrated at the Princeton 250th Anniversary cosmology counterpart to this meeting, entitled "Critical Dialogues in Cosmology." In this paper, the focus will be on a few pivotal issues likely to dominate the 21st century, shaping the future of astrophysics and cosmology and influencing our understanding of fundamental physics.

To gain a perspective on the challenges of the 21st century, it is instructive to recall the successes of the past century, many of which have been pioneered by theorists and experimentalists at Princeton. Cosmology in the 20th century has undergone a remarkable metamorphosis from a field of pure speculation to a field of hard science due to a series of technological advances that have made it possible to probe the distant universe and to test our ideas through observations and experiments. We have seen the first definitive observational evidence that the sun is not the center of the universe; the first evidence that most nebulae are galaxies of stars far away rather than clouds of gas close by; and the first evidence that the universe has been expanding and cooling, explaining the motion of galaxies and the origin of the elements. These discoveries have forced us away from the strongly preferred notion that we live in a static, time-invariant universe. We have also discovered that the early universe had a slightly inhomogeneous distribution of matter and energy that may account for the large-scale structure seen today.

All of these observational breakthroughs have shaped our modern view of the universe, leading us towards a standard model known as the *hot big bang picture*. According to this model, the universe began as a infinitesimal patch of space filled with gas of nearly infinite temperature and nearly infinite density. Suddenly, for reasons not understood, the universe began to expand and cool. The universe observed today is simply the result of fifteen billion years of expanding and cooling and the consequent evolution of matter subject to known physical laws.

Although the hot big bang model is consistent with all observations to date, it leaves important aspects of the universe unexplained. For example, the hot big bang model does not enable us to understand how much matter and energy there is in the universe, or what forms it takes, or from where it originated. It does not explain the geometry of the universe. It does not explain why the universe is homogeneous and isotropic on large scales and, yet, highly inhomogeneous on smaller scales. Finally, the hot big bang model fails to explain the ultimate question: why is the universe the way it is?

In order to address these questions, the universe must be observed to much farther distances and much further back in time. What is so exciting about the coming century is that a whole host of new technologies are providing these very capabilities. Mention of just a few of these new windows on the universe can give some impression of the remarkable potential of the new century:

Over the next ten years, there will be red shift surveys of millions of galaxies that will produce a full, three-dimensional map of the arrangement of galaxies in the universe. Gravitational lensing, the bending of light of distance sources by foreground galaxies and clusters of galaxies, will be used to measure the amount and distribution of matter in the universe, both ordinary baryonic matter and dark matter of unknown kinds. It will become possible to measure with precision the x-ray luminosity of gas in large clusters of galaxies. This makes it possible to measure the amount of baryonic matter which lies in the gas between galaxies. When x-ray and gravitational lensing measurements are combined, it becomes possible to dissect clustered matter into its baryonic and its dark matter components. Observations of the absorption of quasar radiation by Lyman-α clouds, foreground clouds of primordial gas in intergalactic regions where there has been little or no star formation, are another promising source of new information. The absorption of the background quasar radiation may shed light on the first stages of structure formation and on the abundance of elements in the primordial universe. Studies of the Sunyaev-Zel'dovich effect, the rescattering of cosmic microwave background radiation from hot gas and clusters, and the systematic searches for supernova at cosmic distances are providing new standard candles for accurately measuring distance in the universe and determining the expansion, deceleration, and equation-of-state of the universe. And, perhaps the most powerful observational tool of all is the cosmic microwave background anisotropy. The all-sky map of the anisotropy anticipated in the next decade will be a snapshot of the distribution of matter and energy in the universe when it was only 100,000 years old, a decisive test of cosmological models and a new method to constrain key cosmic parameters.

With the wealth of new data pouring in, one should anticipate that the 21st century will bring some startling, unanticipated surprises in cosmology. Hence, the following speculations on the key challenges of the 21st century are offered with some trepidation. The reader is warned: these views may be made obsolete in only a few short years due to a discovery which will dramatically change the the theoretical view and preoccupy cosmologists for at least a century to come.

7.2 IS THE UNIVERSE FLAT?

The common view today is that determining the spatial curvature of the universe is important because it is one of the key tests of inflationary cosmology. Indeed, the inflationary model of the universe does explain how the universe became flat due to extraordinary expansion during the first instants after the big bang and does predict that the universe should remain flat today [1]. However, the issue of the flatness of the universe goes beyond inflation; even if inflation does not survive as an explanation of the observable universe, flatness is a critical issue for the future science of cosmology because it cuts to the heart of a basic assumption of nearly all models, the cosmological principle.

The cosmological principle is the notion that the universe observed from our limited vantage point is representative of the universe entire. That is to say, there should exist a length scale such that a coarse-grain average produces a nearly homogeneous picture. It is important that this length scale be well within an observationally accessible range in order for us to be able to determine the average properties of the universe.

When the cosmological principle was introduced, the dominant idea was that the universe is static and time-invariant. This would mean that there has been a semi-infinite amount of time for distant regions of the universe to interact and reach an equilibrium. Also, there is no limit to the range of vision since there is sufficient time for light to have propagated an unbounded distance. In this scenario, the notion of a coarse-grain length scale and an approach to homogeneity seems plausible.

However, our concept of the universe has changed due to the success of the hot big bang picture. Apparently, our patch of universe has only existed for fifteen billion years; consequently, the maximum distance that can be observed is about fifteen billion light-years, referred to as the "horizon distance." Most likely, the space within our horizon is only a tiny, infinitesimal corner of a much larger universe. This is an essential but seldom emphasized aspect of the big bang picture. Furthermore, regions separated by fifteen billion light-years or more have not had a chance to interact since the beginning of time. Given these conditions, the plausibility of the cosmological principle seems less certain. Is it really possible to to understand the universe entire when we are constrained by causality from observing most of it? The only hope is that there exists a coarse-grain length scale which is smaller than the horizon distance such that the properties of the entire universe can be determined by coarse-grain averaging observations within our limiting horizon. Whether this is so is an issue that must be tested, rather than accepted as assumption.

How might we test the cosmological principle? One test is to measure the distribution of galaxies and determine if there exists a length scale over which we can coarse-grain average and approach homogeneity. Before this year, cosmologists would have to confess that there is no evidence of this coarse-grain length scale. Red shift surveys, such as the Center for Astrophysics Survey [1], show an inhomogeneous distribution of large filamentary and wall-like concentrations of galaxies separated by large voids. In a survey that stretches out several hundred megaparsecs (several hundreds of millions of light-years), the structures themselves stretch several hundreds of megaparsecs, and there is not any indication of approaching homogeneity. In recent months, though, the situation has begun to change as red shift surveys probing deeper distances have begun to report. For example, the Las Campanas Survey of Schechtman et al. [2] has produced maps which reach a distance almost an order of magnitude farther than the Center for Astrophysics study. What that Las Campanas Survey shows is that hundred-metaparsec structures similar to those found in the Center for Astrophysics Survey are found throughout the Las Campanas map, too, but there do not seem to be yet larger structures. If further deep surveys support this conclusion, it will be historic. At the beginning of this century, the predominant view was that the universe consist of a uniform distribution of stars; instead, a hierarchy of increasingly large structures has been found: galaxies, galaxy clusters, superclusters, filaments, walls, and voids. The Las Campanas Survey suggests that the end of this hierarchy has finally been reached and that there truly is a coarse-grain length scale for the galaxy distribution of some few hundred megaparsecs.

Another kind of evidence can be obtained by measuring various electromagnetic radiation backgrounds, such as the x-ray background or the cosmic microwave background. The cosmic microwave background can be interpreted as a snapshot of the distribution of matter and radiation at a time when the universe was only a few hundred thousand years old. As the radiation decoupled from matter and began to stream towards us, the radiation was red-shifted or blue-shifted depending on the gravitational potential in the region from which it last scattered (the "last scattering surface"). Inhomogeneities in the distribution of the radiation can, therefore, be used to determine the homogeneity of matter and energy at this early epoch. The Cosmic Background Explorer (COBE) satellite [4] has measured the root-mean-square inhomogeneity to be a few parts in 10^5, where the microwave antenna produce a coarse-grain average over 10°. Subsequent ground and balloon based experiments suggest similar inhomogeneities at smaller angular scales ranging down to 0.5 degrees. Hence, the cosmic microwave background is strong evidence that the distribution of matter and energy in the large-scale universe was highly homogeneous at early epochs.

For more subtle reasons, the cosmological principle is also tied to the flatness of the universe. Suppose that future empirical evidence were to point to finite spatial curvature based on observations made within our horizon. One possible interpretation is that the entire universe is homogeneously curved (globally). Most discussions of cosmology suggest this interpretation. However, this explanation requires that we live at a very special epoch: although the horizon distance and the curvature vary at different rates as a function of time, it would have to be that we live at the particular epoch when their magnitudes are comparable. (According to the big bang picture, the horizon distance would have to have been negligibly small compared to the radius of curvature at earlier times, and the universe would appear flat; at later times, the curvature would become much smaller than the horizon distance, dominating the expansion of the universe and causing large optical distortions.) It is a coincidence sometimes referred to as the "flatness problem," so-called because it would be an important feature of the universe which had no natural explanation. An embarrassing coincidence is not the only possibility, though.

If future experiments point to non-zero spatial curvature, a second interpretation is that we have stumbled upon a spectrum of large scale spatial distortions spanning scales from subhorizon to superhorizon. By measuring the curvature within our horizon, we have sampled the distribution over some random, horizon-sized patch. A similar patch in a different region of space would have a different spatial curvature.

One explanation, the homogeneously curved universe, is consistent with the cosmological principle. But, the second explanation means that conditions within our horizon are not the same as conditions elsewhere in the universe. The problem is that no observation can distinguish the two possibilities. Consequently, if non-zero curvature becomes established observationally, the cosmological principle and any attempt to explain the universe beyond our horizon must be regarded suspiciously.

Given the significance of the issue, it would be good to report that the curvature can be reliably measured in the near future. Many recent papers have suggested that it is likely in the near future. However, an important warning is due: a careful analysis shows that methods cited as testing the flatness of the universe do not measure the curvature directly, but only in combination with uncertain, model-dependent assumptions. In particular, assumptions must be made about the matter-energy content of the universe, or the spectrum of primordial fluctuations, or other properties of the universe. Consequently, a discussion of measuring the flatness of the universe will be omitted in this section. It will appear in the subsequent section where it is tied to the testing of cosmological models, as is logically appropriate.

7.3 DO WE LIVE IN AN INFLATIONARY UNIVERSE?

Determining the validity of the inflationary model of the universe [1, 5] is one of the key challenges for the 21st century. Inflation is our best hope for answering many of the questions left open by the hot big bang picture [6, 7]. In a single stroke, inflation explains the amount of matter and energy in the universe; how the matter and energy in the universe originated; how the universe became spatially flat; why the universe is homogeneous on large scales; and how the inhomogeneities arose on smaller scales which eventually gave rise to the formation of galaxies and larger-scale structures.

It is noteworthy that the inflationary model creates dynamically the remarkable conditions suggested by the cosmological principle: namely, a universe in which there exists a coarse-grain length scale within the observable horizon over which the universe appears homogeneous and isotropic. Averaged properties within our universe are representative of the greater universe. The cosmological principle is derived rather than assumed.

How will inflation be tested? Two of its generic predictions are a spatially flat universe and a scale invariant spectrum of gaussian, adiabatic energy density fluctuations [8]. The first, spatial flatness, is equivalent to the prediction that the total energy density of the universe is equal to the critical density; that is, $\Omega_{\rm total}$ is equal to one. Throughout the remainder of the paper, Ω_i is used to represent the ratio of the energy density of type i to the critical density needed to close the universe.

Note that I have emphasized here the term "total energy" referring to the sum of all forms of energy including matter energy, radiation, and any other contributions. In particular, inflation does not predict that $\Omega_{\rm matter} = 1$ necessarily, but only that the total energy density is equal to the critical density. $\Omega_{\rm matter} = 1$, as assumed in the standard cold dark matter model of structure formation, is only a special, simple case. Hence, recent evidence suggesting that $\Omega_{\rm matter} < 1$ is inconsistent with the standard cold dark matter model, but it is not inconsistent with inflation.

The second prediction of inflation is a scale invariant spectrum of fluctuations that should have left a mark on the cosmic microwave background anisotropy and may have been responsible for the formation of large-scale structure in the universe [8]. The spectrum is generated by quantum fluctuations which distort the distribution of energy when the universe is microscopically small during the first stages of inflation. Inflation freezes the fluctuation amplitude and stretches the wavelength to cosmic scales. The details of this process can be computed from first principles. The resulting spectrum is scale invariant in the sense that, if one expresses the density as a sum of Fourier modes, the amplitudes are nearly the same from wavelength to wavelength when averaged over the entire universe. Of course, our observations are restricted to a finite horizon, and so measurements in this restricted range of space will produce some deviations from the cosmic average. The spectrum is Gaussian in the sense that the deviations from the cosmic mean are Gaussian-distributed. The spectrum is adiabatic in the sense that all forms of matter and radiation undergo the same fluctuations.

Although there are many types of inflationary scenario in the literature, the predictions described above rely on their common features and, hence, are generic. It is a strong feature of inflationary cosmology that its predictions do not depend sensitively on the details. (As in theoretical models of nearly any kind, it is possible to push on parameters and add contrivances to violate one or more generic predictions in any type of inflationary scenario. However, the conditions are so extreme that the models have little or no predictive power and are physically implausible.)

Even if inflation should pass these initial tests, some would not find them to be a convincing proof of inflation, arguing that the predictions that the universe must be spatially flat [9] and that the primordial perturbations spectrum must be scale invariant (Harrison-Zel'dovich-Peebles) [10] predate the invention of the inflationary model by several years. The argument is questionable. For, while it was discussed that flatness and scale invariant spectra are attractive for a viable cosmological model, there was no dynamic explanation of how they came to be prior to inflation. Symmetry arguments are not compelling. Flatness, while symmetrical, is highly unstable since curvature grows rapidly in big bang cosmology. Scale invariance, while symmetrical, is hard to explain without inflation since it requires primordial fluctuations and scales that exceed the causal horizon. Nevertheless, given that some find the aforementioned inflationary tests unpersuasive, it is important to emphasize more refined predictions of inflation that were not anticipated.

The first refinement is that the predicted perturbation spectrum is not perfectly scale invariant; instead, there is a small but measurable "tilt" [14, 15]. The spectrum can be characterized by a spectral index, n, where the energy perturbation amplitude is defined as $\delta\rho/\rho \propto \lambda^{(1-n)/2}$. Then, $n = 1$ corresponds to perfect scale-invariance. Inflation predicts that the spectral index will deviate from $n = 1$ by a few percent to several tens of a percent, depending on the details of the model. Inflation would predict a perfectly scale invariant ($n = 1$) spectrum if the expansion rate were uniform. However, the expansion rate cannot remain uniform since eventually inflation must slow down to return the universe to the standard big bang expansion rate. In many models, the expansion rate is changing slowly throughout inflation. The tilt is limited by the fact that, if the expansion rate changes too rapidly, inflation ends prematurely, before there is sufficient inflation to solve the cosmological horizon and flatness problems. The allowed range is $n \sim 1.0 \pm 0.3$ [16, 17].

A second refinement is that the perturbations predicted by inflation are a combination of fluctuations in the energy density, which can be sources for the formation of structure, plus fluctuations in the space-time metric which will evolve into gravitational waves [11, 12, 13]. Furthermore, inflation predicts a relationship between the ratio of gravitational waves to energy density fluctuations and the tilt. These more refined predictions may require more than ten years to test, but it is rather likely that they will be tested well before end of the next century. The discovery of these effects should certainly be convincing to the last skeptics, since these are new predictions stemming specifically from analysis of inflationary models.

The critical test of these predictions will come from measurements of the cosmic microwave background anisotropy, which provides a detailed fingerprint of conditions in the early universe. The cosmic fingerprint [16] is obtained from a temperature difference map (Fig. 7.1) which displays the fractional fluctuation in the cosmic background temperature, $\Delta T(\mathbf{x})/T$, as a function of sky direction $\mathbf{x}$. The map represents the deviations in temperature from the mean value, $T = 2.726\pm.010$, after foreground sources of radiation are subtracted. Testing inflation and other cosmological models entails comparing statistical properties of this map to the theoretical predictions. The simplest and most decisive statistical test is the two-point or temperature autocorrelation function. See Fig. 7.1. The temperature autocorrelation function, $C(\theta)$, compares the temperature at points in the sky separated by angle θ:

$$\begin{aligned} C(\theta) &= \langle \tfrac{\Delta T}{T}(\mathbf{x})\tfrac{\Delta T}{T}(\mathbf{x}')\rangle \\ &= \tfrac{1}{4\pi}\textstyle\sum_\ell (2\ell+1)C_\ell P_\ell(\cos\,\theta), \end{aligned} \tag{7.1}$$

where $\langle\rangle$ represents an average over the sky and $\mathbf{x}\cdot\mathbf{x}' = \cos\,\theta$. The coefficients, C_ℓ, are the *multipole moments* (for example, C_2 is the quadrupole, C_3 is the octopole, *etc.*). Roughly speaking, the value of C_ℓ is determined by fluctuations on angular scales $\theta \sim \pi/\ell$. A plot of $\ell(\ell+1)C_\ell$ is referred to as the *cosmic microwave background (CMB) power spectrum.*

There is valuable information in the cosmic microwave background anisotropy in addition to the CMB power spectrum that will be extracted some day. Higher-point temperature correlation functions (entailing three or more factors of $\Delta T/T$) could be obtained from the temperature difference map and be used to test if the fluctuation spectrum is Gaussian, as predicted by inflation. However, the fact that statistical and systematic errors in $\Delta T/T$ compound for higher-point correlations makes precise measurements very challenging. Polarization of the microwave background by the last scattering of photons from the anisotropic electron distribution is another sky signal that provides quantitative data that can be used to test models. However, for known models, the predicted polarization requires more than two orders of magnitude better accuracy than anisotropy measurements alone in order to discriminate models [18]. Although forthcoming satellite experiments will attempt to detect polarization or measure higher-order correlation functions, the most reliable information in the near future will be the CMB power spectrum. Fortunately, the CMB power spectrum is packed with information that can be used, by itself, to discriminate inflation from alternative models.

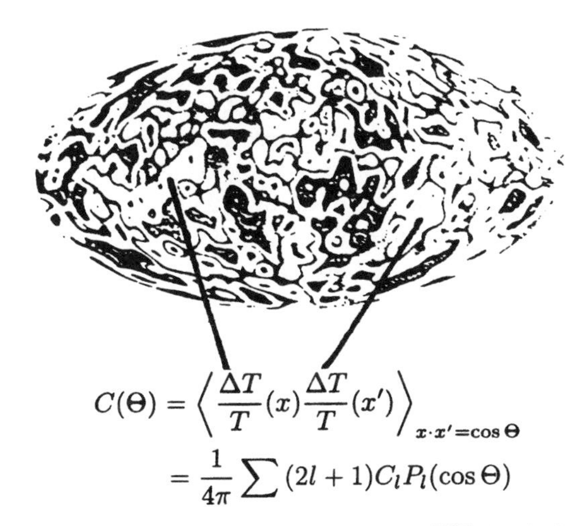

Figure 7.1: The temperature autocorrelation function, $C(\theta)$, is obtained from a map of the sky (here represented by the oval) displaying the difference in the microwave background temperature from the average value, $\Delta T/T$. $C(\theta)$ is computed by taking the map-average of the product of $\Delta T/T$ measured from any two points in the sky separated by angle θ. If $C(\theta)$ is expanded in Legendre polynomials, $P_\ell(\cos\,\theta)$, the coefficients C_ℓ are the *multipole moments*.

Figure 7.2 displays a representative CMB power spectrum for an inflationary model [19]. For this example, the spectral index is $n = 0.85$, with equal contributions of energy density and gravitational wave fluctuations to the large-angular scale anisotropy. To the left of $\ell \approx 100$ are multipoles dominated by fluctuations over distances much larger than the size of the Hubble horizon at the time of last scattering, corresponding to angles $> 1° - 2°$ on the sky. According to the inflationary model, these wavelengths did not have a chance to evolve before last scattering and the beginning of the photon trek towards our detectors. Hence, these fluctuations preserve the imprint of whatever fluctuations were set by inflation. If the fluctuations are remnants of a nearly scale-invariant inflationary spectrum, then the low-ℓ part of the CMB power spectrum should be featureless, just as shown in the figure. If the spectrum is tilted, as in this case, there should be a small downward (or upward) slope over the low multipoles, as shown.

The spectrum includes, in general, both energy density and gravitational wave contributions, as indicated in the example of Fig. 7.2. For inflationary models, the two contributions are predicted to be statistically independent and simply sum to give the total power. The fluctuations in $\Delta T/T$ are also predicted to be Gaussian-distributed for inflationary models. Hence, the C_ℓ's, which are an average over $2\ell + 1$ Gaussian-distributed variables, have a χ^2-distribution.

To the right of $\ell \approx 100$ are multipoles dominated by fluctuations with wavelengths smaller than the horizon at last scattering [16, 17]. The right side of the power spectrum figure appears different from the left because inhomogeneities spanning scales smaller than the Hubble horizon have time to evolve causally. Gravitational waves inside the horizon red-shift away. For energy density fluctuations, the baryon and photon begin to collapse and oscillate acoustically about the centers of high and low energy density, adding to the net microwave background perturbation. Each wavelength laid down by inflation initiates its acoustic oscillation shortly after entering the Hubble horizon. Hence, there is a well-defined phase-relation between the acoustic oscillations on different wavelengths. Waves just entering the horizon and smaller-wavelength waves which have completed a half-integral number of oscillations by last scattering will be at maximum amplitude. Wavelengths in between are mid-phase and will have smaller amplitudes. In a plot of C_ℓ's, increasing ℓ corresponds to multipoles dominated by decreasing wavelengths. The variations of the oscillation phase with wavelength results in a sequence of peaks as a function of ℓ. These peaks are sometimes referred to as Doppler peaks or acoustic peaks.

The position of the first Doppler (or acoustic) peak is of particular interest. Its position along the ℓ-axis, left or right, is most sensitive to the value of $\Omega_{\rm total}$: the peak moves to the right in proportion to $1/\sqrt{\Omega_{\rm total}}$, for large $\Omega_{\rm total}$ [20]. There is only weak dependence on the Hubble constant and other cosmological parameters [21, 22]. Decreasing $\Omega_{\rm total}$ to 0.1, say, causes the first Doppler peak to shift to $\ell \approx 600$ instead of $\ell \approx 200$, a dramatic and decisive difference. As a test of $\Omega_{\rm total}$, the first Doppler peak has the advantage that it is relatively insensitive to the form of the energy density, whether it be radiation, matter, or cosmological constant, and it is relatively difficult to mimic using other physical effects.

In sum, Fig. 7.2 illustrates how all three key features of inflation can be tested by the microwave background power spectrum. Large-angular scale fluctuations are consequences of approximate scale-invariance and the combination of energy density and gravitational wave perturbations. The presence of a combination of energy density and gravitational wave fluctuations can be detected from more subtle features, such as the ratios of the Doppler peaks to the plateau at small ℓ. A gently sloped CMB power spectrum at small ℓ is the signature of being slightly tilted from scale-invariance. Small-angular scale fluctuations, especially the position of the first Doppler peak, are consequences of $\Omega_{\rm total}$ being unity. A sequence of subsequent Doppler peaks is a check that the perturbations are adiabatic [23].

The inflationary prediction for the CMB power spectrum is not unique, since there are undetermined, free parameters having to do both with inflation and with basic cosmic parameters. Figure 7.3 is a representative band of predicted curves for different values of the spectral index. Each example lies within the parameter space achievable in inflationary models. Although the band is wide, there are common features among the curves which can be used as the critical tests of inflation. All have a plateau at small ℓ, a large first Doppler peak at $\ell \approx 200$, and then smaller Doppler hills at larger ℓ.

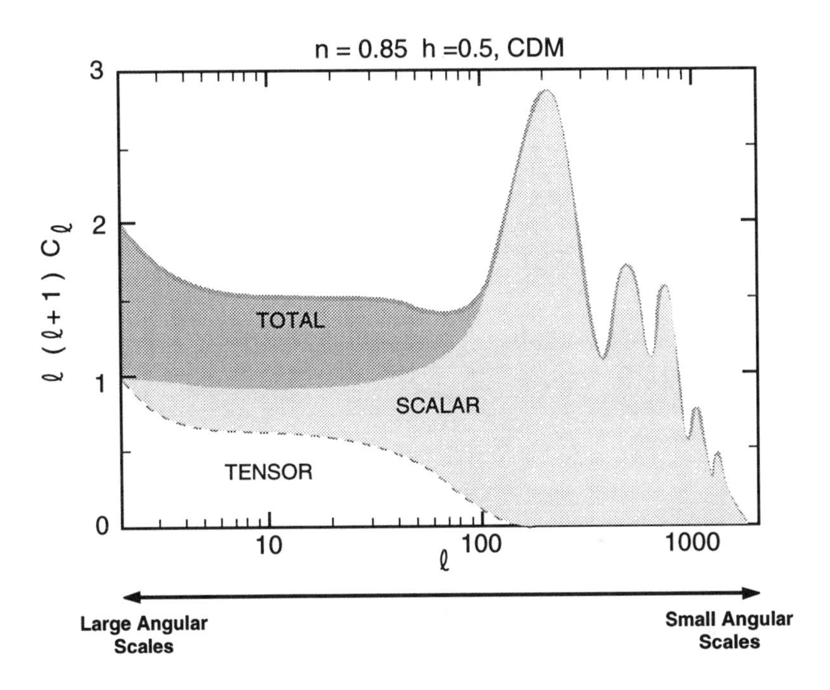

Figure 7.2: The CMB Power Spectrum: A plot of $\ell(\ell+1)C_\ell$ vs. multipole moment number ℓ is the cosmic microwave background power spectrum. For a given ℓ, C_ℓ is dominated by fluctuations on angular scale $\theta \sim \pi/\ell$. In inflation, the power spectrum is the sum of two independent, scalar and tensor contributions.

At this point, the best available CMB anisotropy data is in rough agreement with inflationary predictions [16], but is rather imprecise. In the next ten years, however, there will be dramatic improvements due to a combination of land, air, and space-based experiments, including the MAP (Microwave Anisotropy Probe) satellite to be launched by NASA and the COBRAS/SAMBA satellite to be launched by ESA. These experiments will have an uncertainty comparable to the line thickness at small angular scales, detecting every bump and wiggle.

If inflation or another recognized model (such as cosmic textures) is verified by these measurements, then it will also be possible to determine the flatness, as well, to very high precision. However, it is also important to appreciate that if the measured spectrum does not conform to one of the known patterns, then the value of Ω_{total} will be unconstrained, remaining a key, unsettled issue in the field.

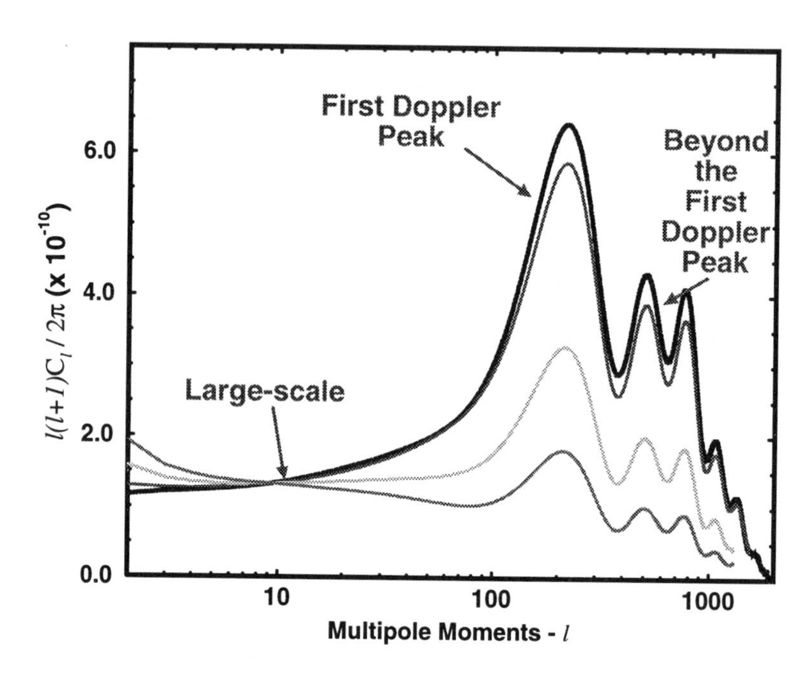

Figure 7.3: A band of microwave background power spectra allowed by inflation. The uppermost curve is a pure-scalar, scale-invariant spectrum, and the lower curves have tilt ($n < 1$) and gravitational waves. Inflationary models with spectra somewhat higher than the uppermost curve are also possible. The common features among these curves—the prime targets for experimental tests of inflation—are a plateau at large angular scales, a prominent first Doppler (or acoustic) peak, and subsequent acoustic oscillation peaks at small angular scales.

7.4 DOES $\Omega_{\text{matter}} = 1$?

In order to speculate further, we shall suppose that during the next ten years inflation passes the tests described in the previous section during the next ten years and that the universe is proven to be flat. The next critical issue for cosmology will be whether or not $\Omega_{\text{matter}} = 1$. There are two reasons why this is a critical issue. If $\Omega_{\rm matter} \geq 0.2$, then we know that some form of non-baryonic dark matter exists. On the other hand, if $\Omega_{\rm matter} < 1$ and the universe has been proven to be flat (*e.g.*, according to the position of the first acoustic peak in the CMB spectrum), then there must exist some other form of energy in the universe besides matter which is a significant, and perhaps, even a dominant contributor to the total energy density of the universe.

The claim that $\Omega_{\rm matter} \geq 0.2$ implies non-baryonic matter stems from the constraint on the baryon density derived from primordial nucleosynthesis. A few seconds after the big bang (or inflation), the universe was sufficiently hot to enable fusion of protons and neutrons into light nuclei, and yet cool enough that the nuclei were not destroyed by subsequent collisions. From the knowledge of the expansion rate and temperature history, along with details of nuclear interactions, the relative abundances of the light elements produced in this primordial epoch can be reliably predicted. The predictions depend on one unknown, $\Omega_b h^2$, where Ω_b is the ratio of the baryon density to the critical density and $h = H_0/(100\,{\rm km/s - Mpc})$ is a standard, dimensionless re-expression of the Hubble constant, H_0. Comparison of observations of primordial helium, deuterium, and lithium with the theoretical predictions have been used to constrain $\Omega_b h^2$ to a range of small values, between 0.01 and 0.02 [24]. Since h lies somewhere between 0.5 and 1, according to current observations, Ω_b is less than 10%. Consequently, if $\Omega_{\rm matter}$ is found to be greater than 0.2, then non-baryonic dark matter must exist.

An important breakthrough is the attempt to constrain the primordial abundance of the light nuclear elements by measuring the deuterium abundance in Lyman-α clouds by measuring absorption of background quasar radiation [25]. These clouds occur in regions far from galaxies where little or no stellar nucleosynthesis has taken place. It is reasonable to suppose that the clouds are representative of matter that evolved very little since primordial nucleosynthesis. Hence, measuring the deuterium abundance compared to the hydrogen abundance in the clouds produces a new observational limit. $\Omega_b h^2$ from this technique have yielded a value slightly higher than the previous bound, $\Omega_b h^2 = 0.024 \pm 0.03$ [25]. As more quasar/Lyman-α systems are measured, it will be possible to test the consistency of the method and perhaps improve the constraint further, making it the most precise method for constraining primordial nucleosynthesis and measuring the $\Omega_b h^2$.

Some may worry about the fact that these measurements are in disagreement with what was before the preferred range. More specifically, the measurements are marginally inconsistent with limits on primordial helium abundance. Some have even spoken about there being a potential crisis in big bang cosmology [26]. However, the the apparent, marginal contradiction is more likely due to underestimated systematic errors. In terms of Ω_b, the higher value of $\Omega_b h^2$ obtained from quasar absorption would permit Ω_b as high as 10%, still well below the critical density. It is possible that further improvements will lead to a higher value yet for $\Omega_b h^2$, but it is hard to imagine that it will exceed 0.035, a range ruled out by additional constraints from solar system measurements. Even allowing a value of $\Omega_b h^2$ near this uppermost bound, Ω_b is constrained to be less than 15%. Hence, even the crudest limits justify our remark that $\Omega_{\mathrm{matter}} \geq 0.20$ is convincing evidence of significant non-baryonic matter in the universe.

The upper bound on Ω_{matter} is also extraordinarily important. In a recent study, J. Ostriker and I mapped all of the constraints on Ω_{matter} and the Hubble expansion rate coming from observations [27]. Figure 7.4 is an updated version using the new constraints on $\Omega_b h^2$ obtained from the Lyman-α cloud measurements [25]. The plot illustrates the constraints coming from measurements of the Hubble constant directly, measures of the age of the universe, measurements of large scale structure, measurements of primordial nucleosynthesis, x-ray luminosity in gas clusters, and bounds on the cosmological constant.

The striking feature is that there is a substantial range of the Ω_b-h plane which is in agreement with all known astrophysical constraints. Furthermore, the range of "cosmic concordance" is well above $\Omega_m = 0.2$ but also well below $\Omega_m = 1$. If the state of affairs does not change, then two striking conclusions emerge: significant amounts of non-baryonic dark matter exist in the universe, *and* there is some additional "missing energy" density in the universe accounting for the difference between Ω_m and unity. At this point, the measurements of the cosmic microwave background anisotropy are also consistent with this trend, but the experimental uncertainty is too large to reach any firm conclusions.

Although the case for missing energy is not conclusive, it is important to appreciate that the current indications come from a variety of observations, and that a combination of changes in disparate measurements is needed to change the qualitative conclusions. The importance of non-baryonic matter has been discussed above and is well-known in the physics community. The notion of missing energy is much less well-known, and, given the observational justification, is an idea worth exploring.

7.5 IF $\Omega_{\mathrm{matter}} < 1$ AND $\Omega_{\mathrm{total}} = 1$, WHAT ELSE IS THERE?

If future observations reveal that the universe is flat but also that Ω_{matter} is less than unity, then then there must exist another contribution to the total energy density of the universe. This additional energy, which we have called "missing energy," must have a different equation-of-state from ordinary matter and non-baryonic dark matter. While missing energy is consistent with inflation and other cosmologies, it is not predicted by any model. The existence of missing energy would surely be one of the most surprising discoveries in the history of cosmology and would immediately produce a puzzle: what is the nature of the missing energy? This question emerges as one of the most important cosmological issues of the 21st century.

The first candidate for missing energy likely to be considered is vacuum density or cosmological constant. However, this choice is based on historical familiarity; it is not uniquely dictated by the observational evidence. In fact, all the data that has been described in this paper would only tell us the amount of missing energy and that this energy has some equation-of-state different from that of ordinary matter. It would not tell what the equation-of-state is.

The equation-of-state is defined as the ratio of the pressure to the energy density, α. Vacuum density or cosmological constant corresponds to $\alpha = -1$ and matter density has $\alpha = 0$. The missing energy could have a value in between, and perhaps the equation-of-state is time-dependent. (For the purposes of this discussion, we do not consider $\alpha > 0$ since this would lead to a universe with a shorter lifetime than a flat model with $\Omega_m = 1$, which is already marginally

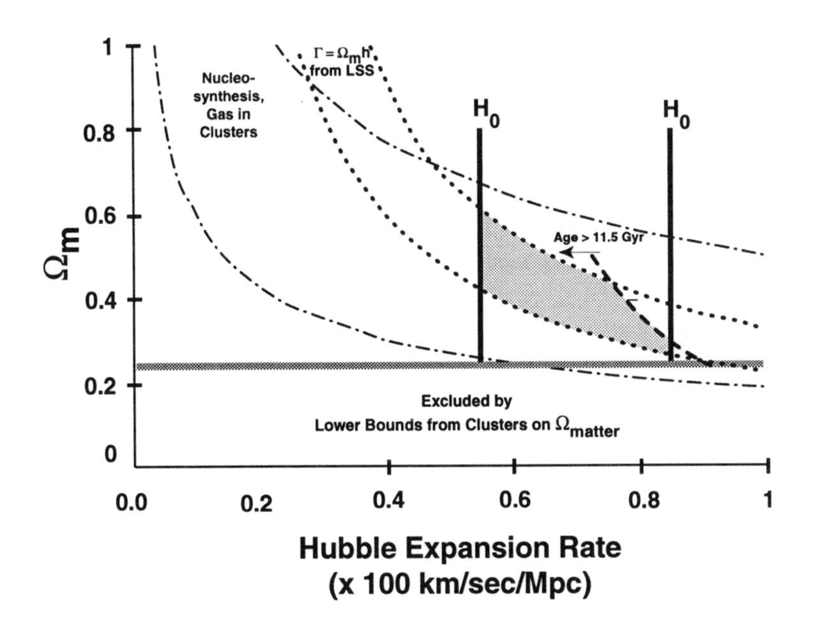

Figure 7.4: A plot in the the Ω_m-h plane showing the range of parameters in concordance with the known astronomical observations. The figure differs from the plot shown in Ostriker and Steinhardt, Ref. [27], in that a higher value of Ω_b is assumed, in accordance with recent limits on deuterium abundance from measurements of Lyman-α absorption of quasar emission [25]. The shaded region (stripes) is the concordance domain for flat CDM + Λ models.

in conflict with lower bounds on the lifetime of the universe.) A simple example of missing energy different from matter or vacuum energy is the field energy associated with a field rolling down an exponential potential, $\exp(-\beta\phi)$, which can have an equation-of-state which lies between $\alpha = 0$ and -1, depending upon the coefficient, β.

The effect of missing energy is to change the expansion rate of the universe at recent epochs. If the missing energy has equation-of-state $\alpha = p/\rho$ =constant, then it will contribute an energy density which scales $1/a^{\mathrm{m}}$, where a is the scale factor which describes the stretching of the universe and m can be simply related to the equation-of-state, $m = 3(1+\alpha)$. For the cosmological constant, $\alpha = -1$ and ρ = constant. For more general forms of missing energy with $\alpha < 0$, the energy density decreases more slowly than matter or radiation energy. In order for the missing energy to be a substantial fraction of the critical density today, it must have been an insignificant fraction of the critical density in the early universe.

If identifying the missing energy emerges as one of the critical questions of the 21st century, the first challenge will be to determine observationally its equation-of-state, α. This may be one of the most difficult tasks in observational cosmology. For example, consider the most powerful tool for measuring cosmological parameters, the cosmic microwave anisotropy. Figure 7.5 illustrates a sequence of curves indicating the predicted microwave background anisotropy. The lowest curve corresponds to the prediction for an inflationary model with $\Omega_{\mathrm{matter}} = 1$. The upper curves correspond to an inflationary model (flat) with a matter density which is 35% of the critical density. The sheath about them indicates the cosmic variance, or theoretical uncertainty in the inflationary prediction. A close look reveals that there is a whole sequence of curves which lie within the cosmic variance sheath and, hence, cannot be distinguished observationally. These correspond to $\Omega_\alpha = 0.65$, with the equation-of-state given values between -0.5 and -1. Hence, the CMB power spectrum does not provide a precise way of determining α. (N.B. The figure also shows that, for $\alpha > -0.5$, the spectrum undergoes more dramatic changes in shape which make it possible to distinguish from $\alpha < -0.5$.) In fact, I have looked for several ways of distinguishing the equation-of-state, including measurements of the luminosity distance-red shift relation for cosmic supernovae and other standard candles in the universe. Thus far, it appears that one cannot determine α precisely by any known method. Should observational evidence establish that there exists missing energy, finding a method for precisely measuring its equation-of-state will emerge as a grand challenge for observational cosmology in the 21st century.

Not only is there the challenge of explaining what the missing energy is, but, in addition, there remains a puzzling "cosmic coincidence." Figure 7.6 shows a plot of the energy density of the universe versus the scale factor. The matter energy density, decreasing as $1/a^3$ and the missing energy density, decreasing as $1/a^m$ with $m < 3$, are indicated. The two lines cross at a time which must correspond to the present epoch, since the matter and missing energy density are

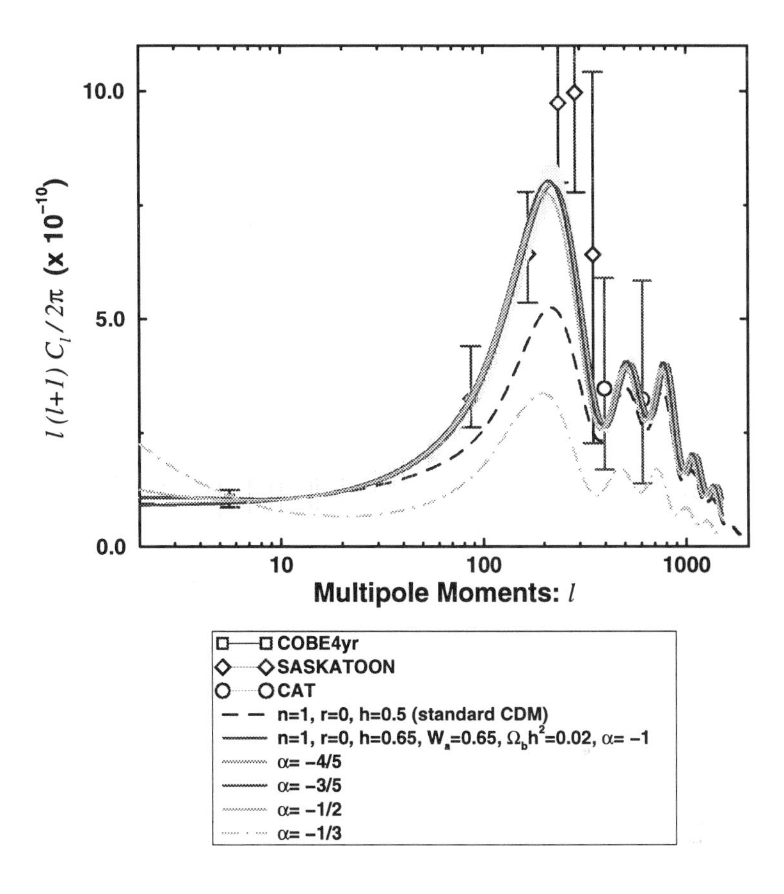

Figure 7.5: A plot of the cosmic microwave background anisotropy power spectrum multipoles vs. multipole moment illustrating the dependence on equation-of-state (α). The middle dashed curve represents standard cold dark matter model with $\Omega_m = 1$. This is easily distinguished from the upper family of (solid) curves corresponding to cosmologically flat models with $\Omega_m = 0.35$ and missing energy $\Omega_{\text{unknown}} = 0.65$. The only difference among the upper curves is the equation-of-state, α, which varies between -1 (vacuum energy density) and -1/2. There is negligible change among the upper curves compared to the cosmic variance (sheath surrounding the curves). If α is decreased further, then the difference in the equation-of-state causes a distinguishable spectrum, *e.g.*, see $\alpha = -1/3$ dot-dashed curve. Some recent data from COBE, Saskatoon, and CAT experiments [16] are superimposed [21].

comparable. Why we should happen to live at this special crossroads? If the missing energy turns out to be the cosmological constant or vacuum density then the missing energy density is strictly constant, a horizontal line in the plot. It is hard to imagine a physical explanation for the coincidence. On the other hand, if the missing energy turns out to have some other equation-of-state, $\alpha > -1$, then it may be due to some field or fields, and one could at least imagine that there are interactions between the missing energy and ordinary matter that might explain naturally why the two energy densities should be nearly equal to one another. Hence, distinguishing whether the missing energy is vacuum energy or not will be essential to theoretical understanding of the apparent cosmic coincidence.

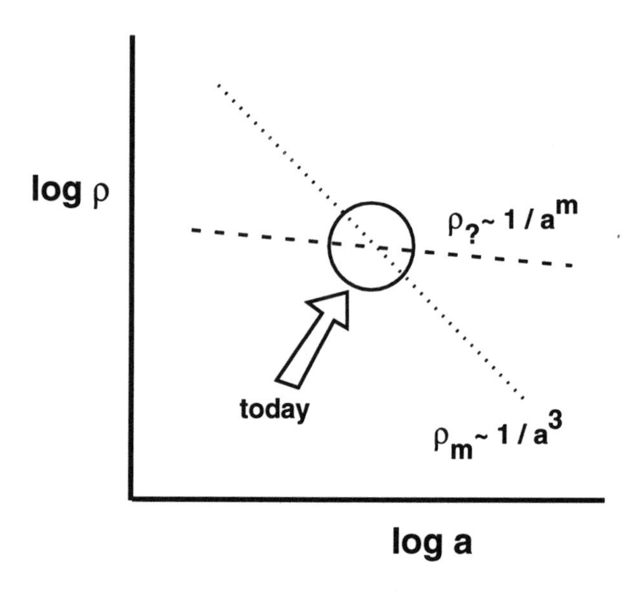

Figure 7.6: A schematic log-log plot of energy density ρ versus scale factor a illustrating the behavior of the matter density ρ_m and putative missing energy density $\rho_?$. The missing energy density decreases more slowly than the matter density. Hence, the missing energy was insignificant in the early universe. The circle represents the cosmic coincidence in which the the missing energy and the matter energy are comparable today.

7.6 ULTIMATE CHALLENGE

Should cosmologists meet all of the challenges described in the previous sections without major changes in the current paradigm, then the field will probably proceed along two different paths. One direction is a deeper understanding of large-scale structure formation and evolution. Large-scale structure already provides some of the most important cosmological tests, and will become an even more powerful constraint as forthcoming red shift surveys are completed. Here, the strategy is to use statistical properties of large-scale structure over the observable universe to constrain cosmological models and parameters. However, the details and variations in large-scale structure formation are interesting as well. What role, for example, does the ionization of the history of the universe have on the formation of the first structures in the universe? In spite of all the progress anticipated in the previous sections, a realm that will remain difficult to probe is below red shift $z = 1000$, where the cosmic microwave background anisotropy provides a snapshot, and above $z = 10$, beyond the reach of optical, x-ray, and infrared measurements. Theorists and observers will struggle to understand how the universe evolved from the tiny ripples seen in the cosmic microwave background to the first stages of large-scale structure formation.

The other path for cosmology will be towards answering its ultimate challenge: why is the universe the way it is? How did the universe we see naturally spring from the fundamental laws which govern the universe? The answers to these questions will not come from observation cosmology so much as from fundamental physics. The issues that fundamental physics must ultimately explain include:

- Nature of dark matter and missing energy
- Initial conditions of the universe
- Value of the cosmological constant
- Mechanism that drives inflationary cosmology

I have listed these issues as topics for future research, but in fact, they are so irresistibly tempting that a significant number of papers have been written on these issues already. There are a large number of candidates proposed for non-baryonic dark matter, fewer proposals for candidates for missing energy. There is the intriguing work of Hartle and Hawking on the wave function of the universe as a proposal for explaining initial conditions [28]. There is also Linde's chaotic, self-reproducing universe picture [29]. Many brands of inflationary models have been proposed which rely on different detailed mechanisms for initializing inflation and bringing inflation to a halt.

However, my own intuition is that some of these speculations may be premature. My special concern is for ideas that depend heavily on processes that take place uncomfortably close to the Planck scale. Many of these ideas rely on field theoretic ideas that may not be valid. For example, recent progress in string theory or M-theory strongly suggests a sequence of compactification scales; this means that field theory may be a bad approximation until the energy density in the universe reduces to 10^{16} GeV or less. In that case, the Hartle-Hawking wave-function and Linde's chaotic, self-reproducing universe may have to be discarded as explanations of the initial conditions since they are based on quantum field theory extrapolated to near the Planck scale. Topological defect models of structure formation are also in jeopardy because they require the Kibble-mechanism [30], a semiclassical field-theoretic notion of bubble nucleation and collision, to produce cosmic strings or textures at masses of order 10^{17} GeV, as required to explain structure formation. Similarly, the monopole problem [31], one of the original motivations for inflationary cosmology, may be obviated since it, too, depends on a Kibble mechanism operating at near Planckian scales. (In supersymmetric models, the monopole mass is above $10^{1}6$ GeV.)

Inflationary cosmology is not necessarily affected directly by physics near the Planck scale. It is certainly possible to construct models in which inflation occurs at energies well below 10^{16} GeV. However, my own intuition is that inflation is enmeshed in strings and Planck-scale physics as well. At present we describe inflation in terms of a scalar field or a scalar condensate of fermions evolving along an effective potential. The effective potential has various features, such as a flat region where the energy density remains nearly constant, a steep region for reheating, and a graceful end to inflation. (The exceptions, chaotic inflation models, rely on near Planck-scale physics.) All of the field theoretic scaffolding is there for one simple purpose: to create a change in the equation-of-state from an epoch in which $\alpha < -1/3$ and the universe inflates, to an epoch where $\alpha \geq 0$ and the universe returns to a decelerating, hot big bang universe. It is conceivable that the same can be achieved without invoking any scalar fields or fermions. For example, some of us are looking at recent progress in understanding stringy properties in the vicinity of black hole horizons to provide insights about stringy properties in a different setting, de Sitter (inflationary) horizons. In the end, it may simply be the behavior of superstrings at high temperatures and high densities that automatically produces this change in the equation-of-state in a manner that cannot be understood from the point-of-view of field theory.

The comments in this section are highly speculative since they are based on highly uncertain guesses about fundamental physics and cosmology. The principal point is that, as each proceeds, fundamental physics and cosmology will have to be reconciled. I have argued that this may not be the time for reconciliation since both are too unsettled. But, based on the current rate of progress, the right time to begin may lie within the 21st-century window. It promises to be one of the most exciting and profound steps in the history of science, drawing together all of our knowledge of space, time, energy, and matter. And, most likely, bringing together detailed knowledge of cosmology and fundamental physics will produce numerous new puzzles and debates that will carry beyond the 21st century. Of all of the predictions I have made, the surest is the last one: that cosmology will once again be a hot topic of discussion one hundred years from now when our great grandchildren meet to celebrate at the Princeton 350th Anniversary Celebration.

ACKNOWLEDGEMENTS

I would like to thank my many collaborators whose work and discussions have shaped the views expressed in the paper.This research was supported by the DOE at Penn (DOE-EY-76-C-02-3071).

REFERENCES

[1] A.H. Guth, Phys. Rev. D **23** 347, (1981).

[2] V. de Lapparent, M.J. Geller, and J. Huchra, Ap. J. **302** L1 (1986); ibid., **332** 44 (1988).

[3] S.A. Shectman, S.D. Landy, A. Oemler, D. Tucker, R.P. Kirshner, H. Lin, and P. Schechter, Ap. J. **470** 172 (1996).

[4] G.F. Smoot et al., Ap. J. **396** L1 (1992); C.L. Bennett et al., submitted to Ap. J. (1994); K.M. Gorski, et al., Ap. J. **430** L89 (1994).

[5] A.D. Linde, Phys. Lett. **108B** 389 (1982); A. Albrecht and P.J. Steinhardt, Phys. Rev. Lett. **48** 1220 (1982).

[6] A.H. Guth and P.J. Steinhardt, "The Inflationary Universe" in *The New Physics*, edited by P. Davies (Cambridge U. Press, Cambridge, 1989), 34-60. Universe, p. 34-60.

[7] A. D. Linde, in *Particle Physics and Inflationary Cosmology*, (Harwood Academic Publishers, New York, 1990).

[8] J. Bardeen, P.J. Steinhardt and M.S. Turner, Phys. Rev. D **28** 679 (1983); A.H. Guth and S.-Y. Pi, Phys. Rev. Lett. **49** 1110 (1982); A.A. Starobinskii, Phys. Lett. B **117** 175 (1982); S.W. Hawking, Phys. Lett. B **115** 295 (1982).

[9] R.H. Dicke and P.J.E. Peebles, in *General Relativity: An Einstein Centenary Survey*; edited by S.W. Hawking and W. Israel (Cambridge: Cambridge U. Press, 1979), 505-517.

[10] E.R. Harrison, Phys. Rev. D **1** 2726 (1970); Ya.B. Zel'dovich, Mon. Not. R. Soc. **160** 1p (1972); P.J.E. Peebles and J.T. Yu, Astrophys. J. **162** 815 (1970).

[11] V.A. Rubakov, M.V. Sazhin, A.V. Veryaskin, Phys. Lett. B **115** 189 (1982).

[12] A.I. Starobinskii, Sov. Astron. Lett. **11** 133 (1985).

[13] L.F. Abbott M. Wise, Nucl. Phys. B **244** 541 (1984).

[14] Tilt in inflationary perturbation spectra was first noted in J. Bardeen, P. J. Steinhardt and M. S. Turner, ibid.

[15] R.L. Davis, H.M. Hodges, G.F. Smoot, P.J. Steinhardt, and M.S. Turner, Phys. Rev. Lett. **69** 1856 (1992).

[16] P.J. Steinhardt, in *Proceedings of the Snowmass Workshop on the Future of Particle Astrophysics and Cosmology*, edited by E.W. Kolb and R. Peccei, (World Scientific, 1995); IJMPA **A10** 1091-1124 (1995).

[17] W. Hu, in *The Universe at High-z, Large Scale Structure and the Cosmic Microwave Background*, edited by E. Martinez-Gonzalez and J.L Sanz (Springer Verlag); W. Hu, N. Sugiyama, and J. Silk, Nature, to appear.

[18] R. Crittenden, R.L. Davis, and P. Steinhardt, Ap. J. **L13** (1993).

[19] R. Crittenden, J.R. Bond, R.L. Davis., G. Efstathiou, and P.J. Steinhardt, Phys. Rev. Lett. **71** 324 (1993).

[20] M. Kamionkowski, D.N. Spergel, N. Sugiyama, Ap. J. Lett. **426** 57 (1994).

[21] J.R. Bond, R. Crittenden, J.R. Bond, R.L. Davis., G. Efstathiou, and P.J. Steinhardt, Phys. Rev. Lett. **72** 13 (1994).

[22] G. Jungman, M. Kamionkowski, A. Kosowsky, D.N. Spergel, Phys. Rev. D **54** 1332 (1996).

[23] W. Hu and M. White, Phys. Rev. Lett. **77** , 1687 (1996).

[24] C. Copi, D.N. Schramm, and M.S. Turner, Science **267** 192 (1995).

[25] D. Tytler, X.-M. Fan, and S. Burles, Nature **381** 207 (1996); D. Tytler, S. Burles, and D. Kirkman, astro-ph/9612121

[26] N. Hata, G. Steigman, S. Bludman, P. Langacker, Phys. Rev. D **55** 540 (1997).

[27] J.P. Ostriker and P.J. Steinhardt, Nature **377** 600 (1995).

[28] J. Hartle and S.W. Hawking, Phys. Rev. D **28** 2960 (1983).

[29] A.D. Linde, Phys. Lett. **129B** 177 (1983); A.D. Linde, Mod. Phys. Lett. **A1** 81 (1986); Phys. Lett. **175B** 395 (1986).

[30] T.W.B. Kibble, J. Phys. **A9** 1387 (1976).

[31] J.P. Presskill, Phys. Rev. Lett. **43** 1365 (1979).

7.7 DISCUSSION

Session Chair: Anthony Hewish
Rapporteur: Andrew Millard

NAPPI: You mentioned "typical inflationary models." Can you tell us if there is a model of inflation favored at the moment and which one it is?

STEINHARDT: A variety of mechanisms of initial conditions have been discovered over the past decade which produce an acceptable inflationary scenario. Although individual theorists have their preferences, there are not reliable guiding principles to determine which variant is correct. Perhaps more specific information will come from what we ultimately learn about fundamental physics close to the Planck scale. There are many detailed ways of accomplishing the same general changes in equations of state and solving the cosmological problems. The predictions discussed in the paper, though, are not sensitive to the particular variant of inflation and, hence, serve as generic tests of inflation. The main lesson is that there is more than one way of doing things.

SPERGEL: In addition to the problems you mentioned, determination of the polarization of the microwave background and observation of high red-shift supernovae are two other potentially powerful probes that may help to distinguish between cosmological parameters.

STEINHARDT: I agree, but the interpretation of these measurements is model-dependent. The parameters can be determined only once the underlying model, such as inflation or cosmic defects, is settled first.

MANN: Suppose it might become possible to measure the neutrino cosmic background radiation, just as the photon background has been measured; what additional constraints would this place on cosmological theories?

STEINHARDT: My initial thought is that the results could be reasonably expected to corroborate and refine existing information. The neutrino background won't have formed at a much higher energy than the photon background (only six orders of magnitude), although something unexpected may have happened between the decouplings which would cause a different between the two backgrounds.

BAHCALL: Several observations limit the cosmological constant with, for instance, measurements of lensing and high red-shift supernovae implying a low value. From the microwave background spectrum, you show a strong peak which may suggest a high cosmological constant value. Are those two inconsistent with each other?

STEINHARDT: I don't think so. All these techniques are in early stages of evolution. Lensing statistics as a function of red shift depend on knowing the effects of evolution and dust. Recent work suggests that the dust effect is large enough to erode the constraint on the cosmological constant to a level where it is consistent with the other data shown in the paper. Cosmic supernovae measurements are a newer and highly promising approach for constraining the cosmological constant. Current limits are tight, but there remains a substantial range of parameter space consistent with the supernovae measurements and the other measurements shown in the paper. It is premature to read too much into microwave background data at the present time. The exciting thing is that all of these techniques will be refined over the next few years.

PARTRIDGE: This morning John Hopfield spoke of "the mark of Cain," the desire to explain complex biological systems in terms of simple fundamental laws. You seem to take the same approach, looking to fundamental physics to answer cosmological questions. Could you say a little bit about the possibility that messy astrophysics—with gamma ray bursts, gravitational lensing, and high red-shift yet elderly galaxies—could provide some interesting cosmological answers in the next century?

STEINHARDT: New phenomena provide added details about evolution of cosmic structures that constrain cosmological scenarios, but their relation to microscopic physics appears to be too remote to use them as a precise tool for testing fundamental physics near the Planck scale.

SCULLI: Several times you said that information concerning fundamental physics was needed, especially knowledge about physics near the Planck scale. Where might this come from? Experimental information? String theory?

STEINHARDT: The information I have in mind relies on progress in our understanding of Planck scale physics, such as superstring theory, and its indirect corroboration in low energy experiment, to be discussed in Saturday's talks.

CHAPTER 8

GRAVITATION AND EXPERIMENT

THIBAULT DAMOUR

Institut des Hautes Etudes Scientifiques and
DARC, CNRS - Observatoire de Paris, France

Talk dedicated to R.H. Dicke and J.H. Taylor

ABSTRACT

The confrontation between Einstein's gravitation theory and experimental results, notably binary pulsar data, is summarized and its significance discussed. Experiment and theory agree at the 10^{-3} level. All the basic structures of Einstein's theory (coupling of gravity to matter; propagation and self-interaction of the gravitational field, including in strong-field conditions) have been verified. However, some recent theoretical findings (cosmological relaxation toward zero scalar couplings) suggest that the present agreement between Einstein's theory and experiment might be naturally compatible with the existence of a long-range scalar contribution to gravity (such as the dilaton, or a moduli field of string theory). This provides a new theoretical paradigm, and new motivations for improving the experimental tests of gravity. Ultra-high precision tests of the Equivalence Principle appear as the most sensitive way to look for possible long-range deviations from General Relativity: they might open a low-energy window on string-scale physics.

8.1 INTRODUCTION

Einstein's gravitation theory can be thought of as defined by two postulates. One postulate states that the action functional describing the propagation and self-interaction of the gravitational field is

$$S_{\text{gravitation}}\,[g_{\mu\nu}] = \frac{c^4}{16\pi\, G} \int \frac{d^4 x}{c}\, \sqrt{g}\, R(g). \tag{8.1}$$

A second postulate states that the action functional describing the coupling of all the (fermionic and bosonic) fields describing matter and its electro-weak and strong interactions is a (minimal) deformation of the special relativistic action functional used by particle physicists (the so called "Standard Model"), obtained by replacing everywhere the flat Minkowski metric $f_{\mu\nu} = \text{diag}(-1, +1, +1, +1)$ by $g_{\mu\nu}(x^\lambda)$ and the partial derivatives $\partial_\mu \equiv \partial/\partial x^\mu$ by g-covariant derivatives ∇_μ. [With the usual subtlety that one must also introduce a field of orthonormal frames, a "vierbein," for writing down the fermionic terms]. Schematically, one has

$$S_{\text{matter}}\,[\psi, A, H, g] = \int \frac{d^4 x}{c}\, \sqrt{g}\, \mathcal{L}_{\text{matter}}, \tag{8.2}$$

$$\begin{aligned} \mathcal{L}_{\text{matter}} \quad = \quad & -\frac{1}{4} \sum \frac{1}{g_*^2}\, \text{tr}(F_{\mu\nu}\, F^{\mu\nu}) - \sum \overline{\psi}\, \gamma^\mu\, D_\mu\, \psi \\ & -\frac{1}{2}\, |D_\mu\, H|^2 - V(H) - \sum y\, \overline{\psi}\, H\, \psi, \end{aligned} \tag{8.3}$$

where $F_{\mu\nu}$ denotes the curvature of a $U(1)$, $SU(2)$, or $SU(3)$ Yang-Mills connection A_μ, $F^{\mu\nu} = g^{\mu\alpha}\, g^{\nu\beta}\, F_{\alpha\beta}$, g_* being a (bare) gauge coupling constant; $D_\mu \equiv \nabla_\mu + A_\mu$; ψ denotes a fermion field (lepton or quark, coming in various flavours and three generations); γ^μ denotes four Dirac matrices such that $\gamma^\mu\, \gamma^\nu + \gamma^\nu\, \gamma^\mu = 2 g^{\mu\nu}\, 1\!\!I_4$, and H denotes the Higgs doublet of scalar fields, with y some (bare Yukawa) coupling constants.

Einstein's theory of gravitation is then defined by finding the extreme of the total action functional,

$$S_{\text{tot}}\,[g, \psi, A, H] = S_{\text{gravitation}}\,[g] + S_{\text{matter}}\,[\psi, A, H, g]. \tag{8.4}$$

Although, seen from a wider perspective, the two postulates (8.1) and (8.2) follow from the unique requirement that the gravitational interaction be mediated only by massless spin-2 excitations [1], the decomposition in two postulates is convenient for discussing the theoretical significance of various tests of General Relativity. Let us discuss in turn the experimental tests of the coupling of matter to gravity (postulate (8.2)), and the experimental tests of the dynamics of the gravitational field (postulate (8.1)). For more details and references we refer the reader to [2] or [3].

8.2 EXPERIMENTAL TESTS OF THE COUPLING BETWEEN MATTER AND GRAVITY

The fact that the matter Lagrangian (8.3) depends only on a symmetric tensor $g_{\mu\nu}(x)$ and its first derivatives (i.e., the postulate of a "metric coupling" between matter and gravity) is a strong assumption (often referred to as the "equivalence principle") which has many observable consequences for the behaviour of localized test systems embedded in given, external gravitational fields. Indeed, using a theorem of Fermi and Cartan [4] (stating the existence of coordinate systems such that, along any given time-like curve, the metric components can be set to their Minkowski values, and their first derivatives made to vanish), one derives from the postulate (8.2) the following observable consequences:

C_1 : Constancy of the "constants" : the outcome of local non-gravitational experiments, referred to local standards, depends only on the values of the coupling constants and mass scales entering the Standard Model. [In particular, the cosmological evolution of the universe at large has no influence on local experiments].

C_2 : Local Lorentz invariance : local non-gravitational experiments exhibit no preferred directions in spacetime [i.e., neither spacelike ones (isotropy), nor timelike ones (boost invariance)].

C_3 : "Principle of geodesics" and universality of free fall : small, electrically neutral, non-self-gravitating bodies follow geodesics of the external spacetime $g_{\mu\nu}(x^\lambda)$. In particular, two test bodies dropped at the same location and with the same velocity in an external gravitational field fall in the same way, independently of their masses and compositions.

C_4 : Universality of gravitational redshift : when intercompared by means of electromagnetic signals, two identically constructed clocks located at two different positions in a static external Newtonian potential $U(\mathbf{x})$ exhibit, independently of their nature and constitution, the (apparent) difference in clock rate:

$$\frac{\tau_1}{\tau_2} = \frac{\nu_2}{\nu_1} = 1 + \frac{1}{c^2}\left[U(\mathbf{x}_1) - U(\mathbf{x}_2)\right] + O\left(\frac{1}{c^4}\right). \tag{8.5}$$

Many experiments or observations have tested the observable consequences $C_1 - C_4$ and found them to hold within the experimental errors. Many sorts of data (from spectral lines in distant galaxies to a natural fission reactor phenomenon which took place at Oklo, Gabon two billion years ago) have been used to set limits on a possible time variation of the basic coupling constants of the Standard Model. The best results concern the electromagnetic coupling, i.e., the fine-structure constant $\alpha_{\rm em}$. A recent reanalysis of the Oklo phenomenon gives a conservative upper bound [5]

$$-6.7 \times 10^{-17}\,\mathrm{yr}^{-1} < \frac{\dot{\alpha}_{\mathrm{em}}}{\alpha_{\mathrm{em}}} < 5.0 \times 10^{-17}\,\mathrm{yr}^{-1}, \tag{8.6}$$

which is much smaller than the cosmological time scale $\sim 10^{-10}\,\mathrm{yr}^{-1}$. It would be interesting to confirm and/or improve the limit (8.6) by direct laboratory measurements comparing clocks based on atomic transitions having different dependences on α_{em}. [Current atomic clock tests of the constancy of α_{em} give the limit $|\dot{\alpha}_{\mathrm{em}}/\alpha_{\mathrm{em}}| < 3.7 \times 10^{-14}\,\mathrm{yr}^{-1}$ [6].]

Any "isotropy of space" having a direct effect on the energy levels of atomic nuclei has been constrained to the impressive 10^{-27} level [6]. The universality of free fall has been verified at the 10^{-12} level both for laboratory bodies [7], e.g. (from the last reference in [7])

$$\left(\frac{\Delta a}{a}\right)_{\mathrm{Be\,Cu}} = (-1.9 \pm 2.5) \times 10^{-12}, \tag{8.7}$$

and for the gravitational accelerations of the Moon and the Earth toward the Sun [8],

$$\left(\frac{\Delta a}{a}\right)_{\mathrm{Moon\,Earth}} = (-3.2 \pm 4.6) \times 10^{-13}. \tag{8.8}$$

The "gravitational redshift" of clock rates given by Eq. (8.5) has been verified at the 10^{-4} level by comparing a hydrogen-maser clock flying on a rocket up to an altitude $\sim 10,000$ km to a similar clock on the ground [9].

In conclusion, the main observable consequences of the Einsteinian postulate (8.2) concerning the coupling between matter and gravity ("equivalence principle") have been verified with high precision by all experiments to date. The traditional paradigm (first put forward by Fierz[10]) is that the extremely high precision of free fall experiments (10^{-12} level) strongly suggests that the coupling between matter and gravity is exactly of the "metric" form (8.2), but leaves open possibilities more general than Eq. (8.1) for the spin-content and dynamics of the fields mediating the gravitational interaction. We shall provisionally adopt this paradigm to discuss the tests of the other Einsteinian postulate, Eq. (8.1). However, we shall emphasize at the end that recent theoretical findings suggest a new paradigm.

8.3 TESTS OF THE DYNAMICS OF THE GRAVITATIONAL FIELD IN THE WEAK FIELD REGIME

Let us now consider the experimental tests of the dynamics of the gravitational field, defined in General Relativity by the action functional (8.1). Following first the traditional paradigm, it is convenient to enlarge our framework by embedding General Relativity within the class of the most natural relativistic theories of gravitation which satisfy exactly the matter-coupling tests discussed above while differing in the description of the degrees of freedom of the gravitational field. This class of theories are the metrically-coupled tensor-scalar theories, first introduced by Fierz [10] in a work where he noticed that the class of non-metrically-coupled tensor-scalar theories previously introduced by Jordan [11] would generically entail unacceptably large violations of the consequence C_1. [The fact that it would, by the same token, entail even larger violations of the consequence C_3 was, probably, first noticed by Dicke in subsequent work]. The metrically-coupled (or equivalence-principle respecting) tensor-scalar theories are defined by keeping the postulate (8.2), but replacing the postulate (8.1) by demanding that the "physical" metric $g_{\mu\nu}$ (coupled to ordinary matter) be a composite object of the form

$$g_{\mu\nu} = A^2(\varphi)\, g^*_{\mu\nu}, \tag{8.9}$$

where the dynamics of the "Einstein" metric $g^*_{\mu\nu}$ is defined by the action functional (8.1) (written with the replacement $g_{\mu\nu} \to g^*_{\mu\nu}$) and where φ is a massless scalar field. [More generally, one can consider several massless scalar fields, with an action functional of the form of a general nonlinear σ model [12]]. In other words, the action functional describing the dynamics of the spin 2 and spin 0 degrees of freedom contained in this generalized theory of gravitation reads

$$S_{\text{gravitational}}\,[g^*_{\mu\nu}, \varphi] = \frac{c^4}{16\pi\, G_*} \int \frac{d^4x}{c}\, \sqrt{g_*}\, [R(g_*) - 2g_*^{\mu\nu}\, \partial_\mu\, \varphi\, \partial_\nu\, \varphi]\,. \tag{8.10}$$

Here, G_* denotes some bare gravitational coupling constant. This class of theories contains an arbitrary function, the "coupling function" $A(\varphi)$. When $A(\varphi) =$ const., the scalar field is not coupled to matter and one falls back (with suitable boundary conditions) on Einstein's theory. The simple, one-parameter subclass $A(\varphi) = \exp(\alpha_0\, \varphi)$ with $\alpha_0 \in R$ is the Jordan-Fierz-Brans-Dicke theory [10, 14, 15]. In the general case, one can define the (field-dependent) coupling strength of φ to matter by

$$\alpha(\varphi) \equiv \frac{\partial \ln A(\varphi)}{\partial \varphi}. \tag{8.11}$$

It is possible to work out in detail the observable consequences of tensor-scalar theories and to contrast them with the general relativistic case (see, e.g., Ref. [12]).

Let us now consider the experimental tests of the dynamics of the gravitational field that can be performed in the solar system. Because the planets move with slow velocities ($v/c \sim 10^{-4}$) in a very weak gravitational potential ($U/c^2 \sim (v/c)^2 \sim 10^{-8}$), solar system tests allow us only to probe the quasi-static, weak-field regime of relativistic gravity (technically described by the so-called "post-Newtonian" expansion). In the limit where one keeps only the first relativistic corrections to Newton's gravity (first post-Newtonian approximation), all solar-system gravitational experiments, interpreted within tensor-scalar theories, differ from Einstein's predictions only through the appearance of two "post-Einstein" parameters $\overline{\gamma}$ and $\overline{\beta}$ (related to the usually considered Eddington parameters γ and β through $\overline{\gamma} \equiv \gamma - 1, \overline{\beta} \equiv \beta - 1$). The parameters $\overline{\gamma}$ and $\overline{\beta}$ vanish in General Relativity, and are given in tensor-scalar theories by

$$\overline{\gamma} = -2 \, \frac{\alpha_0^2}{1 + \alpha_0^2} , \tag{8.12}$$

$$\overline{\beta} = +\frac{1}{2} \, \frac{\beta_0 \, \alpha_0^2}{(1 + \alpha_0^2)^2} , \tag{8.13}$$

where $\alpha_0 \equiv \alpha(\varphi_0)$, $\beta_0 \equiv \partial\alpha(\varphi_0)/\partial\varphi_0$; φ_0 denoting the cosmologically-determined value of the scalar field far away from the solar system. Essentially, the parameter $\overline{\gamma}$ depends only on the linearized structure of the gravitational theory (and is a direct measure of its field content, i.e., whether it is pure spin 2 or contains an admixture of spin 0), while the parameter $\overline{\beta}$ parametrizes some of the quadratic nonlinearities in the field equations (cubic vertex of the gravitational field).

All currently performed gravitational experiments in the solar system, including perihelion advances of planetary orbits, the bending and delay of electromagnetic signals passing near the sun, and very accurate range data to the moon obtained by laser echoes, are compatible with the general relativistic predictions $\overline{\gamma} = 0 = \overline{\beta}$ and give upper bounds on both $|\overline{\gamma}|$ and $|\overline{\beta}|$ (i.e., on possible fractional deviations from General Relativity) of order 10^{-3}. More precisely: (i) the Viking mission measurement of the gravitational time delay of radar signals passing near the sun ("Shapiro effect" [16]) gave [13]

$$|\overline{\gamma}| < 2 \times 10^{-3} , \tag{8.14}$$

with similar limits coming from VLBI measurements of the deflection of radio waves by the Sun [18]; (ii) the Lunar Laser Ranging measurements of a possible polarization of the orbit of the moon toward the sun ("Nordtvedt effect" [19]) give [8]

$$4\overline{\beta} - \overline{\gamma} = -0.0007 \pm 0.0010 , \tag{8.15}$$

which, combined with the above constraint on $\overline{\gamma}$, gives

$$|\overline{\beta}| < 6 \times 10^{-4} ; \tag{8.16}$$

and (iii) measurement of Mercury's orbit through planetary radar ranging gave[20]

$$|\overline{\beta}| < 3 \times 10^{-3} , \tag{8.17}$$

when assuming the above Viking limit on $\overline{\gamma}$ and a value of the sun's quadrupole moment $J_2 \sim 2 \times 10^{-7}$.

Recently, the parametrization of the weak-field deviations between generic tensor-multi-scalar theories and Einstein's theory has been extended to the second post-Newtonian order [14]. Only two post-post-Einstein parameters, ε and ζ, representing a deeper layer of structure of the gravitational interaction, show up. These parameters have been shown to be already significantly constrained by binary-pulsar data: $|\varepsilon| < 7 \times 10^{-2}$, $|\zeta| < 6 \times 10^{-3}$. See [14] for a detailed discussion, including the consequences for the interpretation of future, higher-precision solar-system tests.

8.4 TESTS OF THE DYNAMICS OF THE GRAVITATIONAL FIELD IN THE STRONG FIELD REGIME

In spite of the diversity, number, and often high precision of solar system tests, they have an important qualitative weakness : they probe neither the radiation properties nor the strong-field aspects of relativistic gravity. Fortunately, the discovery [15] and continuous observational study of pulsars in gravitationally bound binary orbits has opened up an entirely new testing ground for relativistic gravity, giving us an experimental handle on the regime of strong and/or radiative gravitational fields.

The fact that binary pulsar data allow one to probe the propagation properties of the gravitational field is well known. This comes directly from the fact that the finite velocity of propagation of the gravitational interaction between the pulsar and its companion generates damping-like terms in the equations of motion, i.e., terms which are directed against the velocities. [This can be understood heuristically by considering that the finite velocity of propagation must cause the gravitational force on the pulsar to make an angle with the instantaneous position of the companion [16], and was verified by a careful derivation of the general relativistic equations of motion of binary systems of compact objects [17]]. These damping forces cause the binary orbit to shrink and its orbital period P_b to decrease. The remarkable stability of the pulsar clock, together with the cleanliness of the binary pulsar system, has allowed Taylor and collaborators to measure the secular orbital period decay $\dot{P}_b \equiv dP_b/dt$ [18], thereby giving us a direct experimental probe of the damping terms present in the equations of motion. Note that, contrary to what is commonly stated, the link between the observed quantity $\dot{P}_b$ and the propagation properties of the gravitational interaction is quite direct. [It appears indirect only when one goes through the common but unnecessary detour of a heuristic reasoning based on the consideration of the energy lost into gravitational waves emitted at infinity].

The fact that binary pulsar data allow one to probe strong-field aspects of relativistic gravity is less well known. The a priori reason for saying that they should is that the surface gravitational potential of a neutron star $Gm/c^2R \simeq 0.2$ is a mere factor 2.5 below the black hole limit (and a factor $\sim 10^8$ above the surface potential of the Earth). Due to the peculiar "effacement" properties of strong-field effects taking place in General Relativity [17], the fact that pulsar data probe the strong-gravitational-field regime can only be seen when contrasting

Einstein's theory with more general theories. In particular, it has been found in tensor-scalar theories [19] that a self-gravity as strong as that of a neutron star can naturally (i.e., without fine tuning of parameters) induce order-unity deviations from general relativistic predictions in the orbital dynamics of a binary pulsar thanks to the existence of nonperturbative strong-field effects. [The adjective "nonperturbative" refers here to the fact that this phenomenon is nonanalytic in the coupling strength of the scalar field, Eq. (8.11), which can be as small as wished in the weak-field limit]. As far as we know, this is the first example where large deviations from General Relativity, induced by strong self-gravity effects, occur in a theory which contains only positive energy excitations and whose post-Newtonian limit can be arbitrarily close to that of General Relativity. [The strong-field deviations considered in previous studies [2, 12] arose in theories containing negative energy excitations.]

A comprehensive account of the use of binary pulsars as laboratories for testing strong-field gravity will be found in ref. [20]. Two complementary approaches can be pursued : a phenomenological one ("Parametrized Post-Keplerian" formalism), or a theory-dependent one [12, 20].

The phenomenological analysis of binary pulsar timing data consists in fitting the observed sequence of pulse arrival times to the generic DD timing formula [21] whose functional form has been shown to be common to the whole class of tensor-multi-scalar theories. The least-squares fit between the timing data and the parameter-dependent DD timing formula allows one to measure, besides some "Keplerian" parameters ("orbital period" P_b, "eccentricity" e,...), a maximum of eight "post-Keplerian" parameters : $k, \gamma, \dot{P}_b, r, s, \delta_\theta, \dot{e}$ and $\dot{x}$. Here, $k \equiv \dot{\omega} P_b / 2\pi$ is the fractional periastron advance per orbit, γ a time dilation parameter (not to be confused with its post-Newtonian namesake), $\dot{P}_b$ the orbital period derivative mentioned above, and r and s the "range" and "shape" parameters of the gravitational time delay caused by the companion. The important point is that the post-Keplerian parameters can be measured without assuming any specific theory of gravity. Now, each specific relativistic theory of gravity predicts that, for instance, $k, \gamma, \dot{P}_b, r$ and s (to quote parameters that have been successfully measured from some binary pulsar data) are some theory-dependent functions of the (unknown) masses m_1, m_2 of the pulsar and its companion. Therefore, in our example, the five simultaneous phenomenological measurements of k, γ, $\dot{P}_b, r$ and s determine, for each given theory, five corresponding theory-dependent curves in the $m_1 - m_2$ plane (through the 5 equations $k^{\text{measured}} = k^{\text{theory}}(m_1, m_2)$, etc...). This yields three $(3 = 5 - 2)$ tests of the specified theory, according to whether the five curves meet at one point in the mass plane, as they should. In the most general (and optimistic) case, discussed in [20], one can phenomenologically analyze both timing data and pulse-structure data (pulse shape and polarization) to extract up to nineteen post-Keplerian parameters. Simultaneous measurement of these 19 parameters in one binary pulsar system would yield 15 tests of relativistic gravity (where one must subtract 4 because, besides the two unknown masses

m_1, m_2, generic post-Keplerian parameters can depend upon the two unknown Euler angles determining the direction of the spin of the pulsar). The theoretical significance of these tests depends upon the physics lying behind the post-Keplerian parameters involved in the tests. For instance, as we said above, a test involving $\dot{P}_b$ probes the propagation (and helicity) properties of the gravitational interaction. But a test involving, say, k, γ, r or s probes (as shown by combining the results of [12] and [19]) strong self-gravity effects independently of radiative effects.

Besides the phenomenological analysis of binary pulsar data, one can also adopt a theory-dependent methodology [12, 20]. The idea here is to work from the start within a certain finite-dimensional "space of theories," i.e., within a specific class of gravitational theories labelled by some theory parameters. Then, by fitting the raw pulsar data to the predictions of the considered class of theories, one can determine which regions of theory-space are compatible (at say the 90% confidence level) with the available experimental data. This method can be viewed as a strong-field generalization of the parametrized post-Newtonian formalism [2] used to analyze solar-system experiments. In fact, under the assumption that strong-gravity effects in neutron stars can be expanded in powers of the "compactness" $c_A \equiv -2\, \partial \ln m_A / \partial \ln G \sim G\, m_A / c^2\, R_A$, Ref. [12] has shown that the observable predictions of generic tensor-multi-scalar theories could be parametrized by a sequence of "theory parameters."

$$\overline{\gamma}\,,\, \overline{\beta}\,,\, \beta_2\,,\, \beta'\,,\, \beta''\,,\, \beta_3\,,\, (\beta\beta')\ldots \tag{8.18}$$

representing deeper and deeper layers of structure of the relativistic gravitational interaction beyond the first-order post-Newtonian level parametrized by $\overline{\gamma}$ and $\overline{\beta}$ (the second layer β_2, β' being equivalent to the parameters ζ, ε describing the second-order post-Newtonian level [14], etc.). When non-perturbative strong-field effects develop, one cannot use the multi-parameter approach just mentioned, based on expansions in powers of the "compactnesses." A useful alternative approach is then to work within specific, low-dimensional "mini-spaces of theories." Of particular interest is the two-dimensional mini-space of tensor-scalar theories defined by the coupling function $A(\varphi) = \exp\left(\alpha_0\, \varphi + \frac{1}{2}\, \beta_0\, \varphi^2\right)$. The predictions of this family of theories (parametrized by α_0 and β_0) are analytically described, in weak-field contexts, by the post-Einstein parameter (8.12), and can be studied in strong-field contexts by combining analytical and numerical methods [22].

After having reviewed the theory of pulsar tests, let us briefly summarize the current experimental situation. Concerning the first discovered binary pulsar PSR1913 + 16 [15], it has been possible to measure with accuracy the three post-Keplerian parameters k, γ, and $\dot{P}_b$. From what was said above, these three simultaneous measurements yield *one* test of gravitation theories. After subtracting a small ($\sim 10^{-14}$ level in $\dot{P}_b$), but significant, perturbing effect caused by the Galaxy [23], one finds that General Relativity passes this $(k - \gamma - \dot{P}_b)_{1913+16}$ test with complete success at the 10^{-3} level. More precisely, one finds [24, 18],

$$\begin{aligned}\left[\frac{\dot{P}_b^{\rm obs}-\dot{P}_b^{\rm galactic}}{\dot{P}_b^{\rm GR}[k^{\rm obs},\gamma^{\rm obs}]}\right]_{1913+16} &= 1.0032 \pm 0.0023({\rm obs}) \pm 0.0026({\rm galactic}) \\ &= 1.0032 \pm 0.0035\,, \end{aligned} \tag{8.19}$$

where $\dot{P}_b^{\rm GR}[k^{\rm obs},\gamma^{\rm obs}]$ is the GR prediction for the orbital period decay computed from the observed values of the other two post-Keplerian parameters k and γ. [More explicitly, this means that the two measurements $k^{\rm obs}$ and $\gamma^{\rm obs}$ are used, together with the corresponding general relativistic predictions $k^{\rm obs} = k^{\rm GR}(m_1,m_2)$, $\gamma^{\rm obs} = \gamma^{\rm GR}(m_1,m_2)$, to compute the two masses m_1 and m_2 that enter the theoretical prediction for $\dot{P}_b$.]

This beautiful confirmation of General Relativity is an embarrassment of riches in that it probes, at the same time, the propagation *and* strong-field properties of relativistic gravity! If the timing accuracy of PSR1913 + 16 could improve by a significant factor two more post-Keplerian parameters (r and s) would become measurable and would allow one to probe separately the propagation and strong-field aspects [24]. Fortunately, the discovery of the binary pulsar PSR1534 + 12 [25] (which is significantly stronger than PSR1913 + 16 and has a more favourably oriented orbit) has opened a new testing ground, in which it has been possible to probe strong-field gravity independently of radiative effects. A phenomenological analysis of the timing data of PSR1534 + 12 has allowed one to measure the four post-Keplerian parameters k, γ, r, and s [24]. From what was said above, these four simultaneous measurements yield *two* tests of strong-field gravity, without mixing of radiative effects. General Relativity is found to pass these tests with complete success within the measurement accuracy [24, 18]. The most precise of these new, pure, strong-field tests is the one obtained by combining the measurements of k, γ, and s. Using the data reported in [33] (with, following [14], doubled statistical uncertainties to take care of systematic errors) one finds agreement at the 1% level:

$$\left[\frac{s^{\rm obs}}{s^{\rm GR[k^{obs},\gamma^{obs}]}}\right]_{1534+12} = 1.010 \pm 0.008\,. \tag{8.20}$$

More recently, it has been possible to extract also the "radiative" parameter $\dot{P}_b$ from the timing data of PSR1534 + 12. Again, General Relativity is found to be fully consistent (at the current $\sim 20\%$ level) with the additional test provided by the $\dot{P}_b$ measurement [26, 33]. Note that this gives our second direct experimental confirmation that the gravitational interaction propagates as predicted by Einstein's theory. Moreover, an analysis of the pulse shape of PSR1534 + 12 has shown that the misalignment between the spin vector of the pulsar and the orbital angular momentum was greater than 8° [20]. This opens the possibility that this system will soon allow one to test the spin precession induced by gravitational spin-orbit coupling.

To end this brief summary, let us mention that several other binary pulsar systems (of a different class than that of $1913 + 16$ and $1534 + 12$) can be used to test relativistic gravity. We have here in mind nearly circular systems made of a neutron star and a white dwarf. Such dissymetric systems are useful probes of the possible existence of dipolar gravitational waves [36] and/or of a possible violation of the universality of free fall linked to the strong self-gravity of the neutron star [37]. A theory-dependent analysis of the published pulsar data on PSRs $1913+16$, $1534+12$, and $0655+64$ (a dissymetric system constraining the existence of dipolar radiation) has been recently performed within the (α_0, β_0)-space of tensor-scalar theories introduced above [22]. This analysis proves that binary-pulsar data exclude large regions of theory-space which are compatible with solar-system experiments. This is illustrated in Fig. 8.1 below (reproduced from Fig. 9 of Ref. [22]) which shows that β_0 must be larger than about -5, while any value of β_0 is compatible with weak-field tests as long as α_0 is small enough. Note that Fig. 8.1 is drawn in the framework of tensor-scalar theories respecting the equivalence principle. In the more general (and more plausible; see below) framework of theories where the scalar couplings violate the equivalence principle one gets much stronger constraints on the coupling parameter α_0 of order $\alpha_0^2 \lesssim 10^{-7}$ [34].

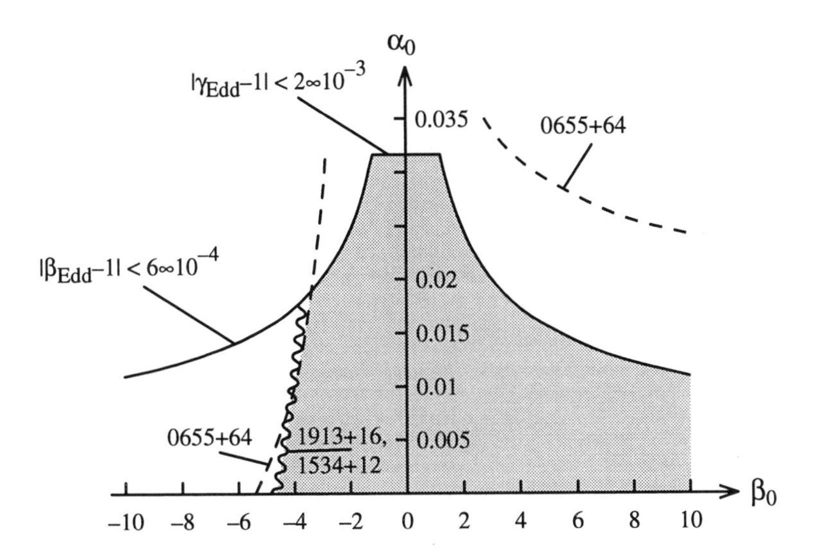

Figure 8.1: Regions of the (α_0, β_0)-plane allowed by (composition-independent) solar-system experiments and three binary pulsar experiments. The region simultaneously allowed by all the tests is shaded. Note that binary pulsar tests exclude a large portion of the region (below the solid line) allowed by solar-system tests. (Figure taken from Ref. [22].)

For a general review of the use of pulsars as physics laboratories the reader can consult Ref. [27].

8.5 WAS EINSTEIN 100% RIGHT?

Summarizing the experimental evidence discussed above, we can say that Einstein's postulate of a pure metric coupling between matter and gravity ("equivalence principle") appears to be, at least, 99.9999999999% right (because of universality-of-free-fall experiments), while Einstein's postulate (8.1) for the field content and dynamics of the gravitational field appears to be, at least, 99.9% correct both in the quasi-static-weak-field limit appropriate to solar-system experiments, and in the radiative-strong-field regime explored by binary pulsar experiments. Should one apply Occam's razor and decide that Einstein must have been 100% right, and then stop testing General Relativity? My answer is definitely, no!

First, one should continue testing a basic physical theory such as General Relativity to the utmost precision available simply because it is one of the essential pillars of the framework of physics. This is the fundamental justification of an experiment such as Gravity Probe B (the Stanford gyroscope experiment), which will advance by two orders of magnitude our experimental knowledge of post-Newtonian gravity. Second, some very crucial qualitative features of General Relativity have not yet been verified : in particular the existence of black holes, and the direct detection on Earth of gravitational waves. Hopefully, the LIGO/VIRGO network of interferometric detectors will observe gravitational waves early in the next century. [See the contribution of Kip Thorne to these proceedings.]

Last, some recent theoretical findings suggest that the current level of precision of the experimental tests of gravity might be naturally (i.e., without fine tuning of parameters) compatible with Einstein being actually only 50% right! By this we mean that the correct theory of gravity could involve, on the same fundamental level as the Einsteinian tensor field $g^*_{\mu\nu}$, a massless scalar field φ.

Let us first question the traditional paradigm (initiated by Fierz [10] and enshrined by Dicke [15], Nordtvedt and Will [2]) according to which special attention should be given to tensor-scalar theories respecting the equivalence principle. This class of theories was, in fact, introduced in a purely *ad hoc* way so as to prevent too violent a contradiction with experiment. However, it is important to notice that the scalar couplings which arise naturally in theories unifying gravity with the other interactions systematically violate the equivalence principle. This is true both in Kaluza-Klein theories (which were the starting point of Jordan's theory) and in string theories. In particular, it is striking that (as first noted by Scherk and Schwarz [40]) the dilaton field Φ, which plays an essential role in string theory, appears as a necessary partner of the graviton field $g_{\mu\nu}$ in all string models. Let us recall that $g_s = e^{\Phi}$ is the basic string coupling constant (measuring the weight of successive string loop contributions) which determines, together with other scalar fields (the moduli), the values of all the coupling constants of the low-energy world. This means, for instance, that the fine-structure constant $\alpha_{\rm em}$ is a function of Φ (and possibly of other moduli fields). This spatiotemporal variability of coupling constants entails a clear violation of the equivalence principle. In particular, $\alpha_{\rm em}$ would be expected to vary on the Hubble time scale (in contradiction with the limit (8.6) above), and materials of different compositions would be expected to fall with different accelerations (in contradiction with the limits (8.7), (8.8) above).

The most popular idea for reconciling gravitational experiments with the existence, at a fundamental level, of scalar partners of $g_{\mu\nu}$ is to assume that all these scalar fields (which are massless before supersymmetry breaking) will acquire a mass after supersymmetry breaking. Typically one expects this mass m to be in the TeV range [41]. This would ensure that scalar exchange brings only negligible, exponentially small corrections $\propto \exp(-mr/\hbar c)$ to the general relativistic predictions concerning low-energy gravitational effects.

However, this idea is fraught with many cosmological difficulties. A first difficulty is that, the dilaton being protected from getting a mass to all orders of perturbation theory, any putative non-perturbative potential $V(\Phi)$ will be extremely shallow, which makes it difficult to fix the VEV of Φ without fine-tuning the initial conditions [42]. A second difficulty is that additional fine-tuning (or some new mechanism) is needed to ensure that the value of the potential $V(\Phi)$ at its minimum is zero, or at least 120 orders of magnitude smaller than the Planck density ("cosmological constant problem"). A third problem is that one generically expects a lot of potential energy to be stored initially in $V(\Phi)$. The cosmological decay of this energy is either too slow or leads to an overproduction of entropy ("Polonyi problem" [29]). Moreover, if cosmological strings exist they tend to radiate a lot of dilatons thereby causing a problem similar to the usual Polonyi problem [44].

Though these cosmological difficulties might be solved by a combination of ad hoc solutions (e.g., introducing a secondary stage of inflation to dilute previously produced dilatons [45, 46], a more radical solution to the problem of reconciling the existence of the dilaton (or any moduli field) with experimental tests and cosmological data has been proposed [47] (see also [28] which considered an equivalence-principle-respecting scalar field). The main idea of Ref. [47] is that string-loop effects (i.e., corrections depending upon $g_s = e^{\Phi}$ induced by worldsheets of arbitrary genus in intermediate string states) may modify the low-energy, Kaluza-Klein type matter couplings ($\propto e^{-2\Phi} F_{\mu\nu} F^{\mu\nu}$) of the dilaton (or moduli) in such a manner that the VEV of Φ be cosmologically driven toward a finite value Φ_m where it decouples from matter. For such a "least coupling principle" to hold, the loop-modified coupling functions of the dilaton, $B_i(\Phi) = e^{-2\Phi} + c_0 + c_1 e^{2\Phi} + \cdots +$ (nonperturbative terms), must exhibit extrema for finite values of Φ, and these extrema must have certain universality properties. More precisely, the most general low-energy couplings induced by string-loop effects will be such that the various terms on the right-hand side of Eq. (8.3) will be multiplied by several different functions of the scalar field(s): say a factor $B_F(\varphi)$ in factor of the kinetic terms of the gauge fields, a factor $B_\psi(\varphi)$ in factor of the Dirac kinetic terms, etc. We work here in the Einstein frame, and with a canonically normalized scalar field φ, i.e., the Lagrangian density has the form

$$\mathcal{L} = \frac{1}{16\pi G_*} \left[R(g_*) - 2 g_*^{\mu\nu} \, \partial_\mu \varphi \, \partial_\nu \varphi \right] - \frac{1}{4} \, B_F(\varphi) \, F_{\mu\nu} \, F^{\mu\nu} + \cdots \tag{8.21}$$

It has been shown in [47] that if the various coupling functions $B_i(\varphi)$, $i = F, \psi, \ldots$, all admit an extremum (which must be a maximum for the "leading" B_i) at some common value φ_m of φ, the cosmological evolution of the coupled tensor-scalar-matter system will drive φ towards the value φ_m, at which φ decouples from matter. As suggested in [47] a natural way in which the required conditions could be satisfied is through the existence of a discrete symmetry in scalar space. [For instance, a symmetry under $\varphi \to -\varphi$ would guarantee that all the scalar coupling functions reach an extremum at the self-dual point $\varphi_m = 0$]. The existence of such symmetries have been proven for some of the scalar fields appearing in string theory (target-space duality for the moduli fields) and conjectured for others (S-duality for the dilaton). This gives us some hope that the mechanism of [47] could apply and thereby naturally reconcile the existence of massless scalar fields with experiment.

A study of the efficiency of attraction of φ towards φ_m estimates that the present vacuum expectation value φ_0 of the scalar field would differ (in a rms sense) from φ_m by

$$\varphi_0 - \varphi_m \sim 2.75 \times 10^{-9} \times \kappa^{-3} \, \Omega^{-3/4} \, \Delta\varphi \tag{8.22}$$

where κ denotes the curvature of $\ln B_F(\varphi)$ around the maximum φ_m and $\Delta\varphi$ the deviation $\varphi - \varphi_m$ at the beginning of the (classical) radiation era. Equation (8.22) predicts (when $\Delta\varphi$ is of order unity[1]) the existence, at the present cosmological epoch, of many small, but not unmeasurably small, deviations from General Relativity proportional to the *square* of $\varphi_0 - \varphi_m$. This provides a new incentive for trying to improve by several orders of magnitude the various experimental tests of Einstein's equivalence principle, i.e., of the consequences $C_1 - C_4$ recalled above. The most sensitive way to look for a small residual violation of the equivalence principle is to perform improved tests of the universality of free fall. The mechanism of Ref. [47] suggests a specific composition-dependence of the residual differential acceleration of free fall and estimates that a non-zero signal could exist at very small level as illustrated in Fig. 8.2 (taken from Ref. [47]). The dashed line in this figure is (as Eq. (8.22) above) a rough analytical estimate

[1]However, $\Delta\varphi$ could be $\ll 1$ if the attractor mechanism already applies during an early stage of potential-driven inflation [49].

(assuming random phases) which reads

$$\left(\frac{\Delta a}{a}\right)^{\max}_{\text{rms}} \sim 1.36 \times 10^{-18}\, \kappa^{-4}\, \Omega^{-3/2}\, (\Delta\varphi)^2, \tag{8.23}$$

where κ is expected to be of order unity (or smaller, leading to a larger signal, in the case where φ is a modulus rather than the dilaton). Let us emphasize again that the strength of the cosmological scenario considered here as counterargument to applying Occam's razor lies in the fact that the very small number on the right-hand side of Eq. (8.23) has been derived without any fine tuning or use of small parameters, and turns out to be naturally smaller than the 10^{-12} level presently tested by equivalence-principle experiments (see Eqs. (8.7), (8.8)). The estimate (8.23) gives added significance to the project of a Satellite Test of the Equivalence Principle (nicknamed STEP, and currently studied by NASA, ESA, and CNES) which aims at probing the universality of free fall of pairs of test masses orbiting the Earth at the 10^{-17} or 10^{-18} level [50].

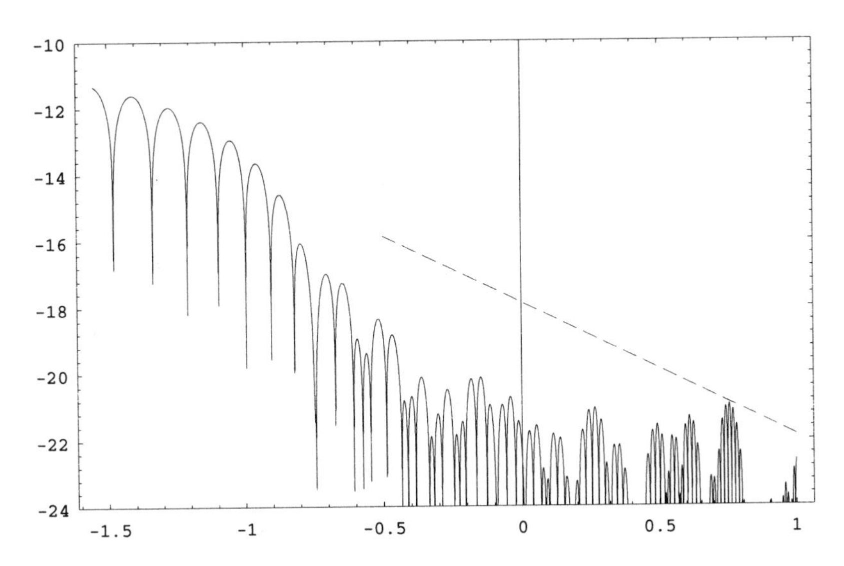

Figure 8.2: The solid line represents $\log_{10}(\Delta a/a)_{\max}$ as a function of $\log_{10}\kappa$, i.e., the expected present level of violation of the universality of free fall as a function of the curvature κ of the (string-loop induced) coupling function $\ln B_F^{-1}(\varphi)$ near a minimum φ. The dashed line is a rough analytical estimate (assuming random phases of oscillations). (Figure taken from Ref. [47].)

REFERENCES

[1] R.P. Feynman, F.B. Morinigo, and W.G. Wagner, *Feynman Lectures on Gravitation*, edited by Brian Hatfield (Addison-Wesley, Reading, 1995); S. Weinberg, Phys. Rev. **138** B988 (1965), V.I. Ogievetsky and I.V. Polubarinov, Ann. Phys. N.Y. **35** 167 (1965); W. Wyss, Helv. Phys. Acta **38** 469 (1965); S. Deser, Gen. Rel. Grav. **1** 9 (1970); D.G. Boulware and S. Deser, Ann. Phys. N.Y. **89** 193 (1975); J. Fang and C. Fronsdal, J. Math. Phys. **20** 2264 (1979); R.M. Wald, Phys. Rev. D **33** 3613 (1986); C. Cutler and R.M. Wald, Class. Quantum Grav. **4** 1267 (1987); R.M. Wald, Class. Quantum Grav. **4** 1279 (1987).

[2] C.M. Will, *Theory and Experiment in Gravitational Physics*, 2nd edition (Cambridge University Press, Cambridge, 1993); and Int. J. Mod. Phys. D **1** 13 (1992).

[3] T. Damour, "Gravitation and Experiment" in *Gravitation and Quantizations*, edited by B. Julia and J. Zinn-Justin, Les Houches, Session LVII (Elsevier, Amsterdam, 1995), pp 1-61.

[4] E. Fermi, Atti Accad. Naz. Lincei Cl. Sci. Fis. Mat. & Nat. **31** 184 and 306 (1922); E. Cartan, *Leçons sur la Géométrie des Espaces de Riemann* (Gauthier-Villars, Paris, 1963).

[5] T. Damour and F. Dyson, Nucl. Phys. B, in press; hep-ph/9606486.

[6] J.D. Prestage, R.L. Tjoelker and L. Maleki, Phys. Rev. Lett. **74** 3511 (1995).

[7] J.D. Prestage et al., Phys. Rev. Lett. **54** 2387 (1985); S.K. Lamoreaux et al., Phys. Rev. Lett. **57** 3125 (1986); T.E. Chupp et al., Phys. Rev. Lett. **63** 1541 (1989).

[8] P.G. Roll, R. Krotkov and R.H. Dicke, Ann. Phys. N.Y. **26** 442 (1964); V.B. Braginsky and V.I. Panov, Sov. Phys. JETP **34** 463 (1972); Y. Su et al., Phys. Rev. D **50** 3614 (1994).

[9] J.O. Dickey et al., Science **265** 482 (1994); J.G. Williams, X.X. Newhall and J.O. Dickey, Phys. Rev. D **53** 6730 (1996).

[10] R.F.C. Vessot and M.W. Levine, Gen. Rel. Grav. **10** 181 (1978); R.F.C. Vessot et al., Phys. Rev. Lett. **45** 2081 (1980).

[11] M. Fierz, Helv. Phys. Acta **29** 128 (1956).

[12] P. Jordan, Nature **164** 637 (1949); *Schwerkraft und Weltall* (Vieweg, Braunschweig, 1955).

[13] T. Damour and G. Esposito-Farèse, Class. Quant. Grav. **9** 2093 (1992).

[14] P. Jordan, Z. Phys. **157** 112 (1959).

[15] C. Brans and R.H. Dicke, Phys. Rev. **124** 925 (1961).

[16] I.I. Shapiro, Phys. Rev. Lett. **13** 789 (1964).

[17] R.D. Reasenberg et al., Astrophys. J. **234** L219 (1979).

[18] D.S. Robertson, W.E. Carter and W.H. Dillinger, Nature **349** 768 (1991); D.E. Lebach et al., Phys. Rev. Lett. **75** 1439 (1995).
[19] K. Nordtvedt, Phys. Rev. **170** 1186 (1968).
[20] I.I. Shapiro, in *General Relativity and Gravitation 1989*, edited by N. Ashby, D.F. Bartlett and W. Wyss (Cambridge University Press, Cambridge, 1990), p. 313.
[21] T. Damour and G. Esposito-Farèse, Phys. Rev. D **53** 5541 (1996).
[22] R.A. Hulse and J.H. Taylor, Astrophys. J. Lett. **195** L51 (1975); see also the 1993 Nobel lectures in physics of Hulse (pp. 699-710) and Taylor (pp. 711-719) in Rev. Mod. Phys. **66** n°3 (1994).
[23] P.S. Laplace, *Traité de Mécanique Céleste*, (Courcier, Paris, 1798-1825), Second part: book 10, chapter 7.
[24] T. Damour and N. Deruelle, Phys. Lett. **A87** 81 (1981); T. Damour, C.R. Acad. Sci. Paris **294** 1335 (1982); T. Damour, in *Gravitational Radiation*, edited by N. Deruelle and T. Piran (North-Holland, Amsterdam, 1983), p. 59.
[25] J.H. Taylor, Class. Quant. Grav. **10** S167 (1993), (Supplement 1993) and references therein; see also J.H. Taylor's Nobel lecture quoted in [15].
[26] T. Damour and G. Esposito-Farèse, Phys. Rev. Lett. **70** 2220 (1993).
[27] T. Damour and J.H. Taylor, Phys. Rev. D. **45** 1840 (1992).
[28] T. Damour and N. Deruelle, Ann. Inst. H. Poincaré **43** 107 (1985) and **44** 263 (1986).
[29] T. Damour and G. Esposito-Farèse, Phys. Rev. D **54** 1474 (1996).
[30] T. Damour and J.H. Taylor, Astrophys. J. **366** 501 (1991).
[31] J.H. Taylor, A. Wolszczan, T. Damour and J.M. Weisberg, Nature **355** 132 (1992).
[32] A. Wolszczan, Nature **350** 688 (1991).
[33] Z. Arzoumanian, Ph.D. thesis, Princeton University, 1995.
[34] T. Damour and D. Vokrouhlicky, Phys. Rev. D **53** 4177 (1996).
[35] A. Wolszczan and J.H. Taylor, to be published (quoted in Taylor's Nobel lecture [15]).
[36] C.M. Will and H.W. Zaglauer, Astrophys. J. **346** 366 (1989).
[37] T. Damour and G. Schäfer, Phys. Rev. Lett. **66** 2549 (1991).
[38] T. Damour and D. Vokrouhlicky, Phys. Rev. D **53** 4177 (1996).
[39] R.D. Blandford et al., editors, *Pulsars as physics laboratories*, Phil. Trans. R. Soc. London A **341** 1-192 (1992); see notably the contributions by J.H. Taylor (117-134) and by T. Damour (135-149).
[40] J. Scherk and J.H. Schwarz, Nucl. Phys. **B81** 118 (1974); Phys. Lett. **B52** 347 (1974).
[41] B. de Carlos, J.A. Casas, F. Quevedo and E. Roulet, Phys. Lett. **B318** 447 (1993).

[42] R. Brustein and P.J. Steinhardt, Phys. Lett. **B302** 196 (1993).

[43] G.D. Coughlan et al., Phys. Lett. **B131** 59 (1983); J. Ellis, D.V. Nanopoulos and M. Quiros, Phys. Lett. **B174** 176 (1986); T. Banks, D.B. Kaplan and A.E. Nelson, Phys. Rev. D **49** 779 (1994).

[44] T. Damour and A. Vilenkin, gr-qc/9610005, submitted to Phys. Rev. Lett.

[45] L. Randall and S. Thomas, Nucl. Phys. **B449** 229 (1995).

[46] D.H. Lyth and E.D. Stewart, Phys. Rev. Lett. **75** 201 (1995); Phys. Rev. D **53** 1784 (1996).

[47] T. Damour and A.M. Polyakov, Nucl. Phys. **B423** 532 (1994); Gen. Rel. Grav. **26** 1171 (1994).

[48] T. Damour and K. Nordtvedt, Phys. Rev. Lett. **70** 2217 (1993); Phys. Rev. D **48** 3436 (1993).

[49] T. Damour and A. Vilenkin, Phys. Rev. D **53** 2981 (1996).

[50] P.W. Worden, in *Near Zero: New Frontiers of Physics*, edited by J.D. Fairbank et al. (Freeman, San Francisco, 1988), p. 766; J.P. Blaser et al., *STEP, Report on the Phase A Study*, ESA document SCI (96)5, March 1996; GEOSTEP Project, CNES report DPI/SC/FJC-N°96/058, April 1996; *Fundamental Physics in Space*, special issue of Class. Quant. Grav. **13** (1996); MiniSTEP, NASA-ESA report, Decembre 1996 (second issue).

8.6 DISCUSSION

Session Chair: Anthony Hewish
Rapporteur: Maulik Parikh

HEWISH: I would like to thank the speaker for his comprehensive survey of tests of Einsteinian gravity. I'd like to mention, in the context of alternate theories of gravity, that there is some recent work by my colleagues Anthony Lasenby, Steve Gull, and Chris Doran at the Cavendish Laboratory using a geometrical language based on Clifford algebras. This work, amounting to about one hundred pages and soon to be published by the Royal Society, presents a new gauge theory of gravitation. The theory is fully consistent with existing experimental tests of Einstein's general relativity, but differs with respect to its topology. For example, wormholes are forbidden and the theory gives new insight into time reversal. It also provides a derivation of the Hawking temperature when the Dirac equation is considered in a black hole background. Thus there are some conflicts with work by Hawking, Kerr, and others. I think this work should be taken seriously; I have faith in it because it was conceived by radio astronomers with a respected track record who have their feet firmly on the ground. It was not dreamed up by some theoretical cranks!

KRUSKAL: How can one ignore the boundary conditions at null infinity in the binary pulsar problem? Without such boundary conditions, with only near-zone physics, one cannot correctly obtain a time-asymmetric result.

DAMOUR: There are indeed boundary conditions at null infinity, namely that there are no incoming gravitational waves. One can then use retarded potentials in the neighborhood of the binary system.

MANN: There is another test of the weak equivalence principle on the elementary particle level. This test comes from the observations of neutrinos and photons from supernova SN 1987A and is good to one part in a thousand. It is the first and only test so far of whether general relativity holds for fermions as well as bosons.

DAMOUR: I agree that this is a qualitatively important test. There are, however, no theories of gravity that predict a difference between fermions and bosons.

CRONIN: One has to be careful with the conclusions reached from tests of the equivalence principle. In comparing the gravitational effects on beryllium and on copper, one is looking at potentially different gravitational interactions for the neutron and the proton. What if these are simply the wrong kind of matter? When one quotes a precision in an equivalence principle test, one must have some idea of why the interaction should be different for the two or more test bodies. I think this proviso has to be added to the conclusions made from such tests.

DAMOUR: This is a subtle issue. The dilaton, or other moduli from string theory, changes the coupling constants, such as α. Now the neutron and proton may have different couplings to the dilaton, but there is also a Coulomb energy, from QED, and the coupling of that to the dilaton dominates. One needs four different pairs of elements to distinguish between the various effects. The separate effects may be accessible to future experiments such as STEP.

SAVVIDI: Almost any string theory one can imagine violates the equivalence principle at high energies. On the other hand, if one could construct quantum gravity directly from an action principle, such as by quantizing the Einstein-Hilbert action, then the theory would obviously satisfy the equivalence principle at high energy.

DAMOUR: Quantum gravity per se has not been successful. On the other hand, one can introduce more fields and a cutoff at the Planck scale; in this way, string loops are calculable. If quantum gravity should ever be consistent at this level, then one would have something to compare. However, one would still want to unify gravity with the other forces, and so string theory leads in a better direction than simply quantizing gravity. This is a good reason for one to hope that string theory is correct.

CHAPTER 9

GRAVITATIONAL WAVES: A NEW WINDOW ONTO THE UNIVERSE

KIP S. THORNE
Theoretical Astrophysics
California Institute of Technology, Pasadena, CA

ABSTRACT

A summary is given of the current status and plans for gravitational-wave searches at all plausible wavelengths, from the size of the observable universe to a few kilometers. The anticipated scientific payoff from these searches is described, including expectations for detailed studies of black holes and neutron stars, high-accuracy tests of general relativity, and hopes for the discovery of exotic new kinds of objects.

9.1 INTRODUCTION

There is an enormous difference between gravitational waves, and the electromagnetic waves on which our present knowledge of the Universe is based:

- Electromagnetic waves are oscillations of the electromagnetic field that propagate through spacetime; gravitational waves are oscillations of the "fabric" of spacetime itself.

- Astronomical electromagnetic waves are almost always incoherent superpositions of emission from individual electrons, atoms, or molecules. Cosmic gravitational waves are produced by coherent, bulk motions of huge amounts of mass-energy—either material mass, or the energy of vibrating, nonlinear spacetime curvature.

- Since the wavelengths of electromagnetic waves are small compared to their sources (gas clouds, stellar atmospheres, accretion disks, ...), from the waves we can make pictures of the sources. The wavelengths of cosmic gravitational waves are comparable to, or larger than, their coherent, bulk-moving sources, so we cannot make pictures from them. Instead, the gravitational waves are like sound; they carry, in two independent waveforms, a stereophonic, symphony-like description of their sources.

- Electromagnetic waves are easily absorbed, scattered, and dispersed by matter. Gravitational waves travel nearly unscathed through all forms and amounts of intervening matter [1, 2].
- Astronomical electromagnetic waves have frequencies that begin at $f \sim 10^7$ Hz and extend on *upward* by roughly 20 orders of magnitude. Astronomical gravitational waves should begin at $\sim 10^4$ Hz (1000-fold lower than the lowest-frequency astronomical electromagnetic waves), and should extend on *downward* from there by roughly 20 orders of magnitude.

These enormous differences make it likely that:

- The information brought to us by gravitational waves will be very different from (almost "orthogonal to") that carried by electromagnetic waves; gravitational waves will show us details of the bulk motion of dense concentrations of energy, whereas electromagnetic waves show us the thermodynamic state of optically thin concentrations of matter.
- Most (but not all) gravitational-wave sources that our instruments detect will not be seen electromagnetically, and conversely, most objects observed electromagnetically will never be seen gravitationally. Typical electromagnetic sources are stellar atmospheres, accretion disks, and clouds of interstellar gas—none of which emit significant gravitational waves, while typical gravitational-wave sources are black holes and other exotic objects which emit no electromagnetic wave at all, and the cores of supernovae which are hidden from electromagnetic view by dense layers of surrounding stellar gas.
- Gravitational waves may bring us great surprises. In the past, when a radically new window has been opened onto the Universe, the resulting surprises have had a profound, indeed revolutionary, impact. For example, the radio universe, as discovered in the 1940s, 50s and 60s, turned out to be far more violent than the optical universe; radio waves brought us quasars, pulsars, and the cosmic microwave radiation, and with them our first direct observational evidence for black holes, neutron stars, and the heat of the big bang [3]. It is reasonable to hope that gravitational waves will bring a similar "revolution."

Gravitational-wave detectors and detection techniques have been under vigorous development since about 1960, when Joseph Weber [4] pioneered the field. These efforts have led to promising sensitivities in four frequency bands:

- The Extremely Low Frequency Band (ELF), 10^{-15} to 10^{-18} Hz, in which the measured anisotropy of the cosmic microwave background radiation places strong limits on gravitational wave strengths, and may, in fact, have detected waves [5, 6]. The only waves expected in this band are relics of the big bang.

- The Very Low Frequency Band (VLF), 10^{-7} to 10^{-9} Hz, in which Joseph Taylor and others have achieved remarkable gravity-wave sensitivities by the timing of millisecond pulsars [7]. The only expected strong sources in this band are processes in the very early universe—the big bang, phase transitions of the vacuum states of quantum fields, and vibrating or colliding defects in the structure of spacetime, such as monopoles, cosmic strings, domain walls, textures, and combinations thereof [8, 9, 10].

- The Low-Frequency Band (LF), 10^{-4} to 1 Hz, in which will operate the Laser Interferometer Space Antenna, LISA; see Sec. 9.3 below. This is the band of massive black holes ($M \sim 1000$ to $10^8\ M_\odot$) in the distant universe, and of other hypothetical massive exotic objects (naked singularities, soliton stars), as well of as binary stars (ordinary, white dwarf, neutron star, and black hole) in our galaxy. Early universe processes should also have produced waves at these frequencies, as in the ELF and VLF bands.

- The High-Frequency Band (HF), 1 to 10^4 Hz, in which operate earth-based gravitational-wave detectors such as LIGO; see Sec. 9.2 below. This is the band of stellar-mass black holes ($M \sim 1$ to $1000\ M_\odot$) and of other conceivable stellar-mass exotic objects (naked singularities and boson stars) in the distant universe, as well as of supernovae, pulsars, and coalescing and colliding neutron stars. Early universe processes should also have produced waves at these frequencies, as in the ELF, VLF, and LF bands.

In this lecture I shall focus primarily on the HF and LF bands, because these are the ones in which observations are likely to bring us the most new information.

9.2 GROUND-BASED LASER INTERFEROMETERS: LIGO & VIRGO

9.2.1 WAVE POLARIZATIONS, WAVEFORMS, AND HOW AN INTERFEROMETER WORKS

According to general relativity theory (which I shall assume to be correct), a gravitational wave has two linear polarizations, conventionally called $+$ (plus) and $\times$ (cross). Associated with each polarization there is a gravitational-wave field, h_+ or $h_\times$, which oscillates in time and propagates with the speed of light. Each wave field produces tidal forces (stretching and squeezing forces) on any object or detector through which it passes. If the object is small compared to the waves' wavelength (as is the case for ground-based interferometers and resonant mass antennas), then relative to the object's center, the forces have the quadrupolar patterns shown in Fig. 9.1. The names "plus" and "cross" are derived from the orientations of the axes that characterize the force patterns [1].

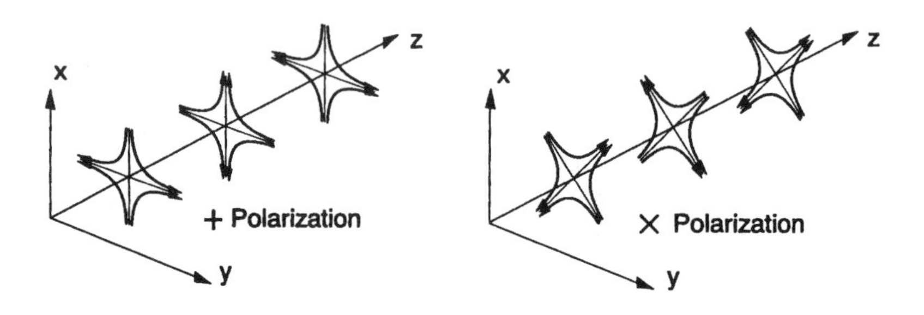

Figure 9.1: The lines of force associated with the two polarizations of a gravitational wave. (From Ref. [11].)

A laser interferometer gravitational wave detector ("interferometer" for short) consists of four masses that hang from vibration-isolated supports as shown in Fig. 9.2, and the indicated optical system for monitoring the separations between the masses [1, 11]. Two masses are near each other, at the corner of an "L," and one mass is at the end of each of the L's long arms. The arm lengths are nearly equal, $L_1 \simeq L_2 = L$. When a gravitational wave, with frequencies high compared to the masses' ~ 1 Hz pendulum frequency, passes through the detector, it pushes the masses back and forth relative to each other as though they were free from their suspension wires, thereby changing the arm-length difference, $\Delta L \equiv L_1 - L_2$. That change is monitored by laser interferometry in such a way that the variations in the output of the photodiode (the interferometer's output) are directly proportional to $\Delta L(t)$.

If the waves are coming from overhead or underfoot and the axes of the $+$ polarization coincide with the arms' directions, then it is the waves' $+$ polarization that drives the masses, and $\Delta L(t)/L = h_+(t)$. More generally, the interferometer's output is a linear combination of the two wave fields:

$$\frac{\Delta L(t)}{L} = F_+ h_+(t) + F_\times h_\times(t) \equiv h(t) \, . \tag{9.1}$$

The coefficients F_+ and $F_\times$ are of order unity and depend in a quadrupolar manner on the direction to the source and the orientation of the detector [1]. The combination $h(t)$ of the two h's is called the *gravitational-wave strain* that acts on the detector; and the time evolutions of $h(t)$, $h_+(t)$, and $h_\times(t)$ are sometimes called *waveforms*.

9.2.2 WAVE STRENGTHS AND INTERFEROMETER ARM LENGTHS

The strengths of the waves from a gravitational-wave source can be estimated using the "Newtonian/quadrupole" approximation to the Einstein field equations. This approximation says that $h \simeq (G/c^4)\ddot{Q}/r$, where $\ddot{Q}$ is the second time derivative of the source's quadrupole moment, r is the distance of the source from earth (and G and c are Newton's gravitation constant and the speed of light). The strongest sources will be highly nonspherical and thus will have $Q \simeq ML^2$, where M is their mass and L their size, and correspondingly will have $\ddot{Q} \simeq 2Mv^2 \simeq 4E_{\rm kin}^{\rm ns}$, where v is their internal velocity and $E_{\rm kin}^{\rm ns}$ is the nonspherical part of their internal kinetic energy. This provides us with the estimate

$$h \sim \frac{1}{c^2}\frac{4G(E_{\rm kin}^{\rm ns}/c^2)}{r}\,; \tag{9.2}$$

i.e., h is about 4 times the gravitational potential produced at earth by the mass-equivalent of the source's nonspherical, internal kinetic energy—made dimensionless by dividing by c^2. Thus, in order to radiate strongly, the source must have a very large, nonspherical, internal kinetic energy.

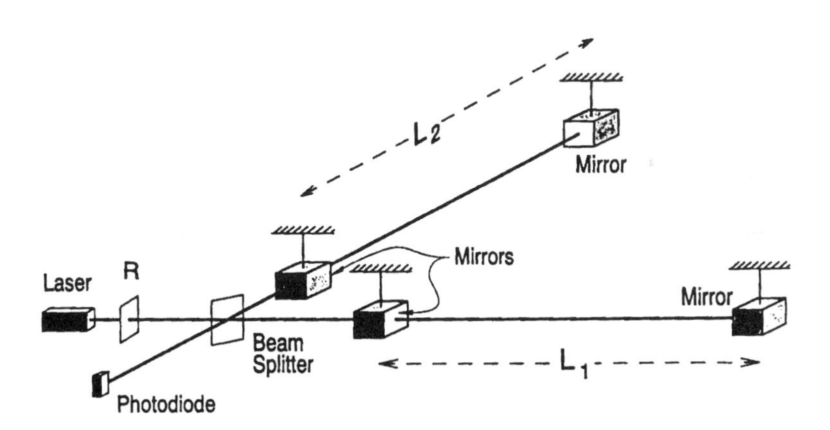

Figure 9.2: Schematic diagram of a laser interferometer gravitational wave detector. (From Ref. [11].)

The best known way to achieve a huge internal kinetic energy is via gravity; and by energy conservation (or the virial theorem), any gravitationally-induced kinetic energy must be of the order of the source's gravitational potential energy. A huge potential energy, in turn, requires that the source be very compact, not much larger than its own gravitational radius. Thus, the strongest gravity-wave sources must be highly compact, dynamical concentrations of large amounts of mass (e.g., colliding and coalescing black holes and neutron stars).

Such sources cannot remain highly dynamical for long; their motions will be stopped by energy loss to gravitational waves and/or the formation of an all-encompassing black hole. Thus, the strongest sources should be transient. Moreover, they should be very rare—so rare that to see a reasonable event rate will require reaching out through a substantial fraction of the universe. Thus, just as the strongest radio waves arriving at earth tend to be extragalactic, so also the strongest gravitational waves are likely to be extragalactic.

For highly compact, dynamical objects that radiate in the high-frequency band, e.g., colliding and coalescing neutron stars and stellar-mass black holes, the internal, nonspherical kinetic energy $E_{\rm kin}^{\rm ns}/c^2$ is of the order of the mass of the sun; and, correspondingly, Eq. (9.2) gives $h \sim 10^{-22}$ for such sources at the Hubble distance (3000 Mpc, i.e., 10^{10} light years); $h \sim 10^{-21}$ at 200 Mpc (a best-guess distance for several neutron-star coalescences per year; see Section 9.4.2), $h \sim 10^{-20}$ at the Virgo cluster of galaxies (15 Mpc); and $h \sim 10^{-17}$ in the outer reaches of our own Milky Way galaxy (20 kpc). These numbers set the scale of sensitivities that ground-based interferometers seek to achieve: $h \sim 10^{-21}$ to 10^{-22}.

When one examines the technology of laser interferometry, one sees good prospects to achieve measurement accuracies $\Delta L \sim 10^{-16}$ cm (1/1000 the diameter of the nucleus of an atom). With such an accuracy, an interferometer must have an arm length $L = \Delta L/h \sim 1$ to 10 km, in order to achieve the desired wave sensitivities, 10^{-21} to 10^{-22}. This sets the scale of the interferometers that are now under construction.

9.2.3 LIGO, VIRGO, AND THE INTERNATIONAL GRAVITATIONAL WAVE NETWORK

Interferometers are plagued by non-Gaussian noise, such as that due to sudden strain releases in the wires that suspend the masses. This noise prevents a single interferometer, by itself, from detecting with confidence short-duration gravitational-wave bursts (though it might be possible for a single interferometer to search for the periodic waves from known pulsars). The non-Gaussian noise can be removed by cross correlating two, or preferably three or more, interferometers that are networked together at widely separated sites.

The technology and techniques for such interferometers have been under development for 25 years, and plans for km-scale interferometers have been developed over the past 15 years. An international network consisting of three km-scale interferometers, at three widely separated sites, is now in the early stages of construction. It includes two sites of the American LIGO Project ("Laser Interferometer Gravitational Wave Observatory") [11], and one site of the French/Italian VIRGO Project (named after the Virgo cluster of galaxies) [12].

LIGO will consist of two vacuum facilities with 4-kilometer-long arms, one in Hanford, Washington (in the northwestern United States; Fig. 9.3) and the other

Figure 9.3: Artist's conception of one of the LIGO interferometers. [Courtesy the LIGO Project.]

in Livingston, Louisiana (in the southeastern United States). These facilities are designed to house many successive generations of interferometers without the necessity of any major facilities upgrade; and after a planned future expansion, they will be able to house several interferometers at once, each with a different optical configuration optimized for a different type of wave (e.g., broad-band burst, or narrow-band periodic wave, or stochastic wave).

The LIGO facilities and their first interferometers are being constructed by a team of about 80 physicists and engineers at Caltech and MIT, led by Barry Barish (the PI), Gary Sanders (the Project Manager), Albert Lazzarini, Rai Weiss, Stan Whitcomb, and Robbie Vogt (who directed the project during the pre-construction phase). Other research groups from many different universities are contributing to R&D for *enhancements* of the first interferometers, or are computing theoretical waveforms for use in data analysis, or are developing data analysis techniques. These groups are linked together by an organization called the *Ligo Research Community*. For further details, see the LIGO World Wide Web Site, http://www.ligo.caltech.edu/.

The VIRGO Project is building one vacuum facility in Pisa, Italy, with 3-kilometer-long arms. This facility and its first interferometers are a collaboration of more than a hundred physicists and engineers at the INFN (Frascati, Napoli, Perugia, Pisa), LAL (Orsay), LAPP (Annecy), LOA (Palaiseau), IPN (Lyon),

ESPCI (Paris), and the University of Illinois (Urbana), under the leadership of Alain Brillet and Adalberto Giazotto.

Both LIGO and VIRGO are scheduled for completion in the late 1990s, and their first gravitational-wave searches are likely to be performed in 2001. Figure 9.4 shows the design sensitivities for LIGO's first interferometers (which are expected to be achieved in 2001) [11] and for enhanced versions of those interferometers (which is expected to be operating five years or so later) [13], along with a benchmark sensitivity goal for a more advanced interferometer that may operate in the LIGO vacuum system some years later [11, 13].

For each interferometer, the quantity shown is the "sensitivity to bursts" that come from a random direction, $h_{\rm SB}(f)$ [11]. This $h_{\rm SB}$ is about 5 times worse than the rms noise level in a bandwidth $\Delta f \simeq f$ for waves with a random direction and polarization, and about $5\sqrt{5} \simeq 11$ worse than the the rms noise level $h_{\rm rms}$ for optimally directed and polarized waves. (In much of the literature, the quantity plotted is $h_{\rm rms} \simeq h_{\rm SB}/11$.)

This $h_{\rm SB}$ is to be compared with the "characteristic amplitude" $h_c(f) = h\sqrt{n}$ of the waves from a source; here, h is the waves' amplitude when they have frequency f, and n is the number of cycles the waves spend in a bandwidth $\Delta f \simeq f$ near frequency f [1, 11]. Any source with $h_c > h_{\rm SB}$ should be detectable with high confidence, even if it arrives only once per year. As examples, Fig. 9.4 shows the characteristic amplitudes h_c of several binary systems made of 1.4 $M_\odot$ neutron stars ("NS") or 10, 25, or 30 $M_\odot$ black holes ("BH"), which spiral together and collide under the driving force of gravitational radiation reaction. As the bodies spiral inward, their waves sweep upward in frequency (rightward across the figure along the dashed lines). From the figure we see that LIGO's first interferometers should be able to detect waves from the inspiral of a NS/NS binary out to a distance of 30 Mpc and from the final collision and merger of a 25 $M_\odot$/25 $M_\odot$ BH/BH binary out to about 300 Mpc.

LIGO alone, with its two sites which have parallel arms, will be able to detect an incoming gravitational wave, measure one of its two waveforms, and (from the time delay between the two sites) locate its source to within a $\sim 1^\circ$ wide annulus on the sky. LIGO and VIRGO together, operating as a *coordinated international network*, will be able to locate the source (via time delays plus the interferometers' beam patterns) to within a 2-dimensional error box with a size between several tens of arcminutes and several degrees, depending on the source direction and on the amount of high-frequency structure in the waveforms. They will also be able to monitor both waveforms $h_+(t)$ and $h_\times(t)$ (except for frequency components above about 1 kHz and below about 10 Hz, where the interferometers' noise becomes severe).

A British/German group is constructing a 600-meter interferometer called GEO600 near Hanover Germany [14], and Japanese groups, a 300-meter interferometer called TAMA near Tokyo. GEO600 may be a significant player in the interferometric network in its early years (by virtue of cleverness and speed of construction), but because of its short arms it cannot compete in the long run. GEO600 and TAMA will both be important development centers and testbeds for interferometer techniques and technology, and in due course they may give rise to kilometer-scale interferometers like LIGO and VIRGO, which could significantly enhance the network's all-sky coverage and ability to extract information from the waves.

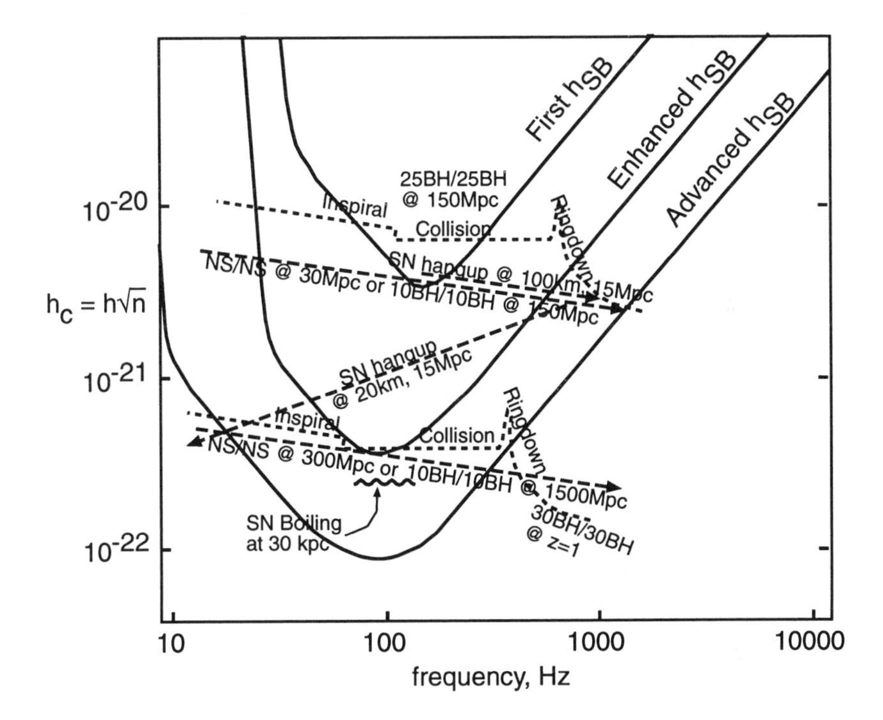

Figure 9.4: LIGO's projected broad-band noise sensitivity to bursts h_{SB} (Refs. [11, 13]) compared with the strengths of the waves from several hypothesized sources. The signal to noise ratios are $\sqrt{2}$ higher than in Ref. [11] because of a factor 2 error in Eq. (29) of Ref. [1].

9.2.4 NARROW-BAND DETECTORS

At frequencies $f \gtrsim 1000$ Hz, the interferometers' photon shot noise becomes a serious obstacle to wave detection. However, narrow-band detectors specially optimized for kHz frequencies show considerable promise. These include a special "dual recycled interferometer" [15], and huge spherical or icosahedral resonant-mass detectors [16] that are modern variants of Joseph Weber's original "bar" detector [4].

9.3 LISA: THE LASER INTERFEROMETER SPACE ANTENNA

The *Laser Interferometer Space Antenna* (LISA) is the most promising detector for gravitational-waves in the low-frequency band, 10^{-4}–1 Hz (10,000 times lower than the LIGO/VIRGO high-frequency band).

LISA was originally conceived (under a different name) by Peter Bender of the University of Colorado, and is being developed by an international team led by Karsten Danzmann of the University of Hanover (Germany) and James Hough of Glasgow University (UK). The European Space Agency tentatively plans to fly it sometime in the 2014–2018 time frame as part of ESA's Horizon 2000+ Program of large space missions. If NASA contributes significantly to LISA, then the flight could be sooner. As presently conceived, LISA will consist of six compact, drag-free spacecraft (i.e., spacecraft that are shielded from buffeting by solar wind and radiation pressure, and that thus move very nearly on geodesics of spacetime). All six spacecraft would be launched simultaneously by a single Ariane rocket. They would be placed into the same heliocentric orbit that the earth occupies, but would follow 20° behind the earth; cf. Fig. 9.5. The spacecraft would fly in pairs, with each pair at the vertex of an equilateral triangle that is inclined at an angle of 60° to the earth's orbital plane. The triangle's arm length would be 5 million km (10^6 times larger than LIGO's arms!). The six spacecraft would track each other optically, using one-Watt YAG laser beams. Because of diffraction losses over the 5×10^6 km arm length, it is not feasible to reflect the beams back and forth between mirrors as is done with LIGO. Instead, each spacecraft will have its own laser; and the lasers will be phase locked to each other, thereby achieving the same kind of phase-coherent out-and-back light travel as LIGO achieves with mirrors. The six-laser, six-spacecraft configuration thereby functions as three, partially independent but partially redundant, gravitational-wave interferometers.

Figure 9.6 depicts the expected noise and sensitivity of LISA in the same language as we have used for LIGO (Fig. 9.4): $h_{\rm SB} = 5\sqrt{5}h_{\rm rms}$ is the sensitivity for high-confidence detection ($S/N = 5$) of a broad-band burst coming from a random direction, assuming Gaussian noise.

At frequencies $f \gtrsim 10^{-3}$ Hz, LISA's noise is due to photon counting statistics (shot noise). The sensitivity curve steepens at $f \sim 3 \times 10^{-2}$ Hz because at larger f than that, the waves' period is shorter than the round-trip light travel time in one of LISA's arms. Below 10^{-3} Hz, the noise is due to buffeting-induced random motions of the spacecraft that are not being properly removed by the drag-compensation system. Notice that, in terms of dimensionless amplitude, LISA's sensitivity is roughly the same as that of LIGO's first interferometers (Fig. 9.4), but at 100,000 times lower frequency. Since the waves' energy flux scales as f^2h^2, this corresponds to 10^{10} better energy sensitivity than LIGO.

LISA can detect and study, simultaneously, a wide variety of different sources scattered over all directions on the sky. The key to distinguishing the different sources is the different time evolution of their waveforms. The key to determining

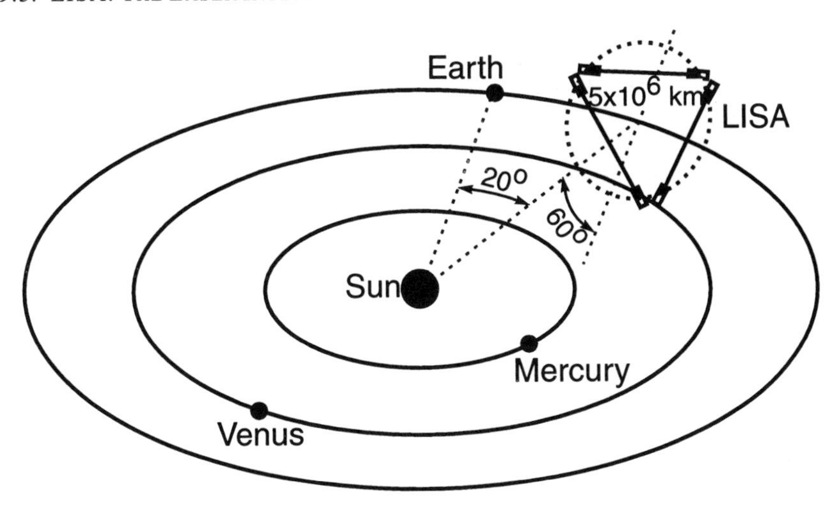

Figure 9.5: LISA's orbital configuration, with LISA magnified in arm length by a factor ~ 10 relative to the solar system.

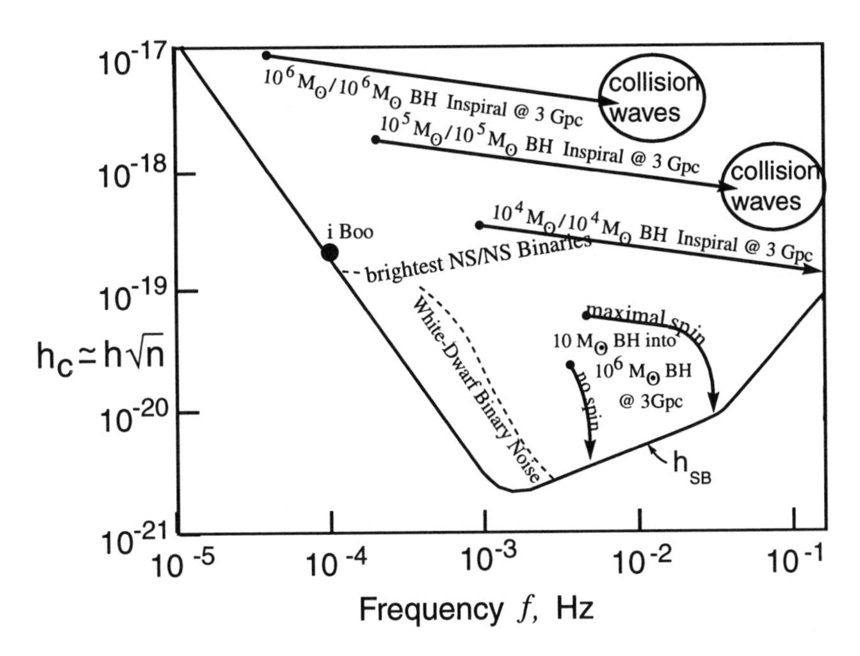

Figure 9.6: LISA's projected sensitivity to bursts $h_{\rm SB}$, compared with the strengths of the waves from several low-frequency sources.

each source's direction, and confirming that it is real and not just noise, is the manner in which its waves' amplitude and frequency are modulated by LISA's complicated orbital motion—a motion in which the interferometer triangle rotates around its center once per year, and the interferometer plane precesses around the normal to the earth's orbit once per year. Most sources will be observed for a year or longer, thereby making full use of these modulations.

9.4 BINARIES MADE OF NEUTRON STARS, BLACK HOLES, AND OTHER EXOTIC OBJECTS: INSPIRAL, COLLISION, AND COALESCENCE IN THE HF BAND

The best understood of all gravitational-wave sources are coalescing, compact binaries composed of neutron stars (NS) and black holes (BH). These NS/NS, NS/BH, and BH/BH binaries may well become the "bread and butter" of the LIGO/VIRGO and LISA diets.

The Hulse-Taylor [17, 18] binary pulsar, PSR 1913+16, is an example of a NS/NS binary whose waves could be measured by LIGO/VIRGO, if we were to wait long enough.

At present PSR 1913+16 has an orbital frequency of about 1/(8 hours) and emits its waves predominantly at twice this frequency, roughly 10^{-4} Hz, which is in LISA's low-frequency band (cf. Fig. 9.6). As a result of their loss of orbital energy to gravitational waves, the PSR 1913+16 NS's are gradually spiraling inward at a rate that agrees with general relativity's prediction to within the measurement accuracy (a fraction of a percent) [18]—a remarkable but indirect confirmation that gravitational waves do exist and are correctly described by general relativity. For this discovery and other aspects of PSR 1913+16, Princeton's Russell Hulse and Joseph Taylor have been awarded the Nobel Prize.

If we wait roughly 10^8 years, this inspiral will bring the waves into the LIGO/VIRGO high-frequency band. As the NS's continue their inspiral, the waves will then sweep upward in frequency, over a time of about 15 minutes, from 10 Hz to $\sim 10^3$ Hz, at which point the NS's will collide and merge. It is this last 15 minutes of inspiral, with $\sim 16,000$ cycles of waveform oscillation, and the final merger, that LIGO/VIRGO seeks to monitor.

9.4.1 WAVE STRENGTHS COMPARED TO LIGO SENSITIVITIES

Figure 9.4 compares the projected sensitivities of interferometers in LIGO [11] with the wave strengths from the last few minutes of inspiral of BH/BH, NS/BH, and NS/NS binaries at various distances from earth. Notice the signal strengths in Fig. 9.4 are in good accord with our rough estimates based on Eq. (9.2); at the endpoint (right end) of each inspiral, the number of cycles n spent near that frequency is of order unity, so the quantity plotted, $h_c \simeq h\sqrt{n}$, is about equal to h—and for a NS/NS binary at 200 Mpc is roughly 10^{-21}, as we estimated in Section 9.2.2.

9.4.2 COALESCENCE RATES

Such final coalescences are few and far between in our own galaxy: about one every 100,000 years, according to 1991 estimates by Phinney [19] and by Narayan, Piran, and Shemi [20], based on the statistics of binary pulsar searches in our galaxy which found three that will coalesce in less than 10^{10} years. Extrapolating out through the universe, Phinney and Narayan et al., infer that to see several NS/NS coalescences per year, LIGO/VIRGO will have to look out to a distance of about 200 Mpc (give or take a factor ~ 2); cf. Fig. 9.4. Since these estimates were made, the binary pulsar searches have been extended through a significantly larger volume of the galaxy than before, and no new ones with coalescence times $\lesssim 10^{10}$ years have been found. This would drive the estimated event rate downward, except that revisions of other aspects of the Phinney/Narayan analyses have compensated, leaving the binary-pulsar-search-based best estimates unchanged, at several events per year out to 200 Mpc [21].

A rate of one every 100,000 years in our galaxy is $\sim$ 100 times smaller than the birth rate of the NS/NS binaries' progenitors: massive, compact, main-sequence binaries [19, 20]. Therefore, either 99 per cent of progenitors fail to make it to the NS/NS state (e.g., because of binary disruption during a supernova or forming TŻO's), or else they do make it, but they wind up as a class of NS/NS binaries that has not yet been discovered in any of the pulsar searches. Several experts on binary evolution have argued for the latter [22, 23, 24]: most NS/NS binaries, they suggest, may form with such short orbital periods that their lifetimes to coalescence are significantly shorter than normal pulsar lifetimes ($\sim 10^7$ years); and with such short lifetimes, they have been missed in pulsar searches. By modeling the evolution of the galaxy's binary star population, these binary experts arrive at best estimates as high as 3×10^{-4} coalescences per year in our galaxy, corresponding to several per year out to 60 Mpc distance [22], though more recent and more conservative models [25] give results more nearly in accord with the binary pulsar searches, several per year out to 200 Mpc. Phinney [19] describes other plausible populations of NS/NS binaries that could increase the event rate, and he argues for "ultraconservative" lower and upper limits of 23 Mpc and 1000 Mpc for how far one must look to see several coalescences per year.

By comparing these rate estimates with the signal strengths in Fig. 9.4, we see that: (i) The first interferometers in LIGO/VIRGO (ca. 2001) have a possibility, but not a high probability, of seeing NS/NS coalescences. (ii) enhanced interferometers (mid 2000's) can be fairly confident of seeing them; the conservatively estimated event rate is $\sim 3 \times (300/200)^3 \simeq 10/\text{year}$. (iii) advanced interferometers are almost certain of seeing them (the requirement that this be so was one factor that forced the LIGO/VIRGO arm lengths to be so long, several kilometers).

We have no good observational handle on the coalescence rate of NS/BH or BH/BH binaries. However, theory suggests that their progenitors might not disrupt during the stellar collapses that produce the NS's and BH's, so their coalescence rate could be about the same as the birth rate for their progenitors: $\sim 1/100,000$ years in our galaxy. This suggests that within 200 Mpc distance there might be several NS/BH or BH/BH coalescences per year [19, 20, 22, 24]. This estimate should be regarded as a plausible upper limit on the event rate and lower limit on the distance to look [19, 20].

If this estimate is correct, then NS/BH and BH/BH binaries will be seen before NS/NS, and might be seen by the first LIGO/VIRGO interferometers or soon thereafter [26]; cf. Fig. 9.4. However, this estimate is far less certain than the (rather uncertain) NS/NS estimates!

Once coalescence waves have been discovered, each further improvement of sensitivity by a factor 2 will increase the event rate by $2^3 \simeq 10$. Assuming a rate of several NS/NS per year at 200 Mpc, the advanced interferometers of Fig. 9.4 should see ~ 100 per year.

9.4.3 Inspiral Waveforms and the Information They Can Bring

Neutron stars and black holes have such intense self gravity that it is exceedingly difficult to deform them. Correspondingly, as they spiral inward in a compact binary, they do not gravitationally deform each other significantly until several orbits before their final coalescence [27, 28]. This means that the inspiral waveforms are determined to high accuracy by only a few, clean parameters: the masses and spin angular momenta of the bodies, and the initial orbital elements (i.e., the elements when the waves enter the LIGO/VIRGO band).

Though tidal deformations are negligible during inspiral, relativistic effects can be very important. If, for the moment, we ignore the relativistic effects—i.e., if we approximate gravity as Newtonian and the wave generation as due to the binary's oscillating quadrupole moment [1], then the shapes of the inspiral waveforms $h_+(t)$ and $h_\times(t)$ are as shown in Fig. 9.7.

The left-hand graph in Fig. 9.7 shows the waveform increasing in amplitude and sweeping upward in frequency (i.e., undergoing a "chirp") as the binary's bodies spiral closer and closer together. The ratio of the amplitudes of the two polarizations is determined by the inclination ι of the orbit to our line of sight (lower right in Fig. 9.7). The shapes of the individual waves, i.e., the waves' harmonic content, are determined by the orbital eccentricity (upper right). (Binaries produced by normal stellar evolution should be highly circular due to past radiation reaction forces, but compact binaries that form by capture events, in dense star clusters that might reside in galactic nuclei [29], could be quite eccentric.) If, for simplicity, the orbit is circular, then the rate at which the frequency sweeps or "chirps." df/dt [or equivalently the number of cycles spent near a given frequency, $n = f^2(df/dt)^{-1}$] is determined solely, in the Newtonian/quadrupole approximation, by the binary's so-called *chirp mass*, $M_c \equiv (M_1 M_2)^{3/5}/(M_1 + M_2)^{1/5}$ (where M_1 and M_2 are the two bodies' masses). The amplitudes of the two waveforms are determined by the chirp mass, the distance to the source, and the orbital inclination. Thus (in the Newtonian/quadrupole approximation), by measuring the two amplitudes, the frequency sweep, and the harmonic content of the inspiral waves, one can determine as direct, resulting observables, the source's distance, chirp mass, inclination, and eccentricity [30, 31].

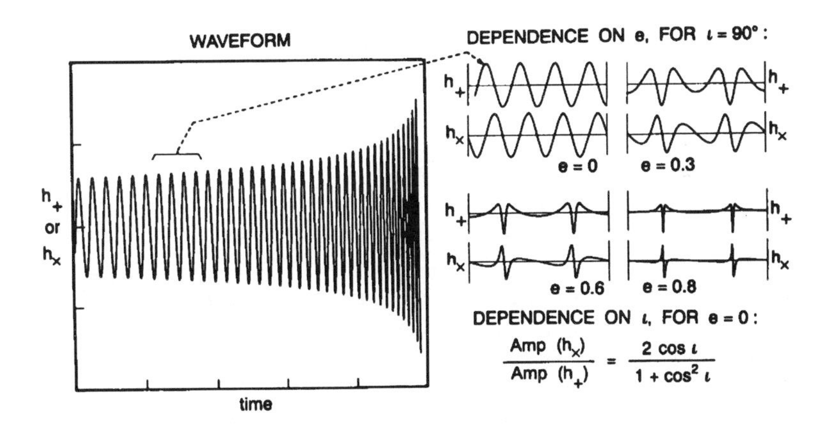

Figure 9.7: Waveforms from the inspiral of a compact binary, computed using Newtonian gravity for the orbital evolution and the quadrupole-moment approximation for the wave generation. (From Ref. [11].)

As in binary pulsar observations [18], so also here, relativistic effects add further information: they influence the rate of frequency sweep and produce waveform modulations in ways that depend on the binary's dimensionless ratio $\eta = \mu/M$ of reduced mass $\mu = M_1 M_2/(M_1 + M_2)$ to total mass $M = M_1 + M_2$, and on the spins of the binary's two bodies. These relativistic effects are reviewed and discussed at length in Refs. [32, 33]. Two deserve special mention: (i) As the waves emerge from the binary, some of them get backscattered one or more times off the binary's spacetime curvature, producing wave *tails*. These tails act back on the binary, modifying its inspiral rate in a measurable way. (ii) If the orbital plane is inclined to one or both of the binary's spins, then the spins drag inertial frames in the binary's vicinity (the "Lense-Thirring effect"); this frame dragging causes the orbit to precess, and the precession modulates the waveforms [32, 34, 35].

Remarkably, the relativistic corrections to the frequency sweep—tails, spin-induced precession and others—will be measurable with rather high accuracy, even though they are typically $\lesssim 10$ per cent of the Newtonian contribution, and even though the typical signal to noise ratio will be only ~ 9. The reason is as follows [36, 37, 32]: The frequency sweep will be monitored by the method of "matched filters"; in other words, the incoming, noisy signal will be cross correlated with theoretical templates. If the signal and the templates gradually get out of phase with each other by more than $\sim 1/10$ cycle as the waves sweep through the LIGO/VIRGO band, their cross correlation will be significantly reduced. Since the total number of cycles spent in the LIGO/VIRGO band will be $\sim 16,000$ for a NS/NS binary, ~ 3500 for NS/BH, and ~ 600 for BH/BH, this means that LIGO/VIRGO should be able to measure the frequency sweep to a fractional precision $\lesssim 10^{-4}$, compared to which the relativistic effects are very large. (This is essentially the same method as Joseph Taylor and colleagues use for high-accuracy radio-wave measurements of relativistic effects in binary pulsars [18].)

Preliminary analyses, using the theory of optimal signal processing, predict the following typical accuracies for LIGO/VIRGO measurements based solely on the frequency sweep (i.e., ignoring modulational information) [38]: (i) The chirp mass M_c will typically be measured, from the Newtonian part of the frequency sweep, to $\sim 0.04\%$ for a NS/NS binary and $\sim 0.3\%$ for a system containing at least one BH. (ii) *If* we are confident (e.g., on a statistical basis from measurements of many previous binaries) that the spins are a few percent or less of the maximum physically allowed, then the reduced mass μ will be measured to $\sim 1\%$ for NS/NS and NS/BH binaries, and $\sim 3\%$ for BH/BH binaries. (Here and below NS means a $\sim 1.4\ M_\odot$ neutron star and BH means a $\sim 10\ M_\odot$ black hole.) (iii) Because the frequency dependences of the (relativistic) μ effects and spin effects are not sufficiently different to give a clean separation between μ and the spins, if we have no prior knowledge of the spins, then the spin/μ correlation will worsen the typical accuracy of μ by a large factor, to $\sim 30\%$ for NS/NS, $\sim 50\%$ for NS/BH, and a factor ~ 2 for BH/BH. These worsened accuracies might be improved somewhat by waveform modulations caused by the spin-induced precession of the orbit [34, 35], and even without modulational information, a certain combination of μ and the spins will be determined to a few per cent. Much additional theoretical work is needed to firm up the measurement accuracies.

To take full advantage of all the information in the inspiral waveforms will require theoretical templates that are accurate, for given masses and spins, to a fraction of a cycle during the entire sweep through the LIGO/VIRGO band. Such templates are being computed by an international consortium of relativity theorists (Blanchet and Damour in France, Iyer in India, Will and Wiseman in the U.S.) [39], using post-Newtonian expansions of the Einstein field equations. This enterprise is rather like computing the Lamb shift to high order in powers of the fine structure constant, for comparison with experiment. Cutler and Flanagan [40] have estimated the order to which the computations must be carried in order that systematic errors in the theoretical templates will not significantly impact the information extracted from the LIGO/VIRGO observational data. The answer appears daunting: radiation-reaction effects must be computed to three full post-Newtonian orders [six orders in v/c =(orbital velocity)/(speed of light)] beyond the leading-order radiation reaction, which itself is 5 orders in v/c beyond the Newtonian theory of gravity—though by judicious use of Padé approximates, these requirements might be relaxed [41].

It is only about ten years since controversies over the leading-order radiation reaction [42] were resolved by a combination of theoretical techniques and binary pulsar observations. Nobody dreamed then that LIGO/VIRGO observations may require pushing post-Newtonian computations onward from $O[(v/c)^5]$ to $O[(v/c)^{11}]$. This requirement epitomizes a major change in the field of relativity research: At last, 80 years after Einstein formulated general relativity, experiment has become a major driver for theoretical analyses.

Remarkably, the goal of $O[(v/c)^{11}]$ is achievable. The most difficult part of the computation, the radiation reaction, has been evaluated to $O[(v/c)^9]$ beyond Newton by the French/Indian/American consortium [39] and $O[(v/c)^{10}]$ is coming under control.

These high-accuracy waveforms are needed only for extracting information from the inspiral waves, after the waves have been discovered; they are not needed for the discovery itself. The discovery is best achieved using a different family of theoretical waveform templates, one that covers the space of potential waveforms in a manner that minimizes computation time instead of a manner that ties quantitatively into general relativity theory [32, 43]. Such templates are in the early stage of development.

9.4.4 TESTING GR, MEASURING THE COSMOLOGICAL UNIVERSE, MAPPING BLACK HOLES, AND SEARCHING FOR EXOTIC OBJECTS

LIGO/VIRGO observations of compact binary inspirals have the potential to bring us far more information than just binary masses and spins:

- They can be used for high-precision tests of general relativity. In scalar-tensor theories (some of which are attractive alternatives to general relativity [44]), radiation reaction due to emission of scalar waves places a unique signature on the gravitational waves that LIGO/VIRGO would detect—a signature that can be searched for with high precision [45].

- They can be used to measure the Universe's Hubble constant, deceleration parameter, and cosmological constant [30, 31, 46, 47]. The keys to such measurements are that: (i) Advanced interferometers in LIGO/VIRGO will be able to see NS/NS out to cosmological redshifts $z \sim 0.3$, and NS/BH out to $z \sim 2$. (ii) The direct observables that can be extracted from the observed waves include the source's luminosity distance $r_{\rm L}$ (measured to accuracy $\sim$ 10 per cent in a large fraction of cases), and its direction on the sky (to accuracy $\sim$ 1 square degree)—accuracies good enough that only one or a few electromagnetically-observed clusters of galaxies should fall within the 3-dimensional gravitational error boxes, thereby giving promise to joint gravitational/electromagnetic statistical studies. (iii) Another direct gravitational observable is $(1 + z)M$ where z is redshift and M is any mass in the system (measured to the accuracies quoted above). Since the masses of NS's in binaries seem to cluster around 1.4 $M_\odot$, measurements of $(1 + z)M$ can provide a handle on the redshift, even in the absence of electromagnetic aid.

- For a NS or small BH spiraling into a massive $\sim$ 50 to 500 $M_\odot$ compact central body, the inspiral waves will carry a "map" of the massive body's spacetime geometry. This map can be used to determine whether the massive body is a black hole (in which case the geometry must be that of Kerr, with all its features uniquely determined by its mass and angular momentum) or is some other kind of exotic compact object, e.g., a soliton star or naked singularity [48, 49]. As we shall see in Section 9.6.2 below, this type of black-hole study and search for exotic objects can be carried out with much higher precision by LISA at large masses and low frequencies, than by LIGO/VIRGO at low masses and high frequencies.

9.4.5 COALESCENCE WAVEFORMS AND THEIR INFORMATION

The waves from the binary's final coalescence can bring us new types of information.

BH/BH Coalescence

In the case of a BH/BH binary, the coalescence will excite large-amplitude, highly nonlinear vibrations of spacetime curvature near the coalescing black-hole horizons—a phenomenon of which we have very little theoretical understanding today. Especially fascinating will be the case of two spinning black holes whose spins are not aligned with each other or with the orbital angular momentum. Each of the three angular momentum vectors (two spins, one orbital) will drag space in its vicinity into a tornado-like swirling motion—the general relativistic "dragging of inertial frames," so the binary is rather like two tornados with orientations skewed to each other, embedded inside a third, larger tornado with a third orientation. The dynamical evolution of such a complex configuration of coalescing spacetime warpage (as revealed by its emitted waves) might bring us surprising new insights into relativistic gravity [11]. Moreover, if the sum of the BH masses is fairly large, $\sim$ 40 to 200 $M_\odot$, then the waves should come off in a frequency range $f \sim$ 40 to 200 Hz where the LIGO/VIRGO broad-band interferometers have their best sensitivity and can best extract the information the waves carry.

To get full value out of such wave observations will require [26] having theoretical computations with which to compare them. There is no hope to perform such computations analytically; they can only be done as supercomputer simulations. The development of such simulations is being pursued by several research groups, including an eight-university American consortium of numerical relativists and computer scientists called the Two-Black-Hole Grand Challenge Alliance [50]. I have a bet with Richard Matzner, the lead PI of this alliance, that LIGO/VIRGO will discover waves from such coalescences with misaligned spins before the Alliance is able to compute them.

NS/NS Coalescence

The final coalescence of NS/NS binaries should produce waves that are sensitive to the equation of state of nuclear matter, so such coalescences have the potential to teach us about the nuclear equation of state [11, 32]. In essence, LIGO/VIRGO will be studying nuclear physics via the collisions of atomic nuclei that have nucleon numbers $A \sim 10^{57}$—somewhat larger than physicists are normally accustomed to. The accelerator used to drive these nuclei up to the speed of light is the binary's self gravity, and the radiation by which the details of the collisions are probed is gravitational.

Unfortunately, the final NS/NS coalescence will emit its gravitational waves in the kHz frequency band (800 Hz $\lesssim f \lesssim$ 2500 Hz) where photon shot noise will prevent them from being studied studied by the standard, "workhorse," broadband interferometers of Fig. 9.4. However, a specially configured ("dual-recycled") interferometer, which could have enhanced sensitivity in the kHz region at the price of reduced sensitivity elsewhere, may be able to measure the waves and extract their equation of state information, as might massive, spherical, resonant-mass detectors [32, 51]. Such measurements will be very difficult and are likely only when the LIGO/VIRGO network has reached a mature stage.

A number of research groups [52] are engaged in numerical astrophysics simulations of NS/NS coalescence, with the goal not only to predict the emitted gravitational waveforms and their dependence on equation of state, but also (more immediately) to learn whether such coalescences might power the γ-ray bursts that have been a major astronomical puzzle since their discovery in the early 1970s.

NS/NS coalescence is a popular explanation for the γ-ray bursts because (i) the bursts are isotropically distributed on the sky, (ii) they have a distribution of number versus intensity that suggests they might lie at near-cosmological distances, and (iii) their event rate is roughly the same as that predicted for NS/NS coalescence ($\sim$ 1000 per year out to cosmological distances, if they are cosmological). If LIGO/VIRGO were now in operation and observing NS/NS inspiral, it could report definitively whether or not the γ-bursts are produced by NS/NS binaries; and if the answer were yes, then the combination of γ-burst data and gravitational-wave data could bring valuable information that neither could bring by itself. For example, it would reveal when, to within a few msec, the γ-burst is emitted relative to the moment the NS's first begin to touch; and by comparing the γ and gravitational times of arrival, we might test whether gravitational waves propagate with the speed of light to a fractional precision of $\sim 0.01\text{sec}/3 \times 10^9\,\text{lyr} = 10^{-19}$.

NS/BH Coalescence

A NS spiraling into a BH of mass $M \gtrsim 10\ M_\odot$ should be swallowed more or less whole. However, if the BH is less massive than roughly $10\ M_\odot$, and especially if it is rapidly rotating, then the NS will tidally disrupt before being swallowed. Little is known about the disruption and accompanying waveforms. To model them with any reliability will likely require full numerical relativity, since the circumferences of the BH and NS will be comparable and their physical separation at the moment of disruption will be of order their separation. As with NS/NS, the coalescence waves should carry equation of state information and will come out in the kHz band, where their detection will require advanced, specialty detectors.

Christodoulou Memory

As the coalescence waves depart from their source, their energy creates (via the nonlinearity of Einstein's field equations) a secondary wave called the "Christodoulou memory" [53, 54, 55]. Whereas the primary waves may have frequencies in the kHz band, the memory builds up on the timescale of the primary energy emission profile, which is likely to be of order 0.01 sec, corresponding to a memory frequency in the optimal band for the LIGO/VIRGO workhorse interferometers, $\sim$ 100 Hz. Unfortunately, the memory is so weak that only very advanced ground-based interferometers have much chance of detecting and studying it—and then, perhaps only for BH/BH coalescences and not for NS/NS or NS/BH [56]. LISA, by contrast, should easily be able to measure the memory from supermassive BH/BH coalescences.

9.5 OTHER HIGH-FREQUENCY SOURCES

9.5.1 STELLAR CORE COLLAPSE AND SUPERNOVAE

When the core of a massive star has exhausted its supply of nuclear fuel, it collapses to form a neutron star or black hole. In some cases, the collapse triggers and powers a subsequent explosion of the star's mantle—a supernova explosion. Despite extensive theoretical efforts for more than 30 years, and despite wonderful observational data from Supernova 1987A, theorists are still far from a definitive understanding of the details of the collapse and explosion. The details are highly complex and may differ greatly from one core collapse to another [57].

Several features of the collapse and the core's subsequent evolution can produce significant gravitational radiation in the high-frequency band. We shall consider these features in turn, the most weakly radiating first.

Boiling of the Newborn Neutron Star

Even if the collapse is spherical, so it cannot radiate any gravitational waves at all, it should produce a convectively unstable neutron star that "boils" vigorously (and nonspherically) for the first $\sim$ 0.1 seconds of its life [58]. The boiling dredges up high-temperature nuclear matter ($T \sim 10^{12}$ K) from the neutron star's central regions, bringing it to the surface (to the "neutrino-sphere"), where it cools by neutrino emission before being swept back downward and reheated. Burrows [59] has pointed out that the boiling should generate $n \sim 10$ cycles of gravitational waves with frequency $f \sim 100$ Hz and amplitude large enough to be detectable by LIGO/VIRGO. Recent, preliminary 3+1 dimensional simulations by Janka, Keil, and Müller [60] (which should be more reliable than Burrows' earlier 2+1 dimensional simulations) suggest an amplitude $h \sim 10^{-22}(30\ \mathrm{kpc}/r)$ (where r is the distance to the source), corresponding to a characteristic amplitude $h_c \simeq h\sqrt{n} \sim 3 \times 10^{-22}(30\ \mathrm{kpc}/r)$; cf. Fig. 9.4. LIGO/VIRGO will be able to detect such waves only in our own galaxy and its satellites, where the supernova rate is probably no larger than $\sim$ 1 each 20 years. However, neutrino detectors have a similar range, and there could be a high scientific payoff from correlated observations of the gravitational waves emitted by the boiling's mass motions and neutrinos emitted from the boiling neutrino-sphere. With neutrinos to trigger on, the sensitivities of LIGO detectors should be about twice as good as shown in Fig. 9.4, so enhanced LIGO interferometers may see such a source anywhere in our galaxy

Axisymmetric Collapse, Bounce, and Oscillations

Rotation will centrifugally flatten the collapsing core, enabling it to radiate as it implodes. If the core's angular momentum is small enough that centrifugal forces do not halt or strongly slow the collapse before it reaches nuclear densities, then the core's collapse, bounce, and subsequent oscillations are likely to be axially symmetric. Numerical simulations [61, 62] show that in this case the waves from collapse, bounce, and oscillation will be quite weak: the total energy radiated as gravitational waves is not likely to exceed $\sim 10^{-7}$ solar masses (about 1 part in a million of the collapse energy) and might often be much less than this; and correspondingly, the waves' characteristic amplitude will be $h_c \lesssim 3 \times 10^{-21}(30\ \mathrm{kpc}/r)$. These collapse-and-bounce waves will come off at frequencies $\sim$ 200 Hz to $\sim$ 1000 Hz, and will precede the boiling waves by a fraction of a second. Though a little stronger than the boiling waves, they probably cannot be seen by LIGO/VIRGO beyond the local group of galaxies and thus will be a very rare occurrence.

Rotation-Induced Bars and Break-Up

If the core's rotation is large enough to strongly flatten the core before or as it reaches nuclear density, then a dynamical and/or secular instability is likely to break the core's axisymmetry. The core will be transformed into a bar-like configuration that spins end-over-end like an American football, and that might even break up into two or more massive pieces. In this case, the radiation from the spinning bar or orbiting pieces *could* be almost as strong as that from a coalescing neutron-star binary, and thus could be seen by the LIGO/VIRGO first interferometers out to the distance of the Virgo cluster (where the supernova rate is several per year), by enhanced interferometers out to $\sim$ 100 Mpc (supernova rate several thousand per year), and by advanced interferometers out to several hundred Mpc (supernova rate $\sim$ (a few)$\times$ $\sim$ 10^4 per year); cf. Fig. 9.4. It is far from clear what fraction of collapsing cores will have enough angular momentum to break their axisymmetry, and what fraction of those will actually radiate at this high rate; but even if only $\sim 1/1000$ or $1/10^4$ do so, this could ultimately be a very interesting source for LIGO/VIRGO.

Several specific scenarios for such non-axisymmetry have been identified:

Centrifugal hangup at $\sim$ 100 km radius: If the pre-collapse core is rapidly spinning (e.g., if it is a white dwarf that has been spun up by accretion from a companion), then the collapse may produce a highly flattened, centrifugally supported disk with most of its mass at radii $R \sim 100$ km, which then (via instability) may transform itself into a bar or may bifurcate. The bar or bifurcated lumps will radiate gravitational waves at twice their rotation frequency, $f \sim 100$ Hz—the optimal frequency for LIGO/VIRGO interferometers. To shrink on down to $\sim$ 10 km size, this configuration must shed most of its angular momentum. *If* a substantial fraction of the angular momentum goes into gravitational waves, then independently of the strength of the bar, the waves will be nearly as strong as those from a coalescing binary. The reason is this: the waves' amplitude h is proportional to the bar's ellipticity e; the number of cycles n of wave emission is proportional to $1/e^2$; and the characteristic amplitude $h_c = h\sqrt{n}$ is thus independent of the ellipticity and is about the same whether the configuration is a bar or is two lumps [31]. The resulting waves will thus have h_c roughly half as large, at $f \sim 100$ Hz, as the h_c from a NS/NS binary (half as large because each lump might be half as massive as a NS), and the waves will chirp upward in frequency in a manner similar to those from a binary.

It is rather likely, however, that most of excess angular momentum does *not* go into gravitational waves, but instead goes largely into hydrodynamic waves as the bar or lumps, acting like a propeller, stir up the surrounding stellar mantle. In this case, the radiation will be correspondingly weaker.

Centrifugal hangup at $\sim$ 20 km radius: Lai and Shapiro [63] have explored the case of centrifugal hangup at radii not much larger than the final neutron star, say $R \sim 20$ km. Using compressible ellipsoidal models, they have deduced that, after a brief period of dynamical bar-mode instability with wave emission at $f \sim 1000$ Hz (explored by Houser, Centrella, and Smith [64]), the star switches to a secular instability in which the bar's angular velocity gradually slows while the material of which it is made retains its high rotation speed and circulates through the slowing bar. The slowing bar emits waves that sweep *downward* in frequency through the LIGO/VIRGO optimal band $f \sim 100$ Hz, toward ~ 10 Hz. The characteristic amplitude (Fig. 9.4) is only modestly smaller than for the upward-sweeping waves from hangup at $R \sim 100$ km, and thus such waves should be detectable near the Virgo Cluster by the first LIGO/VIRGO interferometers, near 100 Mpc by enhanced interferometers, and and at distances of a few 100 Mpc by advanced interferometers.

Successive fragmentations of an accreting, newborn neutron star: Bonnell and Pringle [65] have focused on the evolution of the rapidly spinning, newborn neutron star as it quickly accretes more and more mass from the pre-supernova star's inner mantle. If the accreting material carries high angular momentum, it may trigger a renewed bar formation, lump formation, wave emission, and coalescence, followed by more accretion, bar and lump formation, wave emission, and coalescence. Bonnell and Pringle speculate that hydrodynamics, not wave emission, will drive this evolution, but that the total energy going into gravitational waves might be as large as $\sim 10^{-3}\ M_{\odot}$. This corresponds to $h_c \sim 10^{-21}(10\ \mathrm{Mpc}/r)$.

9.5.2 SPINNING NEUTRON STARS; PULSARS

As the neutron star settles down into its final state, its crust begins to solidify (crystallize). The solid crust will assume nearly the oblate axisymmetric shape that centrifugal forces are trying to maintain, with poloidal ellipticity $\epsilon_p \propto$(angular velocity of rotation)^2. However, the principal axis of the star's moment of inertia tensor may deviate from its spin axis by some small "wobble angle" θ_w, and the star may deviate slightly from axisymmetry about its principal axis; i.e., it may have a slight ellipticity $\epsilon_e \ll \epsilon_p$ in its equatorial plane.

As this slightly imperfect crust spins, it will radiate gravitational waves [66]: ϵ_e radiates at twice the rotation frequency, $f = 2f_{\mathrm{rot}}$ with $h \propto \epsilon_e$, and the wobble angle couples to ϵ_p to produce waves at $f = f_{\mathrm{rot}} + f_{\mathrm{prec}}$ (the precessional side-band of the rotation frequency) with amplitude $h \propto \theta_w \epsilon_p$. For typical neutron-star masses and moments of inertia, the wave amplitudes are

$$h \sim 6 \times 10^{-25} \left(\frac{f_{\mathrm{rot}}}{500\ \mathrm{Hz}}\right)^2 \left(\frac{1\ \mathrm{kpc}}{r}\right) \left(\frac{\epsilon_e \text{ or } \theta_w \epsilon_p}{10^{-6}}\right) . \tag{9.3}$$

The neutron star gradually spins down, due in part to gravitational-wave emission but perhaps more strongly due to electromagnetic torques associated with its spinning magnetic field and pulsar emission. This spin-down reduces the strength of centrifugal forces, and thereby causes the star's poloidal ellipticity

ϵ_p to decrease, with an accompanying breakage and resolidification of its crust's crystal structure (a "starquake") [67]. In each starquake, θ_w, ϵ_e, and ϵ_p will all change suddenly, thereby changing the amplitudes and frequencies of the star's two gravitational "spectral lines" $f = 2f_{\rm rot}$ and $f = f_{\rm rot} + f_{\rm prec}$. After each quake, there should be a healing period in which the star's fluid core and solid crust, now rotating at different speeds, gradually regain synchronism. By monitoring the amplitudes, frequencies, and phases of the two gravitational-wave spectral lines, and by comparing with timing of the electromagnetic pulsar emission, one might learn much about the physics of the neutron-star interior.

How large will the quantities ϵ_e and $\theta_w\epsilon_p$ be? Rough estimates of the crystal shear moduli and breaking strengths suggest an upper limit in the range $\epsilon_{\rm max} \sim 10^{-4}$ to 10^{-6}, and it might be that typical values are far below this. We are extremely ignorant, and correspondingly, there is much to be learned from searches for gravitational waves from spinning neutron stars.

One can estimate the sensitivity of LIGO/VIRGO (or any other broad-band detector) to the periodic waves from such a source by multiplying the waves' amplitude h by the square root of the number of cycles over which one might integrate to find the signal, $n = f\hat{\tau}$ where $\hat{\tau}$ is the integration time. The resulting effective signal strength, $h\sqrt{n}$, is larger than h by

$$\sqrt{n} = \sqrt{f\hat{\tau}} = 10^5 \left(\frac{f}{1000\ \text{Hz}}\right)^{1/2} \left(\frac{\hat{\tau}}{4\ \text{months}}\right)^{1/2} . \tag{9.4}$$

Four months of integration is not unreasonable in targeted searches; but for an all-sky, all-frequency search, a coherent integration might not last longer than a few days because of computational limitations associated with having to apply huge numbers of trial neutron-star spindown corrections and earth-motion doppler corrections [68].

Equation (9.4) for $h\sqrt{n}$ should be compared (i) to the detector's rms broad-band noise level for sources in a random direction, $\sqrt{5}h_{\rm rms}$, to deduce a signal-to-noise ratio, or (ii) to $h_{\rm SB}$ to deduce a sensitivity for high-confidence detection when one does not know the waves' frequency in advance [1]. Such a comparison suggests that the first interferometers in LIGO/VIRGO might possibly see waves from nearby spinning neutron stars, but the odds of success are very unclear.

The deepest searches for these nearly periodic waves will be performed by narrow-band detectors, whose sensitivities are enhanced near some chosen frequency at the price of sensitivity loss elsewhere—e.g., dual-recycled interferometers [15] or resonant-mass antennas (Section 9.2.4). With "advanced-detector technology" and targeted searches, dual-recycled interferometers might be able to detect with confidence spinning neutron stars that have [1]

$$(\epsilon_e \text{ or } \theta_w\epsilon_p) \gtrsim 3 \times 10^{-10} \left(\frac{500\ \text{Hz}}{f_{\rm rot}}\right)^2 \left(\frac{r}{1000\ \text{pc}}\right)^2 . \tag{9.5}$$

There may well be a large number of such neutron stars in our galaxy; but it is also conceivable that there are none. We are extremely ignorant.

Some cause for optimism arises from several physical mechanisms that might generate radiating ellipticities large compared to 3×10^{-10}:

- It may be that, inside the superconducting cores of many neutron stars, there are trapped magnetic fields with mean strength $B_{\rm core} \sim 10^{13}$ G or even 10^{15} G. Because such a field is actually concentrated in flux tubes with $B = B_{\rm crit} \sim 6 \times 10^{14}$ G surrounded by field-free superconductor, its mean pressure is $p_B = B_{\rm core}B_{\rm crit}/8\pi$. This pressure could produce a radiating ellipticity $\epsilon_{\rm e} \sim \theta_w \epsilon_p \sim p_B/p \sim 10^{-8}B_{\rm core}/10^{13}{\rm G}$ (where p is the core's material pressure).

- Accretion onto a spinning neutron star can drive precession (keeping θ_w substantially nonzero), and thereby might produce measurably strong waves [69].

- If a neutron star is born rotating very rapidly, then it may experience a gravitational-radiation-reaction-driven instability. In this "CFS" (Chandrasekhar, [70] Friedman, Schutz [71]) instability, density waves propagate around the star in the opposite direction to its rotation, but are dragged forward by the rotation. These density waves produce gravitational waves that carry positive energy as seen by observers far from the star, but negative energy from the star's viewpoint; and because the star thinks it is losing negative energy, its density waves get amplified. This intriguing mechanism is similar to that by which spiral density waves are produced in galaxies. Although the CFS instability was once thought ubiquitous for spinning stars [71, 72], we now know that neutron-star viscosity will kill it, stabilizing the star and turning off the waves, when the star's temperature is above some limit $\sim 10^{10}$ K [73] and below some limit $\sim 10^9$ K [74]; and correspondingly, the instability should operate only during the first few years of a neutron star's life, when $10^9 \text{ K} \lesssim T \lesssim 10^{10}$ K.

9.6 LOW-FREQUENCY GRAVITATIONAL-WAVE SOURCES

9.6.1 WAVES FROM THE COALESCENCE OF MASSIVE BLACK HOLES IN DISTANT GALAXIES

LISA would be a powerful instrument for studying massive black holes in distant galaxies. Figure 9.6 shows, as examples, the waves from several massive black hole binaries at 3 Gpc distance from earth (a cosmological redshift of unity). The waves sweep upward in frequency (rightward in the diagram) as the holes spiral together. The black dots show the waves' frequency one year before the holes' final collision and coalescence, and the arrowed lines show the sweep of frequency and characteristic amplitude $h_c = h\sqrt{n}$ during that last year. For simplicity, the figure is restricted to binaries with equal-mass black holes: $10^4\ M_\odot/10^4\ M_\odot$, $10^5\ M_\odot/10^5\ M_\odot$, and $10^6\ M_\odot/10^6\ M_\odot$.

By extrapolation from these three examples, we see that LISA can study much of the last year of inspiral, and the waves from the final collision and coalescence, whenever the holes' masses are in the range $3\times10^4\ M_\odot \lesssim M \lesssim 3\times 10^8\ M_\odot$ [26]. Moreover, LISA can study the final coalescences with remarkable signal to noise ratios: $S/N \gtrsim 1000$. Since these are much larger S/N's than LIGO/VIRGO is likely to achieve, we can expect LISA to refine the experimental understanding of black-hole physics, and of highly nonlinear vibrations of warped spacetime, which LIGO/VIRGO initiates—*provided* the rate of massive black-hole coalescences is of order one per year in the Universe or higher. The rate might well be that high, but it also might be much lower.

By extrapolating Fig. 9.6 to lower BH/BH masses, we see that LISA can observe the last few years of inspiral, but not the final collisions, of binary black holes in the range $100\ M_\odot \lesssim M \lesssim 10^4\ M_\odot$, out to cosmological distances [26].

Extrapolating the BH/BH curves to lower frequencies using the formula (time to final coalescence) $\propto f^{-8/3}$, we see that equal-mass BH/BH binaries enter LISA's frequency band roughly 1000 years before their final coalescences, more or less independently of their masses, for the range $100\ M_\odot \lesssim M \lesssim 10^6\ M_\odot$. Thus, if the coalescence rate were to turn out to be one per year, LISA would see roughly 1000 additional massive binaries that are slowly spiraling inward, with inspiral rates df/dt readily measurable. From the inspiral rates, the amplitudes of the two polarizations, and the waves' harmonic content, LISA can determine each such binary's luminosity distance, redshifted chirp mass $(1+z)M_c$, orbital inclination, and eccentricity; and from the waves' modulation by LISA's orbital motion, LISA can learn the direction to the binary with an accuracy of order one degree.

9.6.2 WAVES FROM COMPACT BODIES SPIRALING INTO MASSIVE BLACK HOLES OR EXOTIC OBJECTS IN DISTANT GALAXIES

When a compact body with mass μ spirals into a much more massive black hole with mass M, the body's orbital energy E at fixed frequency f (and correspondingly at fixed orbital radius a) scales as $E \propto \mu$, the gravitational-wave luminosity $\dot{E}$ scales as $\dot{E} \propto \mu^2$, and the time to final coalescence thus scales as $t \sim E/\dot{E} \propto 1/\mu$. This means that the smaller is μ/M, the more orbits are spent in the hole's strong-gravity region, $a \lesssim 10GM/c^2$, and thus the more detailed and accurate will be the map of the hole's spacetime geometry, which is encoded in the emitted waves.

For holes observed by LIGO/VIRGO, the most extreme mass ratio that we can hope for is $\mu/M \sim 1\ M_\odot/300\ M_\odot$, since for $M > 300\ M_\odot$ the inspiral waves are pushed to frequencies below the LIGO/VIRGO band. This limit on

μ/M seriously constrains the accuracy with which LIGO/VIRGO can hope to map out the spacetime geometries of black holes and test the black-hole no-hair theorem [48, 49] (end of Section 9.4.4). By contrast, LISA can observe the final inspiral waves from objects of any mass $M \gtrsim 0.5\ M_\odot$ spiraling into holes of mass $3 \times 10^5\ M_\odot \lesssim M \lesssim 3 \times 10^7\ M_\odot$.

Figure 9.6 shows the example of a 10 $M_\odot$ black hole spiraling into a $10^6\ M_\odot$ hole at 3 Gpc distance. The inspiral orbit and waves are strongly influenced by the hole's spin. Two cases are shown [75]: an inspiraling circular orbit around a non-spinning hole, and a prograde, circular, equatorial orbit around a maximally spinning hole. In each case the dot at the upper left end of the arrowed curve is the frequency and characteristic amplitude one year before the final coalescence. In the nonspinning case, the small hole spends its last year spiraling inward from $r \simeq 7.4GM/c^2$ (3.7 Schwarzschild radii) to its last stable circular orbit at $r = 6GM/c^2$ (3 Schwarzschild radii). In the maximal spin case, the last year is spent traveling from $r = 6GM/c^2$ (3 Schwarzschild radii) to the last stable orbit at $r = GM/c^2$ (half a Schwarzschild radius). The $\sim 10^5$ cycles of waves during this last year should carry, encoded in themselves, rather accurate values for the massive hole's lowest few multipole moments [48] (or, equivalently, a fairly accurate map of the hole's spacetime geometry (or,hole's). If the measured moments satisfy the "no-hair" theorem (i.e., if they are all determined uniquely by the measured mass and spin in the manner of the Kerr metric), then we can be sure the central body is a black hole. If they violate the no-hair theorem, then (assuming general relativity is correct), either the central body was an exotic object (e.g., soliton star or naked singularity) rather than a black hole, or else an accretion disk or other material was perturbing its orbit [76]. From the evolution of the waves one can hope to determine which is the case, and to explore the properties of the central body and its environment [49].

Models of galactic nuclei, where massive holes reside, suggest that inspiraling stars and small holes typically will be in rather eccentric orbits [77]. This is because they get injected into such orbits via gravitational deflections off other stars, and by the time gravitational radiation reaction becomes the dominant orbital driving force, there is not enough inspiral left to fully circularize their orbits. Such orbital eccentricity will complicate the waveforms and complicate the extraction of information from them. Efforts to understand the emitted waveforms are just now getting underway.

The event rates for inspiral into massive black holes are not at all well understood. However, since a significant fraction of all galactic nuclei are thought to contain massive holes, and since white dwarfs and neutron stars, as well as small black holes, can withstand tidal disruption as they plunge toward the massive hole's horizon, and since LISA can see inspiraling bodies as small as $\sim 0.5\ M_\odot$ out to 3 Gpc distance, the event rate is likely to be interestingly large.

9.7 Conclusion

It is now 36 years since Joseph Weber initiated his pioneering development of gravitational-wave detectors [4] and 25 years since Forward [78] and Weiss [79] initiated work on interferometric detectors. Since then, hundreds of talented experimental physicists have struggled to improve the sensitivities of these instruments. At last, success is in sight. If the source estimates described in this lecture are approximately correct, then the planned interferometers should detect the first waves in 2001 or several years thereafter, thereby opening up this rich new window onto the Universe.

Acknowledgments

My group's research on gravitational waves and their relevance to LIGO/VIRGO and LISA is supported in part by NSF grants AST-9417371 and PHY-9424337 and by NASA grant NAGW-4268. This manuscript was largely adapted and updated from my Ref. [10].

References

[1] K. S. Thorne, in *Three Hundred Years of Gravitation*, edited by S. W. Hawking and W. Israel, (Cambridge University Press, 1987), p. 330–458.

[2] K. S. Thorne, in *Gravitational Radiation*, edited by N. Deruelle and T. Piran, (North Holland, 1983), p. 1.

[3] W. J. Sullivan, *The Early Years of Radio Astronomy*, (Cambridge University Press, 1984).

[4] J. Weber, Phys. Rev. **117** 306 (1960).

[5] L. M. Krauss and M. White, Phys. Rev. Lett. **69** 969 (1992).

[6] R. L. Davis, H. M. Hodges, G. F. Smoot, P. J. Steinhardt, and M. S. Turner, Phys. Rev. Lett. **69** 1856 (1992).

[7] V. M. Kaspi, J. H. Taylor, and M. F. Ryba, Astrophys. J. **428** 713 (1994).

[8] Ya. B. Zel'dovich, Mon. Not. Roy. Astron. Soc. **192** 663 (1980).

[9] A. Vilenkin, Phys. Rev. D **24** 2082 (1981).

[10] X. Martin and A. Vilenkin, Phys. Rev. Lett. **77** 2879 (1996).

[11] A. Abramovici, et al., Science **256** 325 (1992).

[12] C. Bradaschia, et al., Nucl. Instrum. & Methods **A289** 518 (1990).

[13] B. Barish and G. Sanders et al., LIGO Advanced R and D Program Proposal, Caltech/MIT, unpublished, 1996.

[14] J. Hough and K. Danzmann et al., GEO600, Proposal for a 600 m Laser-Interferometric Gravitational Wave Antenna, unpublished, 1994.

[15] B. J. Meers, Phys. Rev. D **38** 2317 (1988).

[16] W. W. Johnson and S. M. Merkowitz, Phys. Rev. Lett. **70** 2367 (1993).

[17] R. A. Hulse and J. H. Taylor, Astrophys. J. **324** 355 (1975).
[18] J. H. Taylor, Rev. Mod. Phys. **66** 711 (1994).
[19] E. S. Phinney, Astrophys. J. **380** L17 (1991).
[20] R. Narayan, T. Piran, and A. Shemi, Astrophys. J. **379** L17 (1991).
[21] E. P. J. van den Heuvel and D. R. Lorimer, Mon. Not. Roy. Astron. Soc. (1996), [in press].
[22] A. V. Tutukov and L. R. Yungelson, Mon. Not. Roy. Astron. Soc. **260** 675 (1993).
[23] H. Yamaoka, T. Shigeyama, and K. Nomoto, Astron. Astrophys. **267** 433 (1993).
[24] V. M. Lipunov, K. A. Postnov, and M. E. Prokhorov, Astrophys. J. **423** L121 (1994), and related, unpublished work.
[25] S. F. P. Zwart and H. N. Spreeuw, Astron. Astrophys. **312** 670 (1996).
[26] E. E. Flanagan and S. A. Hughes, Phys. Rev. D. (1997), submitted for publication.
[27] C. Kochanek, Astrophys. J. **398** 234 (1992).
[28] L. Bildsten and C. Cutler, Astrophys. J. **400** 175 (1992).
[29] G. Quinlan and S. L. Shapiro, Astrophys. J. **321** 199 (1987).
[30] B. F. Schutz, Nature **323** 310 (1986).
[31] B. F. Schutz, Class. Quant. Grav. **6** 1761 (1989).
[32] C. Cutler, T. A. Apostolatos, L. Bildsten, L. S. Finn, E. E. Flanagan, D. Kennefick, D. M. Markovic, A. Ori, E. Poisson, G. J. Sussman, and K. S. Thorne, Phys. Rev. Lett. **70** 1984 (1993).
[33] C. M. Will, in *Relativistic Cosmology*, edited by M. Sasaki, (Universal Academy Press, 1994), p. 83.
[34] T. A. Apostolatos, C. Cutler, G. J. Sussman, and K. S. Thorne, Phys. Rev. D **49** 6274 (1994).
[35] L. E. Kidder, Phys. Rev. D **52** 821 (1995).
[36] C. Cutler and E. E. Flanagan, Phys. Rev. D, **49** 2658 (1994).
[37] L. S. Finn and D. F. Chernoff, Phys. Rev. D, **47** 2198 (1993).
[38] E. Poisson and C. M. Will, Phys. Rev. D **52** 848 (1995).
[39] L. Blanchet, T. Damour, B. R. Iyer, C. M. Will, and A. G. Wiseman, Phys. Rev. Lett. **74** 3515 (1995).
[40] C. Cutler and E. E. Flanagan, Phys. Rev. D. (paper in preparation).
[41] T. Damour and B.S. Sathyaprakash, Research in progress, (1997).
[42] A. Ashtekar, in *Gravitational Radiation*, edited by N. Deruelle and T. Piran, (North Holland, Amsterdam, 1983), p. 421.
[43] B. J. Owen, Phys. Rev. D **53** 6749 (1996).
[44] T. Damour and K. Nordtvedt, Phys. Rev. D **48** 3436 (1993).
[45] C. M. Will, Phys. Rev. D **50** 6058 (1994).
[46] D. Markovic, Phys. Rev. D **48** 4738 (1993).

[47] D. F. Chernoff and L. S. Finn, Astrophys. J. Lett. **411** L5 (1993).

[48] F. D. Ryan, Phys. Rev. D **53** 3064 (1995).

[49] F. D. Ryan, L. S. Finn, and K. S. Thorne, Phys. Rev. Lett. (in preparation).

[50] Numerical Relativity Grand Challenge Alliance, 1995. References and information on the World Wide Web, http://jean-luc.ncsa.uiuc.edu/GC.

[51] D. Kennefick, D. Laurence, and K. S. Thorne, Phys. Rev. D. (in preparation).

[52] X. Zhuge, J. M. Centrella, and S. L. W. McMillan, Phys. Rev. D **50** 6247 (1994). Also references therein.

[53] D. Christodoulou, Phys. Rev. Lett. **67** 1486 (1991).

[54] K. S. Thorne, Phys. Rev. D **45** 520–524 (1992).

[55] A. G. Wiseman and C. M. Will, Phys. Rev. D **44** R2945 (1991).

[56] D. Kennefick, Phys. Rev. D **50** 3587 (1994).

[57] *Supernovae*, A. G. Petschek, editor, (Springer Verlag, 1990).

[58] H. A. Bethe, Rev. Mod. Phys. **62** 801 (1990).

[59] A. Burrows, private communication, 1995.

[60] H.-T. Yanka, W. Keil, and E. Müller, 1996.

[61] L. S. Finn, Ann. N. Y. Acad. Sci. **631** 156 (1991).

[62] R. Mönchmeyer, G. Schäfer, E. Müller, and R. E. Kates, Astron. Astrophys. **256** 417 (1991).

[63] D. Lai and S. L. Shapiro, Astrophys. J. **442** 259 (1995).

[64] J. L. Houser, J. M. Centrella, and S. C. Smith, Phys. Rev. Lett. **72** 1314 (1994).

[65] I. A. Bonnell and J. E. Pringle, Mon. Not. Roy. Astron. Soc. **273** L12 (1995).

[66] M. Zimmermann and E. Szedenits, Phys. Rev. D **20** 351 (1979).

[67] S. L. Shapiro and S. A. Teukolsky, Wiley: Interscience, 1983. Section 10.10 and references cited therein.

[68] P Brady, J Creighton, C Cutler, an B. Schutz, submitted to Phys. Rev. D (1997).

[69] B. F. Schutz, private communication, (1995).

[70] S. Chandrasekhar, Phys. Rev. Lett. **24** 611 (1970).

[71] J. L. Friedman and B. F. Schutz, Astrophys. J. **222** 281 (1978).

[72] R. V. Wagoner, Astrophys. J. **278** 345 (1984).

[73] L. Lindblom, Astrophys. J. **438** 265 (1995).

[74] L. Lindblom and G. Mendell, Astrophys. J. **444** 804 (1995).

[75] L. S. Finn and K. S. Thorne, Phys. Rev. D.(in preparation).

[76] D. Molteni, G. Gerardi, and S. K. Chakrabarti, Astrophys. J. **436** 249 (1994).

[77] D. Hils and P. Bender, Astrophys. J. Lett. **445** L7 (1995).

[78] G. E. Moss, L. R. Miller, and R. L. Forward, Applied Optics **10** 2495 (1971).

[79] R. Weiss. Quarterly Progress Report of RLE, MIT **105** 54 (1972).

[80] K. S. Thorne, in *Proceedings of the Snowmass 95 Summer Study on Particle and Nuclear Astrophysics and Cosmology*, edited by E. W. Kolb and R. Peccei, (World Scientific, 1995), p. 398.

9.8 DISCUSSION

Session Chair: Joseph Taylor
Rapporteur: Dean Jens

TAYLOR: Maybe I could begin by asking you from whence comes the great gain in sensitivity below 10 Hz in the lowest curves that you show? i.e., for the sensitivity projections furthest into the future? (Editor's note: Fig. 9.4 does not show the region below 10 Hz, but in the lecture that region was shown; cf. [13].)

THORNE: Optimism about seismic isolation. What has been put into those curves is simply an assumption that over the long haul one has conquered seismic noise—and it should be conquerable, but it's not easy by any means—and what is left is gravity gradient noise. What is shown here is gravity gradient noise produced by Rayleigh waves propagating along the surface of the earth created by wind on trees in the distance and so forth at the two sites where we have measurements of background spectrum. These seismic waves create density fluctuations which create a fluctuating gravity field on the test masses. This is the background level at quiet times when the wind is not blowing much. A second noise region deserves mention: $\sim$ 50 to 100 Hz. There the noise is largely due to the Heisenberg uncertainty principle, the test masses behaving quantum mechanically assuming you're not doing any so-called quantum nondemolition, that you're just limited by the half-width of the Schroedinger wave function of the center of mass of the test masses. There are techniques, even, it appears, practical techniques, of doing quantum nondemolition and doing better than this at least with narrow banded interferometers. That's going to be one of the challenges of this field over the next thirty years: to get to a situation where you can routinely take a test mass like this and manipulate its wave function in the process of monitoring its motion in a way that's better than you might have naively thought possible on the basis of the uncertainty principle.

SAMIOS: Do you think you could give us a cost for going from the first performance curve (Fig. 9.4), to the second, to the third ... ?

THORNE: I think I would like Rai Weiss to comment on that.

WEISS: Right now it's typically costing something like \$30 million per interferometer group. We wouldn't have to change the vacuum system facilities that are in place, but as you make these progressions, each time you make a major change in the configuration of the interferometer you're changing three interferometers, and if you go by what we're spending now it's about \$30 million. The total project cost at this point is about \$300 million for construction, operation, and research for the first three years. ... May I say something?

TAYLOR: Go ahead, please do.

WEISS: Since this is a celebration of Princeton, everyone who is interested in careful and difficult experiments really has to go back to the original 1962 paper that Bob Dicke and colleagues wrote on the Eötves experiment; I give it to all my graduate students to read. Now aside from achieving a remarkably small figure for the difference of the ratio between inertial and gravitational passive mass of two different objects, it also developed a number of techniques without which you couldn't do any of these gravity wave experiments. The fact that LIGO or LISA can even be contemplated involves two ideas that were incorporated in that experiment and thought through *ab initio* by Bob and the guys around him. You and I can estimate the fundamental limits of an experiment—calculate the noise—but to actually achieve them is the hard part, and what that experiment did was to demonstrate the techniques by which you do that. I'll give you two of them; they're deeply ingrained in the whole LIGO project. One: you suffer in every experiment at low frequencies from $\frac{1}{f}$ noise; that's just a rubric for saying you don't know something. One of the major experimental techniques that is used everywhere now is to take the information you want, in this case it happens to be the fringe information—the ability to split a fringe to one part in 10^{11}—you take that information and move it from the $\frac{1}{f}$ dominated region to higher frequencies where the fundamental noise terms dominate. That means a frequency conversion, and that is what you'll see in the Eötvos experiment. It's something we now accept without thinking about it, but it really was first applied in that experiment, and it made it possible; it also makes these gravity detection experiments possible. Two: the ability to damp motion by active control. That sounds very engineering-like, but it turns out that that is the heart of being able to make an experiment as complicated as LIGO work. Now, of course, you and I deal with servo loops all the time; every power supply has those. At the time, it was not standard practice to do the following thing, and this is the thing for which you want to give credit to Bob. You have a controller somewhere, and you very carefully feed back a control signal that makes your signal go to zero. Then you measure your signal by measuring the control signal. Now it sounds like an elementary idea, but it turns out that it has a number of applications: damping a system in a noise-free way, refrigerating a system, taking the noise out of it, the ideas that are central to being able to get down to 10^{-16}cm, and without any of that all of this would be impossible. Bob is the unsung hero behind a lot of this very tricky technology that's being done here.

THORNE: I'd like to expand on that a little bit and say that to me as a theorist there's somebody else in Princeton who is really the inspiration for this field, and that's Johnny Wheeler. He is the person who first understood in a deep way and conveyed to the rest of the community the dynamics of geometry, what John called geometrodynamics, and that is what is going on here in spades. It was really this pair of people for me as a graduate student at Princeton—John Wheeler and Bob Dicke—who turned me on to this whole area from both the theoretical side and the experimental side. I think that Rai, who spent time here as a postdoc when I was a graduate student, was similarly influenced by them. Those were really wonderful years and very exciting in both of those research groups at Princeton.

DAMOUR: In this vein I want to mention that the person who pioneered the idea that binary neutron stars were really very interesting objects, as far back as 1964, three years before the discovery of pulsars, ten years before the discovery of binary neutron stars by Joe Taylor, was Freeman Dyson at the Institute. He foresaw many years in advance that that would be an interesting source of gravitational waves.

CHAPTER 10

NEUTRINO OSCILLATIONS

D.H. PERKINS

Particle and Nuclear Physics Laboratory
University of Oxford, U.K.

10.1 SOME NEUTRINO PROPERTIES

Before discussing neutrino oscillations, let me first remind you of some of the salient properties of neutrinos. They are members of a family of fundamental fermions called leptons, listed in Table 10.1. Antileptons have the opposite charge and magnetic moment to the leptons. The neutral particles, neutrinos and antineutrinos (denoted by a bar), occur in 3 flavor states, designated by the subscript ν_e, ν_μ, ν_τ of the charged leptons. Leptons are assigned a quantum number $L = +1$, antileptons $L = -1$. Lepton conservation means the total lepton number (the difference in the number of leptons and antileptons) is a constant. This is an experimental fact: there appears to be no deep underlying theoretical reason.

Table 10.1: The Lepton Family

Q/e	Leptons (lepton number+1)		
-1	e^-	μ^-	τ^-
0	ν_e	ν_μ	ν_τ
	Antileptons (lepton number-1)		
+1	e^+	μ^+	τ^+
0	$\bar{\nu}_e$	$\bar{\nu}_\mu$	$\bar{\nu}_\tau$

In certain circumstances—and in essentially all laboratory experiments to date—the flavour number L_e, L_μ, L_τ is also conserved. Thus γ-rays, in traversing matter, may produce e^+e^- or $\mu^+\mu^-$ pairs, but not μ^+e^- or $e^+\mu^-$. In the decay of the neutron

$$n \to p + e^- + \bar{\nu}_e$$

the antineutrino $\bar{\nu}_e$, if it had sufficient energy when it interacted with matter, would transform to e^+, never μ^+.

The hypothesis of neutrino flavour oscillations, to be discussed shortly, challenges flavour number conservation, and postulates that, on long enough timescales, one flavour of neutrino may transform into another.

Neutrino masses, as directly measured, are very small in comparison with the charged lepton masses, as shown in Table 10.2. Only upper limits can be given. The ν_e mass limit comes from the measurement of the end-point of the β-spectrum in tritium decay (sensitive because of the low $Q = 18.6$ keV), that of ν_μ from precise measurement of the muon momentum in pion decay, and accurate values of pion and muon masses. In both cases one actually measures m_ν^2, which somewhat embarrassingly comes out negative in most experiments! So, there may be systematic errors which are not at present understood, which limit the precision.

Table 10.2: Direct Neutrino Mass Measurements

Flavour	Mass	Method
ν_e	< 15 eV/c^2	$^3\mathrm{H}_1 \to {}^3\mathrm{He}_2 + e^- + \bar{\nu}_e$
ν_μ	< 170 keV/c^2	$\pi^+ \to \mu^+ + \nu_\mu$
ν_τ	< 24 MeV/c^2	$\tau \to 5\pi + \nu_\tau$

In the Standard Model of particle physics, which makes detailed predictions in impressive and detailed agreement with all laboratory experiments so far, neutrinos are assumed to be massless. Furthermore they are completely polarized. Neutrinos have the projection of the spin vector ($j_z = -1/2$) against the momentum vector, i.e., they are completely left-handed (LH), while antineutrinos are RH. This is, of course, a relativistically invariant description only if $m_\nu = 0$ (otherwise we could switch to a reference frame moving faster than the neutrino, from which the handedness would be reversed).

Grand Unification Theories (GUTs) do postulate very massive RH neutrinos. The massless LH states get very weakly mixed with the RH states, of mass M, and acquire a mass, given by the so-called "seesaw formula"

$$m_\nu \sim m_D^2/M$$

where M is the GUT scale mass and m_D is a typical "Dirac mass," that is of normal charged fermions with both LH and RH states. Purely as an example, if one takes for m_D the masses of charge 2/3 quarks u, c, t and M=10^{12} GeV, one gets

$$m_{\nu_e} = 10^{-5}\ \mathrm{eV}$$
$$m_{\nu_\mu} = 10^{-3}\ \mathrm{eV}$$
$$m_{\nu_\tau} = 10\ \mathrm{eV}$$

Such values for m_{ν_μ} and m_{ν_τ} could for instance be relevant for the solar neutrino problem and for hot dark matter respectively.

10.2 NEUTRINO OSCILLATIONS

In the 1960s, as soon as it became clear that neutrinos occurred in different flavour eigenstates, it was suggested that, in analogy with $K^0 - \bar{K}^0$ mixing, neutrino flavour mixing could occur. To simplify the problem, consider just 2 flavours, ν_e and ν_μ (with these same symbols to represent the particle wavefunctions). The idea is that at the moment of production in a weak interaction or decay process, these neutrinos are flavour eigenstates: but as they propagate through space-time, they have to be considered as a superposition of mass eigenstates, say ν_1 and ν_2. The flavour and mass eigenstates are connected by a unitary transformation

$$\begin{vmatrix} \nu_e \\ \nu_\mu \end{vmatrix} = \begin{vmatrix} \cos\theta & \sin\theta \\ -\sin\theta & \cos\theta \end{vmatrix} \begin{vmatrix} \nu_1 \\ \nu_2 \end{vmatrix} \tag{10.1}$$

specified by a mixing angle θ. Thus, if θ is small, ν_e is mostly ν_1 with a little ν_2, and ν_μ conversely. ν_1 and ν_2 have the same momentum, but with different masses m_1 and m_2, will have slightly different angular frequencies (energies). Hence a phase difference will develop, i.e., $\nu_1(t) = \nu_1(0)\exp(-i\omega_1 t)$, $\nu_2(t) = \nu_2(0)\exp(-i\omega_2 t)$ with $\omega_2 > \omega_1$ if $m_2 > m_1$. Figure 10.1 shows the situation for $\theta = 45°$, where ν_1 and ν_2 successively go in and out of phase, that is, as defined in (1), as pure ν_e or ν_μ eigenstates. At arbitrary times, the superposition of ν_1 and ν_2 corresponds as far as weak interactions are concerned, to a mixture of ν_e and ν_μ.

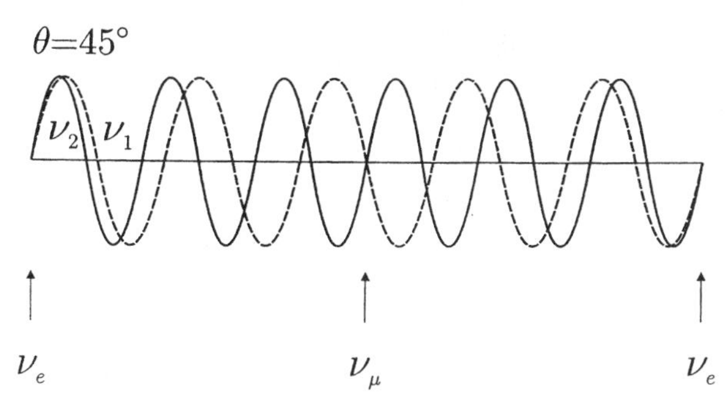

Figure 10.1: Amplitude of ν_1 and ν_2 mass eigenstates as a function of time, for the case $\theta = 45°$. The two are in phase at the beginning and end of the plot and here correspond to a pure ν_e weak-interaction eigenstate. The two amplitudes are out of phase—see Eq.(1)—in the centre of the plot, where they correspond to the ν_μ weak interaction eigenstate.

In general, the expression for the survival probability of an initially pure ν_e beam will be

$$P(\nu_e \to \nu_e) = 1 - \sin^2 2\theta \sin^2 \left(\frac{1.27 \Delta m^2 L}{E} \right) . \tag{10.2}$$

The value for the phase difference $(\omega_2 - \omega_1)t = 1.27\Delta m^2 L/E$ assumes $\Delta m^2 = m_2^2 - m_1^2 << E^2$ where E is the beam energy, and the factor $1.27 = 1/4\hbar c$ when the distance travelled L is in meters, E is in MeV and Δm in eV/c^2.

As soon as neutrino oscillations were suggested, experimental searches started at accelerators and reactors. On several occasions in the past 30 years, neutrino oscillations have been "discovered," only to be "un-discovered" by subsequent experiments. The most recent claim is from the LSND project at Los Alamos [1]. Figure 10.2 shows the probability contours in a plot of $\sin^2 2\theta$ against Δm^2.

The experiment consists of bombarding a target with 800 MeV protons, producing pions which decay to muons. Muons in turn decay at rest; $\mu^+ \to e^+ + \nu_e + \bar{\nu}_\mu$. Here the $\bar{\nu}_\mu$ energy extends up to 55 MeV. In a 170-ton liquid scintillator detector they look for $\bar{\nu}_\mu \to \bar{\nu}_e$ oscillation via the reaction $\bar{\nu}_e + p \to e^+ + n$. The

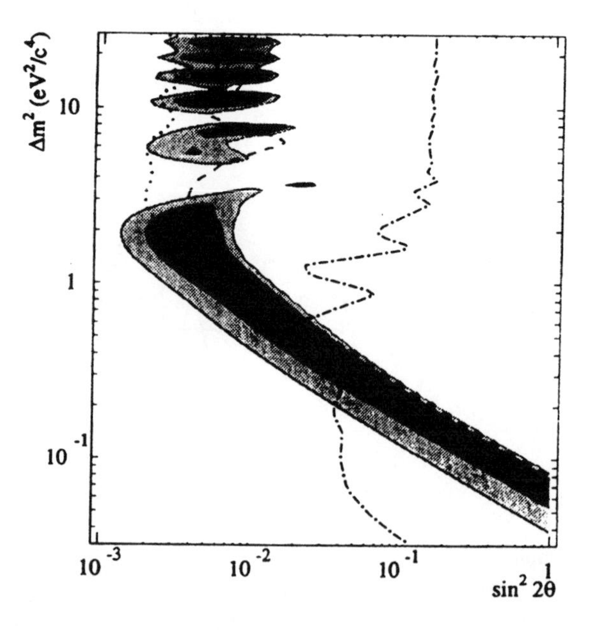

Figure 10.2: Probability contours from LSND experiment in $\sin^2 2\theta$ vs. Δm^2 plot. The dark shaded areas show where these oscillation parameters lie with 90% probability. Also shown are results from the KARMEN experiment at RAL [2] and from the E776 experiment [3] at BNL. They observe no evidence for oscillations, and set upper limits at 90% CL, shown by the dashed and dotted curves respectively.

signal is of two pulses of γ-rays, the first from positron annihilation $e^+e^- \to 2\gamma$, and a further delayed 2.2 MeV γ-ray from neutron capture in hydrogen of the scintillator, to form a deuteron. After two years the experiment finds 22 events, against a computed background of 5 (from pion or muon decays in flight, neutron background, etc.). Similar experiments at RAL and BNL find no effect. Clearly, since the probability that all three experiments are compatible is less than 3%, much more data, with variations in L or E, are needed before the evidence can be taken seriously.

10.3 SOLAR NEUTRINOS

The present best evidence for neutrino oscillations, such as it is, comes from several independent experiments on solar neutrinos and on atmospheric neutrinos.

Figure 10.3 shows the solar neutrino spectrum calculated by Bahcall [4]. The principal source of ν_e from the sun is the *pp* reaction ($p + p \to d + e^+ + \nu_e$) but there are other sources (^{7}Be, *pep* and ^{8}B) associated with side reactions which extend to 15 MeV energy.

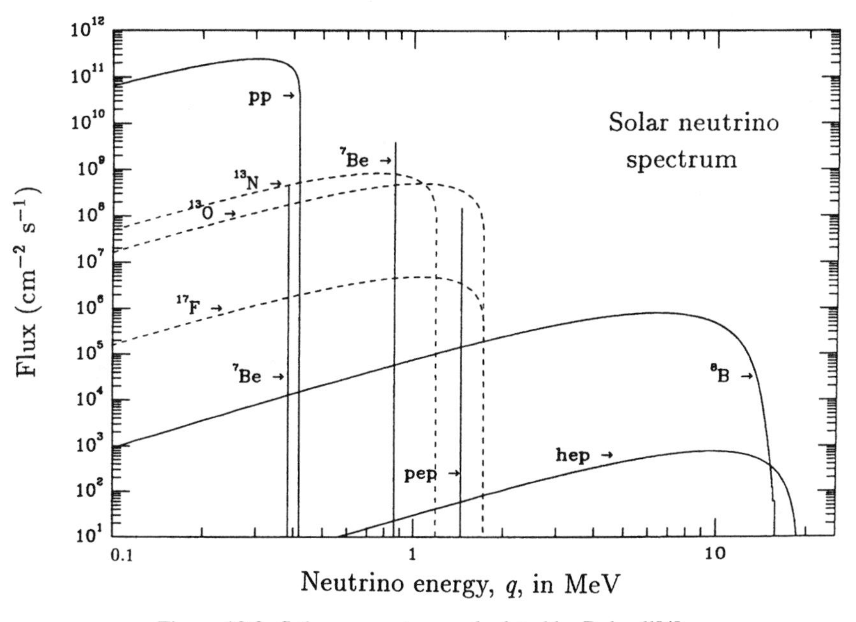

Figure 10.3: Solar ν_e spectrum calculated by Bahcall[4].

The four experiments observing solar neutrinos to date are the radiochemical SAGE [5] and GALLEX [6] projects, observing ^{71}Ge produced by neutrino capture in ^{71}Ga. These have 0.2 MeV threshold energy and are sensitive to *pp* neutrinos (60% of count rate) plus all the rest (40%). Although the flux from these other sources is very small compared with *pp*, the cross-sections vary typically as E_ν^3, so they make significant contributions to the rate. The other radiochemical experiment is HOMESTAKE [7], using a C_2Cl_4 detector and observing the ^{37}Cl $\rightarrow$ ^{37}Ar transition. This has 0.8 MeV threshold energy, so is sensitive to one ^{7}Be line, to *pep* and to ^{8}B sources. Finally the KAMIOKA [8] water detector observes Cerenkov light from electrons above 6 MeV produced by $\nu e \rightarrow \nu e$ scattering. It is sensitive only to ^{8}B neutrinos. The events are observed in real time, and detected as a forward peak in the angular distribution of Cerenkov signals, when measured relative to the solar direction.

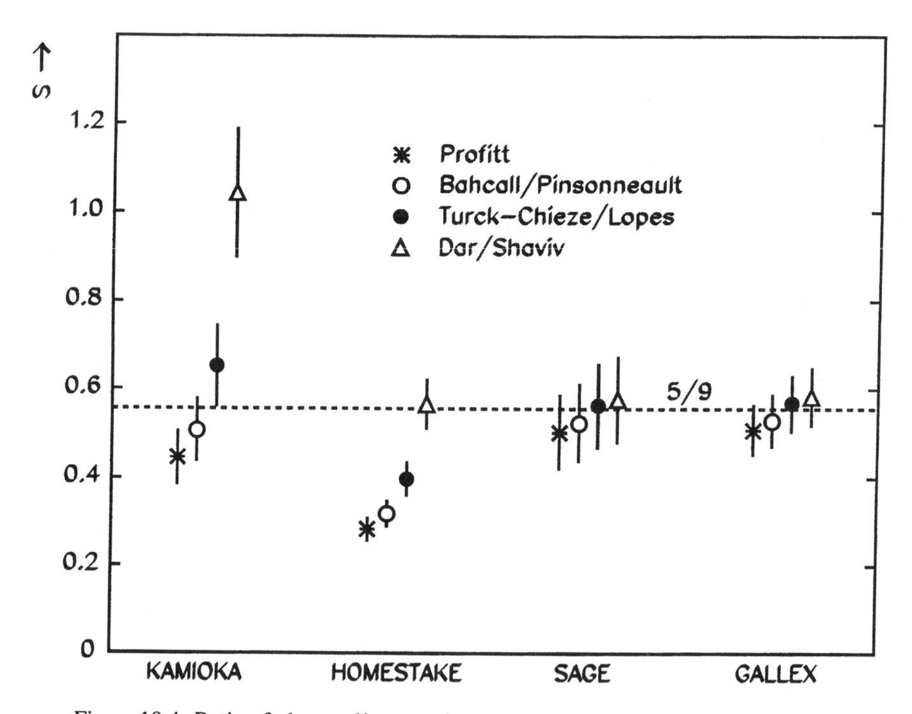

Figure 10.4: Ratio of observed/expected event rates, for each of four predictions for the four solar neutrino experiments.

Figure 10.4 shows the results for the four experiments, plotting the observed/ expected event rates, for each of four predictions of the fluxes. The usually-quoted and most extensive calculations are those of Bahcall/Pinsonneault [9] and Turck-Chieze/Lopez [10]. Other flux calculations shown are those of Profitt [11] and Dar and Shaviv [12]. The latter take extreme values for the solar opacity and some of the relevant cross-sections, and come out with much smaller ^{8}B and ^{7}Be neutrino fluxes. The main points to be made from Fig. 10.4 are as follows:

The two gallium experiments are in good agreement, and the observed/ expected event ratio is also about the same for the various calculated fluxes. Furthermore, the absolute efficiency of both experiments has been directly measured with ^{51}Cr laboratory neutrino sources. So this result, indicating a 40–50% deficit in the expected rate, is very solid.

By contrast, the ratios for the KAMIOKA and HOMESTAKE experiments vary widely for the different calculations. However, the feature that all have in common is that the KAMIOKA ratio (threshold $\sim$ 6 MeV) is larger than that for HOMESTAKE (threshold $\sim$ 0.8 MeV): there is apparently an energy-dependent suppression.

Ten years ago, Mikheyev and Smirnov [12], following the work of Wolfenstein [14], described a matter-enhanced suppression mechanism, now called the MSW effect. The idea is that while all flavours of neutrino scatter from electrons via Z^0 exchange, only ν_e can scatter via $W^{\pm}$ exchange.

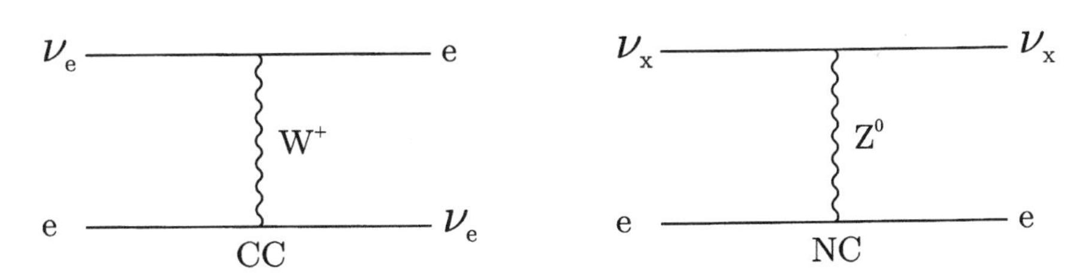

Figure 10.5: ν_e scattering from electrons via $W^{\pm}$ exchange, and $\nu_{e,\nu,\tau}$ scattering via Z^0 exchange.

For a muon neutrino to do this, for example, is impossible because the threshold for $\nu_\mu + e^- \rightarrow \mu + \nu_e$ is 110 MeV and solar neutrinos have energies up to 15 MeV only. Uniquely therefore, in traversing a medium the ν_e suffers an extra potential V_{eff} leading to a change in effective mass with an energy dependence:

$$\begin{aligned} V_{\text{eff}} &= G\sqrt{2}N_e \\ m^2 &= E^2 - p^2 \rightarrow (E + V_{\text{eff}})^2 - p^2 \approx m^2 + 2EV_{\text{eff}} \\ \Delta m_m^2 &= 2\sqrt{2}GN_e.E_\nu \end{aligned} \tag{10.3}$$

where N_e is the electron density, Δm_m^2 is the matter-induced shift in mass squared. In the sun, the density varies exponentially with radius, from 150 gm cm^{-3} ($N_e \simeq 10^{26}$) at the core to 10^{-6} g cm^{-3} at the photosphere. Clearly, if N_e and E in (3) are such that $\Delta m_m^2 = \Delta m_V^2$, where '$m$' stands for matter and 'V' stands for vacuum, and $\Delta m_V^2 = m_2^2 - m_1^2$, a resonant transition can occur. The actual resonant condition is

$$\Delta m_m^2 = \Delta m_V^2 \cos 2\theta_V . \tag{10.4}$$

So, basically what can happen is that a ν_e starts out in the solar core, predominantly (for θ_V small) as the ν_1 eigenstate, and the extra potential—if N_e and E are right—lifts the ν_1 to the state ν_2, of mass m_2. This mass eigenstate then passes out of the sun without change, provided the interaction is adiabatic, that is, the variation of N_e per oscillation length is small. (If not, ν_2 can jump back to ν_1 and one obtains only partial conversion). The ν_2 state emerging into the vacuum now has to be identified as mostly ν_μ, so that a $\nu_e \rightarrow \nu_\mu$ matter-induced oscillation has occurred, regardless of how small the vacuum mixing angle θ_V may be.

Figure 10.6 shows the regions of $\sin^2 2\theta$ and Δm^2 favoured by the experiments. The gallium experiments basically determine Δm^2, while the HOMESTAKE and KAMIOKA experiments determine the vacuum mixing angle. The favoured solution has $\theta \sim 2°$, $\Delta m^2 \sim 10^{-5}$ eV^2. The MSW suppression is largest in the region $E = 1 - 5$ MeV (i.e., for ^{7}Be and *pep* neutrinos). This provides a neat explanation of the solar data.

In the future, efforts are being made to

(i) detect the distortion of the ^{8}B energy spectrum due to the MSW effect, in SUPERKAMIOKA, a large 50,000 ton water Cerenkov detector which came into operation in April 1996;

(ii) detect the "converted" ν_e as ν_μ and ν_τ through "neutral current" interactions: $\nu_\mu + d \rightarrow p + n + \nu_\mu$. This is the goal of the SNO experiment, employing 1 kiloton of heavy water, and hoping to detect the neutron from disintegration of the deuteron;

(iii) investigate the suppression of ^{7}Be neutrinos (which should be maximal) in the Borexino liquid scintillator experiment. All these projects will be very difficult, and a final resolution of the solar neutrino problem is probably several years away.

10.4 ATMOSPHERIC NEUTRINOS

The atmosphere provides a source of neutrinos where anomalies are also found. Incident primary cosmic rays (protons and heavier nuclei) in the GeV energy

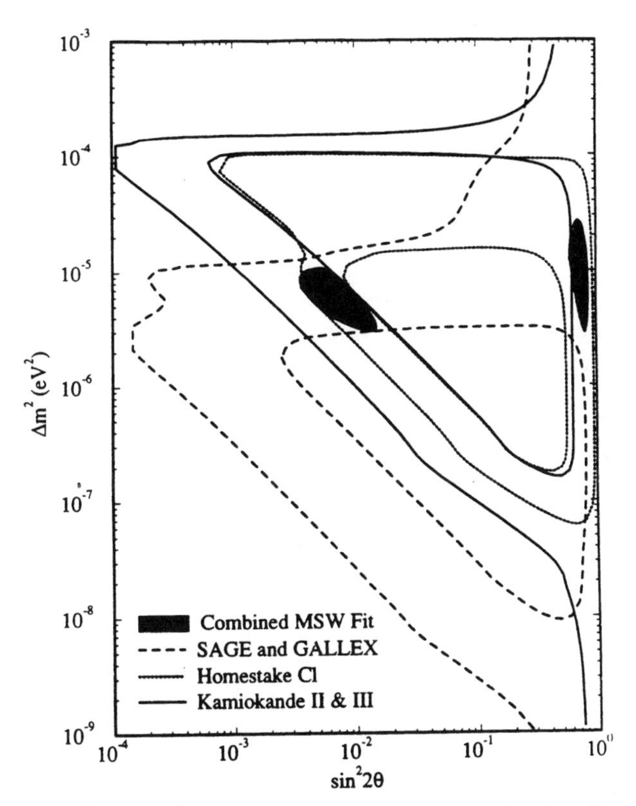

Figure 10.6: $\sin^2 2\theta/\Delta m^2$ exclusion plot, the shaded areas showing where the four solar neutrino experiments define the 90% CL MSW solutions. There is one small angle solution, and one, less probable, large (vacuum) angle solution.

range impinge on the atmosphere and generate secondary pions, which decay to muons and neutrinos, the muons decaying in turn:

$$\begin{array}{ll} \pi^+ \to \mu^+ + \nu_\mu & \pi^- \to \mu^- + \bar{\nu}_\mu \\ \mu^+ \to e^+ + \nu_e + \bar{\nu}_\mu & \mu^- \to e^- + \bar{\nu}_e + \nu_\mu . \end{array} \tag{10.5}$$

At energies of 1 GeV or below, essentially all pions and muons decay in flight, rather than interacting or reaching sea-level. Furthermore, the kinematics in (5) are such that all the neutrinos get, on average, about the same energy—roughly 20–25% of the pion energy—and it is therefore no surprise that MC calculations come out with a flux ratio $\phi(\nu_\mu + \bar{\nu}_\mu)/\phi(\nu_e + \bar{\nu}_e) \approx 2:1$. For

$E_\nu > 1$ GeV, the ratio is greater than 2, because at higher energies the longer-lived muons lose substantial ionization energy before decay (the decay length of a muon is 6.6 km/GeV, and the scale height of the atmosphere is 6.3 km). There is some uncertainty ($\sim$ 20%) in the absolute fluxes, so what is plotted is the ratio of ratios $R = (N_\mu/N_e)_{\text{obs}}/(N_\mu/N_e)_{\text{MC}}$, where the MC calculation of the relative numbers of ν_μ induced events to ν_e induced events includes efficiencies and energy thresholds. Figure 10.7 shows the ratio R for the two water Cerenkov (IMB [15] and KAMIOKA [16]) and the three electronic tracking detectors (NUSEX [17], FREJUS [18] and SOUDAN II [19]). The weighted average of the five experiments is $R_{av} = 0.64 \pm .06$, with a value of $\chi^2 = 1.04$ per degree of freedom, for the dispersion of the results about the sample mean. So the different experiments are consistent with each other (and with $R = 2/3$ predicted by the maximal mixing scheme—see below).

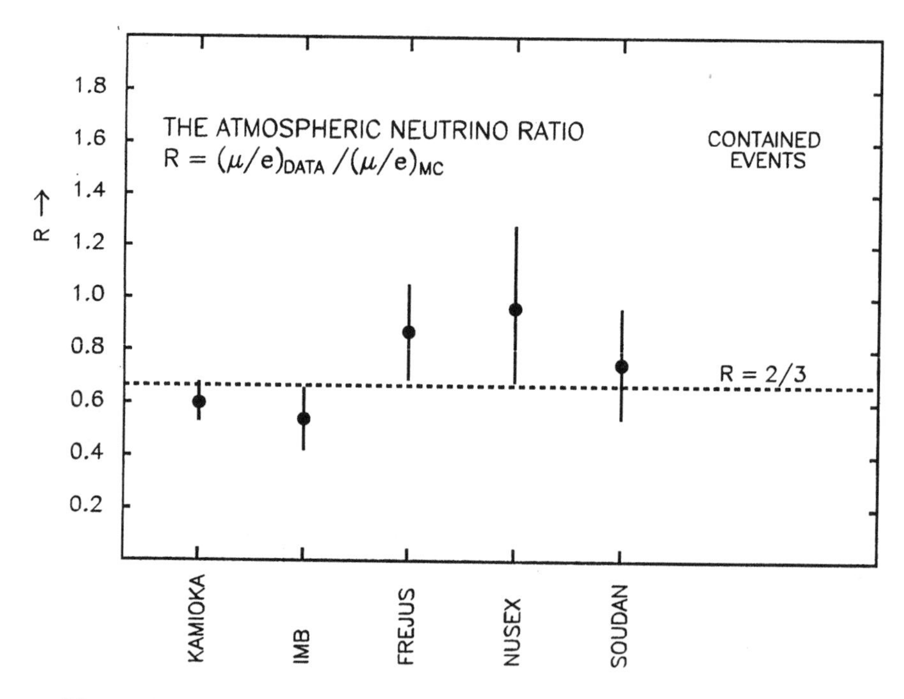

Figure 10.7: The quantity R, which is the ratio of μ/e events observed, to that calculated, for 5 atmospheric neutrino detectors.

In the KAMIOKA experiment there are, in addition to the fully-contained events, so-called "multi-GeV" events, where the interaction is identified but the charged lepton is of high energy and exits the detector. Only a lower limit to the energy is known for these events (all have $E_\nu > 1.5$ GeV): the mean energy will be about 5 GeV, from the calculated spectrum. At such energies, there is a strong angular correlation (within 20° or less) between the neutrino and the secondary charged lepton, so the zenith angle distribution of these leptons reflects that of the neutrinos. Figure 10.8 shows preliminary evidence for a zenith angle dependence.

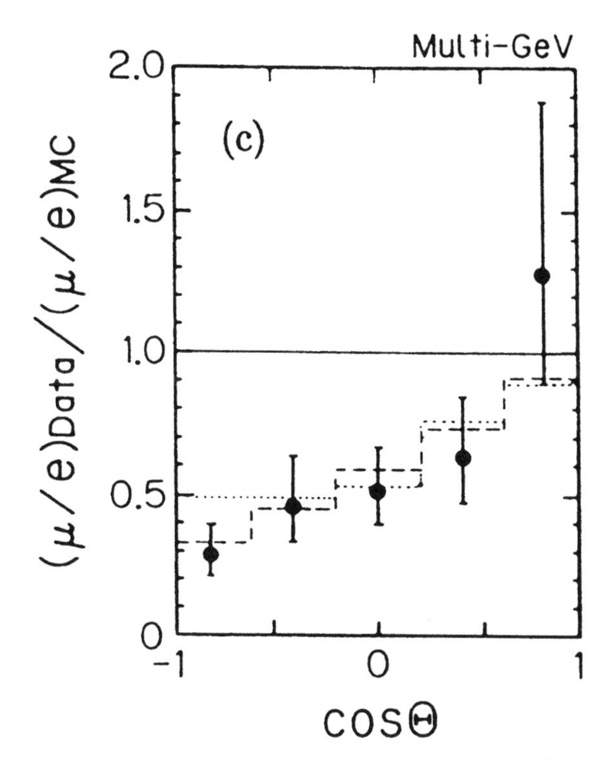

Figure 10.8: Zenith angle dependence of ratio of ratios R for the multi-GeV events in the KAMIOKA detector.

Leptons from interactions of neutrinos coming downwards (and generated in the stratosphere some 20 km distant) seem to show a value of $R \sim 1$, while those from upward-coming neutrinos generated in the atmosphere on the far side of the earth—typically 5,000–10,000 km distant—show a smaller R value. This is exactly the effect expected from neutrino oscillations with a wavelength of the order of the earth radius.

10.5 MAXIMAL MIXING

With 3 neutrino flavours, there is a 3×3 matrix coupling flavour and mass eigenstates. At the very simplest, this matrix will involve 3 Euler angles and 1

CP-violating phase angle: in addition there are 2 independent mass differences, or 6 parameters in all. With present data, simplifying assumptions are necessary. For example, one can assume no CP violation and only 2 non-zero Euler angles. One small angle and one mass difference, together with the MSW mechanism, can account for the solar neutrino suppression: and another mass difference and another large mixing angle can account for the atmospheric anomaly. Essentially this amounts to two independent sets of 2×2 mixings, say $\nu_e \to \nu_\mu$ and $\nu_\mu \to \nu_\tau$.

Of course, there is an infinite number of mixing schemes, and I will just outline one rather unique scheme, that of maximum mixing, proposed many years ago by Gribov and Pontecorvo [20], by Wolfenstein [21], by Cabibbo [22] and by Nussinov [23]. This idea seems to have withered away, when it was found that the off-diagonal elements in the CKM quark mixing matrix were in fact very small. However the idea was revived in 1994 by Harrison and Scott and applied to neutrino mixing [24].

The general formula for threefold mixing gives the probability, starting with a neutrino flavour α to end up with one of flavour β after a time t

$$P(\nu_\alpha \to \nu_\beta) = \left| \sum_{i=1}^{3} U^\dagger_{\alpha i} e^{-iE_i t} U_{\beta i} \right|^2 \qquad (10.6)$$

where $i = 1, 2, 3$ refers to the mass eigenstates, and $E_i \simeq p + m_i^2/2E_i$. The exponential term thus includes a factor $\Delta_m^2 L/2E$, where L is the distance travelled from the source.

For maximum mixing, the unitary matrix U is very simple: all entries are equal in magnitude:

$$U_{\alpha i} = \frac{1}{\sqrt{3}} \begin{vmatrix} \omega_1 & \omega_1 & \omega_1 \\ \omega_1 & \omega_2 & \omega_3 \\ \omega_1 & \omega_3 & \omega_2 \end{vmatrix} \qquad (10.7)$$

where $\omega_1, \omega_2, \omega_3$ are the (complex) cube roots of unity ($\omega_1 = 1; \omega_2 = -\frac{1}{2} + \frac{\sqrt{3}}{2}i; \omega_3 = -\frac{1}{2} - \frac{\sqrt{3}}{2}i$). The various elements, while equal in magnitude, do differ in phase. The Euler angles here are fixed (at $\pi/4, \pi/4$ and $\sin^{-1} 1/\sqrt{3}$) and the CP-violating phase angle is $\pi/2$, that is, CP violation is maximal (although it turns out in fact to be undetectable). There are only two free parameters, namely the mass differences among m_1, m_2 and m_3. Assuming a mass hierarchy as for the charged leptons, we expect one large mass difference squared, say $\Delta m^2 = m_3^2 - m_2^2$, and one small, say $\Delta m'^2 = m_2^2 - m_1^2$.

For values such that $\Delta m'^2 L/E << 1$, the oscillations are determined entirely by the value of Δm^2. Furthermore the survival probability $P(\nu_\alpha \to \nu_\alpha)$ as a function of time is the same for all flavours of neutrino and antineutrino. So, all data from all experiments—$\bar{\nu}_e$ from reactors, ν_e from the sun, ν_μ and $\bar{\nu}_\mu$ from accelerators, ν_μ and ν_e from the atmosphere—can be plotted on one graph, as shown in Fig. 10.9. Under the above assumption the survival probability varies according to

$$P(\nu_\alpha \to \nu_\alpha) = \frac{5}{9} + \frac{4}{9}\cos(\Delta m^2 L/2E). \tag{10.8}$$

So $P(\nu_\alpha \to \nu_\alpha)$ oscillates between 1 and 1/9, with an average of 5/9. In practice, experiments usually involve a range in neutrino energy and, for the accelerator and atmospheric sources, a range in baseline L. Thus when integrated over L and E, the oscillations are rapidly damped down as a function of L/E and P acquires its average value. Figure 10.9 shows the values of P deduced from various experiments. The curve is the best-fit oscillation solution for

$$\Delta m^2 = (0.72 \pm 0.18).10^{-2}\ \text{eV}^2. \tag{10.9}$$

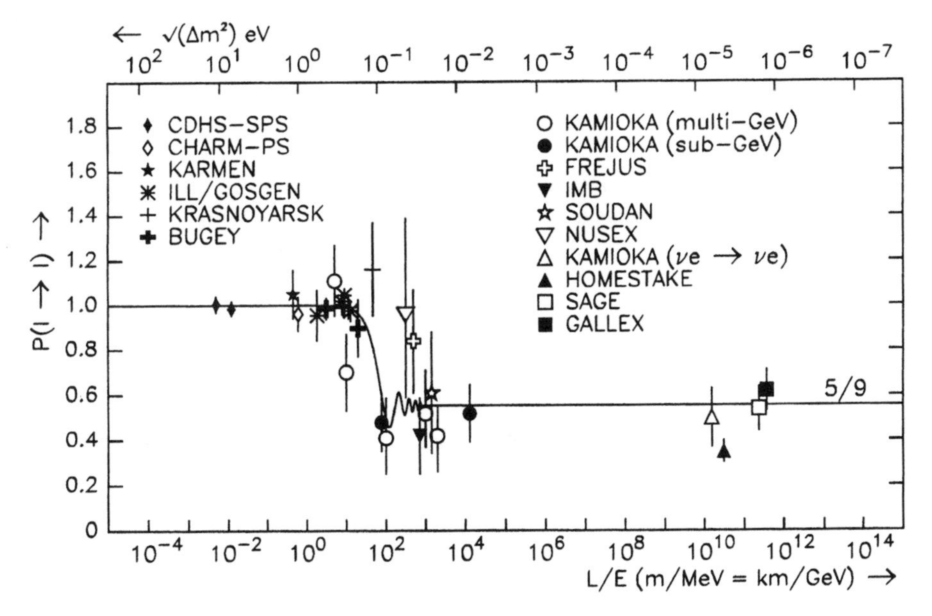

Figure 10.9: Results for the neutrino survival probability $P(l \to l)$ measured in disappearance experiments at accelerators and reactors, together with corrected results from atmospheric and solar experiments.

Excluding the HOMESTAKE point, the fit has $\chi^2 = 19.2$ for 26 degrees of freedom. Including HOMESTAKE, it is $\chi^2 = 36.1/27$. The conclusion is that the hypothesis fits 27 points from 19 experiments, but fails for HOMESTAKE, assuming the Bahcall-Pinsonneault fluxes [16]. It is possible to analyse the solar data differently, by choosing that combination of the 4 flux calculations which gives the best fit for maximal mixing. The relative ^{8}B flux is thereby reduced by 25% and the hypothesis fits all the solar data at 11% CL—while the same procedure for the small angle MSW fit reaches 60% CL.

At much larger values of L/E, such that $\Delta m'^2 L/E \sim 1$, the mean survival probability would drop to 1/3, after a second "threshold": there is however no sign of this in the data, and one concludes that $\Delta m'^2 < 10^{-11}$ eV^2.

It is perhaps worth mentioning here that the MSW hypothesis and maximum mixing are mutually exclusive. In the case of maximal mixing, the MSW mechanism has no effect on the survival probability. on the other hand, if the energy dependence of $P(\nu_e \to \nu_e)$ suggested by the Homestake and KAMIOKA results is taken seriously the MSW effect provides the only reasonable explanation and the maximum mixing hypothesis, with P independent of energy, is excluded.

10.6 APPEARANCE EXPERIMENTS

The appearance probability predicted by maximum mixing is given by

$$P(\nu_\alpha \to \nu_\beta) = \frac{2}{9} - \frac{2}{9}\cos(\Delta m^2 L/2E). \tag{10.10}$$

Inserting Δm^2 from Eq. (10.9), we get the dependence shown in Fig. 10.10. To date, appearance experiments have employed short baselines, that is $\Delta m^2 L/2E$ $<< 1$. The LSND result for $P(\bar{\nu}_\mu \to \bar{\nu}_e)$ is some 3 orders of magnitude larger than the prediction: so this result, if accepted, excludes maximum mixing. The 90% CL upper limits for the BNL (E776) and KARMEN experiments are also given.

The presently-running NOMAD[25] and CHORUS[26] experiments at CERN, searching for $\nu_\mu \to \nu_\tau$ transitions are shown by dashed lines, the upper limit of each line being the 90% CL (upper) limit if no ν_τ events are seen (i.e., it corresponds to 2.3 events). So even the observation of a single ν_τ event would also demolish the maximum mixing hypothesis.

10.7 LONG BASELINE EXPERIMENTS

During the last years, several long baseline experiments at accelerators and reactors have been proposed. Their motivation is to push to smaller values of Δm^2 than hitherto, and they have been stimulated by the atmospheric neutrino results. Basically these new proposals aim at values of $L/E \sim 100$ km GeV^{-1} ($\sim$ 100m MeV^{-1}) so as to mimic the parameters of the atmospheric data. Their importance, in relation to the atmospheric results, is that they can be far more

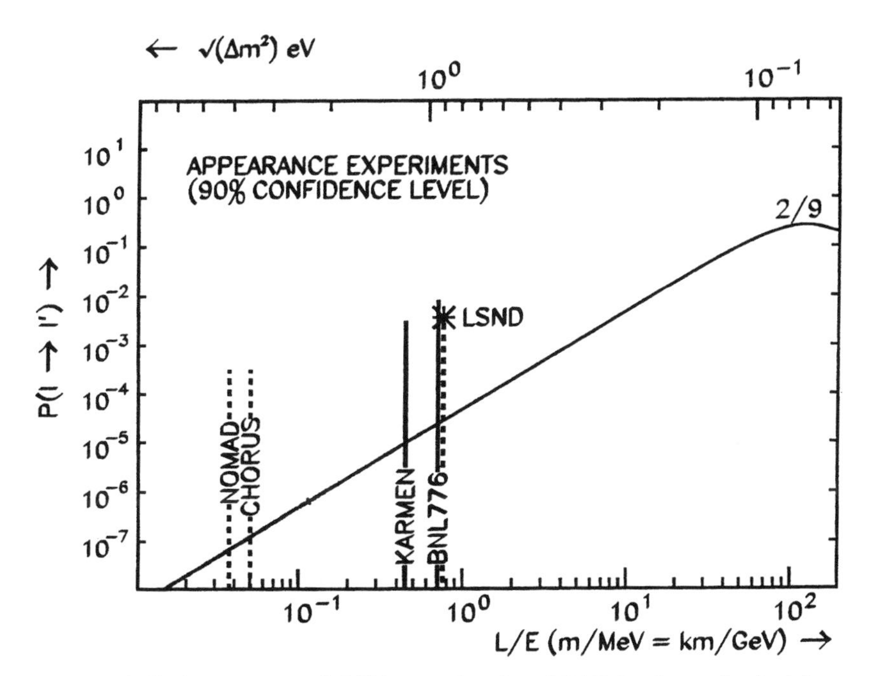

Figure 10.10: Appearance probabilities as a function of L/E, for the maximal mixing hypothesis with $\Delta m^2 = 0.72 \times 10^{-2}$ eV^2, together with limits or values from five experiments.

quantitative: the beam can be turned on and off, the energy E can be varied, and so can L; and the detectors can have good energy resolution (better than 10%).

Figure 10.11. shows results expected for two typical long baseline projects, out of five or six world-wide. In each case, the event rate (assuming no oscillations) is shown at top, and the survival/appearance probability underneath. The CHOOZ [27] project is a reactor experiment in the Ardennes which detects the reaction $\bar{\nu}_e + p \to n + e^+$. The energy in each event is known since both n and e^+ are recorded as a prompt γ-ray pulse due to e^+e^- annihilation, and a delayed γ-ray pulse following neutron capture in gadolinium. The baseline is $L = 1$ km, so L/E is known for every event (the reactor source size ≈ 1 m only). This means that it should be possible to observe a complete oscillation ($P = 1 \to 1/9 \to 1$) of the survival probability. A similar experiment is being mounted in the USA, at PALO-VERDE [29].

For the accelerator projects, the beams are predominantly ν_μ and $\bar{\nu}_\mu$, with only a small (0.5%) $\nu_e, \bar{\nu}_e$ admixture. Both survival $P(\nu_\mu \to \nu_\mu)$ and appearance probabilities $P(\nu_\mu \to \nu_e, \nu_\mu \to \nu_\tau)$ can be measured, via the charged current reactions $\nu_e + N \to e^+ + \ldots, \nu_\tau + N \to \tau^- + \ldots$. Again, Fig. 10.11 shows the expected rate for the MINOS experiment [29], with $L = 730$ km running from the Fermilab accelerator to the Soudan mine in N. Minnesota. A BNL proposal with 4 detectors over a 64 km baseline and lower mean energy ($\sim$ 1 GeV) was unfortunately not funded. Additionally there is a KEK $\rightarrow$ SUPERKAMIOKA project which is funded, and proposals for a CERN $\rightarrow$ GRAN SASSO beam.

Of the four projects which are going ahead, CHOOZ is running now and should present results within one year, with PALO-VERDE following. The KEK and MINOS projects will not start data-taking for four or five years. So, there are good prospects that new regions in the $\Delta^2 m$ v. $\sin^2 2\theta$ plot can be opened up in the near future, and it is quite possible that for the first time, full neutrino oscillations may be detected.

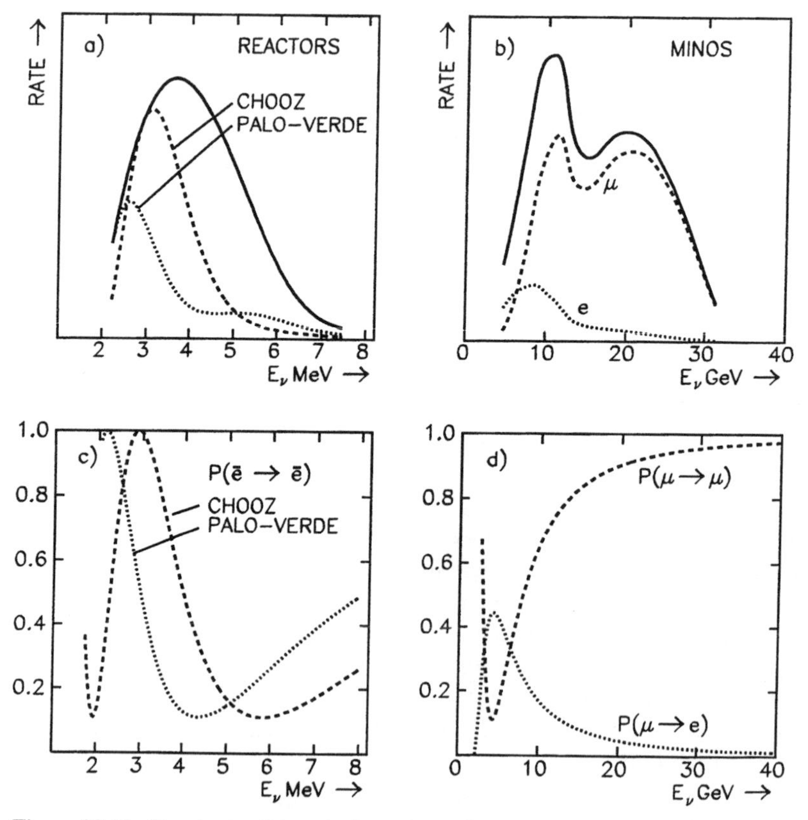

Figure 10.11: Event rates for typical reactor and accelerator long baseline neutrino beams, together with the survival and appearance probabilities according to the maximal mixing fit of Fig. 10.9.

10.8 A NEUTRINO CRISIS?

Some people have suggested that there is a crisis in neutrino physics. If one accepts the LSND, solar and atmospheric results, they appear to require for their explanation, three different values of Δm^2 (~ 1 eV^2, 10^{-5} eV^2 and 10^{-2} eV^2 respectively). With only three mass eigenstates, this is impossible, and people have speculated that one should include a fourth, sterile neutrino state. However, it is far more likely that at least one of these results is incorrect, and it is wiser to wait until experiments specifically designed to try to reproduce the observed effects (from atmospheric neutrinos) narrow down the range of mixing angle and mass difference.

Finally, it cannot be over-emphasized that, if neutrino oscillations are eventually discovered, this will herald a clear departure from the Standard Model (where $m_\nu = 0$) and the onset of new physics. It is generally believed that the very small mass values, in comparison with those of the quarks or charged leptons, is a manifestation, via the see-saw mechanism, of very massive right-handed neutrinos ($M \sim 10^{12} - 10^{15}$ GeV). In turn the decay of such massive states, violating lepton number, in an out-of-equilibrium cosmological scenario, may be connected with the generation of the cosmological baryon asymmetry at an early stage in the Big Bang.

REFERENCES

[1] C. Athanassopoulos et al., Phys. Rev. Lett. **75** 2260 (1995); Phys. Rev. (1996); see also J.E. Hill, Phys. Rev. Lett. **75** 2694 (1995).

[2] B. Armbruster et al., Nucl. Phys. B (Proc. Suppl.) **38**, 235 (1995).

[3] B. Blumenfeld et al., Phys. Rev. Lett. **62** 2237 (1989).

[4] J. Bahcall, *Neutrino Astrophysics* (CUP, 1989), p. 13.

[5] A.I. Abazov et al., Phys. Rev. Lett. **67** 3332 (1991); J.N. Abdurashitov et al., Phys. Lett. **B328** 234 (1994).

[6] P. Anselmann et al., Phys. Lett. **B285** 376 (1992); **B314** 445 (1993); **B342** 440 (1995); **B357** 237 (1995).

[7] B.T. Cleveland et al., Nucl. Phys. B (Proc. Suppl.) **38** 47 (1995).

[8] K.S. Hirata et al., Phys. Rev. Lett. **65** 1297 (1990); **66** 9 (1991); Phys. Rev. D **44** 146 (1992).

[9] J.N. Bahcall and M.H. Pinsonneault, Rev. Mod. Phys. **64** 885 (1992); **67** 781 (1995).

[10] S. Turck-Chieze et al., Phys. Rep. **230** 57 (1993).

[11] C.R. Profitt, Ap. J. **425** 849 (1994).

[12] A. Dar and G. Shaviv, Nucl. Phys. B (Proc. Suppl.) **38** 235 (1995).

[13] S.P. Mikheyev and A. Yu Smirnov, Il. Nuov. Cim **C9** 17 (1986).

[14] L. Wolfenstein, Phys. Rev. D **17** 2369 (1978); **20** 2634 (1979).

[15] R. Becky-Szandy et al., Phys. Rev. D **46** 3720 (1992); D. Casper et al., Phys. Rev. Lett. **66** 2561 (1992).

[16] K.S. Hirata et al., Phys. Lett. B **205** 416 (1988); **280** 146 (1992); Y. Fukuda et al., Phys. Lett. B **335** 237 (1994).

[17] M. Aglieatta et al., Europhys. Lett. **15** 559 (1991).

[18] C. Berger et al., Phys. Lett. B **227** 489 (1989); **245** 305 (1990).

[19] P.J. Litchfield, in *Proceedings International Europhysics Conference High Energy Physics, Marseille, 1993*, edited by J. Carr and M. Perrottet (Editions Frontières), p. 557.

[20] V. Gribov and B. Pontecorvo, Phys. Lett. **28B** 493 (1969).

[21] L. Wolfenstein, Phys. Rev. D **18** 958 (1978).

[22] N. Cabibbo, Phys. Lett. B **72** 333 (1978).

[23] S. Nussinov, Phys. Lett. B **63** 201 (1976).

[24] P.F. Harrison et al., Phys. Lett. B **349** 137 (1995); **374** 111 (1996).

[25] P. Astier et al., CERN SPSC/P261.

[26] N. Armenise et al., CERN SPSC/P254.

[27] Y. Declais et al., "Search for Neutrino Oscillations at 1 Km distance from two power reactors at CHOOZ". Letter of Intent (1992).

[28] F. Boehm et al., Palo-Verde Neutrino Oscn. Expt. (1995) (CHOOZ).

[29] E. Ables et al., FNAL proposal P–875 (1995).

10.9 DISCUSSION

Session Chair: Leon Lederman
Rapporteur: Keir Neuman

MANN: I want to echo the very last comment that Don made and give a minor justification. Consider the very first equation that he showed, which is the probability for an oscillation. This equation has an amplitude and it has an oscillating term. If we assume, for simplicity, oscillations between two neutrino types, then the oscillating term has four factors in it. The first factor is a constant, the second is the factor that we have been trying to measure, and the last two are length and energy. Don is certainly correct—we have spent 30 years trying to measure an effect which, ultimately, we will have to demonstrate possesses the correct length and energy dependence and has a non-zero amplitude. The elusive nature of this measurement can be seen either as a commentary on our inabilities, or as indicating some natural effect that has contrived to make the measurement so difficult. Regardless of the exact cause of the measurement difficulty, this is really a most elementary problem to solve as compared with some that have been discussed here, and yet, for various reasons, we have been unable to solve it so far. I believe, though, that Don's predictions are correct. In the next few years, by having devoted ourselves in this direction, we are likely to solve this problem and we will know something about neutrino masses.

SULAK: In your beautiful and comprehensive presentation, you neglected to mention the most direct test, to my mind, of the suggestions from the IMB and Kamioka data, of a deficit of muon neutrinos in the atmospheric cosmic ray showers. The KEK experiment will expose the super Kamiokande detector at 100 km as well as a nearby normalizing detector to 1 GeV neutrinos. This experiment should eliminate the last unknown, the atmospheric flux, from the test for muon neutrino transformations. By the year 2000 we should know.

PERKINS: True, and the similar proposed experiment at CERN, aiming the neutrino beam from the accelerator to the grand SASSO laboratory, may be abandoned due to the financial problems in Europe.

CHAPTER 11

THE TEVATRON FROM TOP TO BOTTOM (AND I'M DREAMING OF A MUON COLLIDER)

ALVIN TOLLESTRUP
Fermi National Accelerator Laboratory
Batavia, IL

A Prologue Addressed to the Conference on the 300th Birthday of Princeton

The 250th Birthday Conference is over, and I am writing up the presentation I delivered. Professor Sam Treiman reviewed the conference of 50 years ago. I would like to direct my gaze in the other direction, toward the future. I am unable to resist the opportunity to address some remarks to the proceedings 50 years hence. At the 250th Professors Witten and Wilczek gave concluding lectures about how the world must surely be in the energy region beyond our present reach. Regardless of how sure the theorists may be of this vision, we experimentalists have a different set of problems to face. By the next celebration, the Standard Model will certainly have become a thing of the past, and graduate students may not even recognize the phrase. Suffice it to say, that the model has been incredibly successful in the energy region that we can explore at the end of the 20th century. We now face a future that can only be compared with the state of physics at the beginning of the century when the firm foundations of classical physics were just beginning to crumble.

We have indications that there are exciting discoveries to be made in the energy region beyond our present reach. However, unlike the earlier period, there is a plethora of theories that have been proposed to fill the space between our presently available energies and the Planck scale. It is thus far from clear exactly what configuration of machine will be appropriate for the next step. In the past, machines were relatively inexpensive, and each of the large laboratories could build machines of various types and of their own choosing to explore the physics that appealed. At this moment we hope that the LHC will be the appropriate machine to explore the next level of interesting energies. But what is the complement to, or even more important, the extension beyond the LHC? This question, of course, is no different from that faced in the past. In fact, we have built machines that have been exceedingly productive, as well as others that have been barren in their physics results simply because their energy was in the wrong range. The complicating issue this time is that the cost of any useful machine has far outstripped the available budget of any single laboratory, or indeed, country.

In this paper, I will review some aspects of the history of the Tevatron with the hope that we can learn from our past. It was the first superconducting synchrotron and has formed the basis for the design of future machines. It was also fortunate in having sufficient energy to complete the quark structure required by the Standard Model, and complete enough in its instrumentation to permit the study of b-physics. For your amusement, I have included some photographs of components. A photograph is a marvelous invention in that it will give you a precise view of some of the things that have happened at the Tevatron. However, the equivalent invention of something that would allow us to view the future equally well would be much more useful at this stage. I only can dream of a muon collider as being (perhaps!) a possible next step. At your 300th birthday party, there may well be photographs of such a machine along with accounts of its discoveries. Yet other technology, undiscovered now, may prevail and the "dream" may have been only a dream. It would be fun to be there!

11.1 THE TEVATRON

The Tevatron is unique in many ways. It was the first superconducting synchrotron, and it has become the model for all of the subsequent machines. It was also the last machine in the era of the great machine builders such as E.O. Lawrence and R.R. Wilson when the machine's conception and construction was dominated by a single physicist. See Fig. 11.1. At its start there were no DOE construction reviews; there were no WBS forms to fill out; the bureaucracy that is so familiar now was absent. In the end, of course, bureaucracy caught up. But the early stages of the Tevatron were firmly controlled by the foresight, genius, and direction of one man, R.R. Wilson.

The Tevatron required developing ways to apply existing superconducting technology and cryogenics on a scale never tried before. Wire strand had to be developed in industry and methods found to make it into cable. The cable had to be fabricated into coils and then into magnets whose field was accurate to about one part in 10,000. These magnets were mounted in cryostats cooled with liquid helium. Finally, a cryogenic system had to be constructed that extended over many kilometers in order to cool the magnets to 4 K. The basic knowledge was available, but the technology base had to be developed both in industry and at the site for the project to succeed.

Figure 11.1: R.R. Wilson winding a one foot long model magnet in early 1976. This may be the last picture ever taken that will show a laboratory director getting his hands dirty!

11.2 THE FACTORY CONCEPT

Wilson, from the outset, was intent on building a factory that could make full-scale magnets for the final machine. This required mastering mass production techniques. The factory technology was developed hand-in-hand with the magnet technology. There was never a case of hand crafting magnets with old-world craftsmanship. The factory was both the R&D laboratory for the magnet development and the ultimate manufacturing center. Model magnets were used to develop both the tooling technology and the magnet configuration. This approach made it possible to introduce and test new magnet designs quickly, and ultimately, nearly a hundred one-foot model magnets were constructed and measured.

We realized that the results obtained from these short model magnets might not apply to full-length magnets. So from the outset full-scale magnets were also constructed as soon as sufficient cable became available. Many of these magnets were not of accelerator quality, and some even destroyed themselves. Many wound up in secondary beam lines as their field was adequate for that purpose. Their operation in beam lines also spread the technology base in the laboratory and resulted in training the technicians in the new technology. In fact, 20 percent of the magnets constructed were not suitable for use in the accelerator. However, the end result was a factory well tuned to turning out the high quality magnets required by the accelerator, and a large base of highly trained personnel within the laboratory that could operate these magnets.

The close integration of the magnet R&D program with the development of the factory resulted in a number of innovations in the fabrication. The magnet coils were formed in 22-foot long presses, and magnetic field calculation had indicated that the tolerance on the dimensions of the coil cross section should be less than .001 inch. Machining very long steel forms for the presses to such high precision proved to be difficult, expensive, and very time consuming.

The problem was solved through the innovation of "laminated tooling." The forms were constructed by stacking precision steel stampings .06" thick onto a precision flat steel bed. Dies for making such stampings were readily available from industry and had a precision greater than we required. The turn-around time for procuring either dies or stampings was short, and the cross section was easily modified. Ultimately, all of the tooling for coil fabrication was made by using this technique.

Many other inventions contributed to the eventual success of the factory. Precision presses were developed for forming and curing the coils. Quality control of all of the components was necessary in order to obtain the ultimate necessary precision. Gauges to measure complex cross sectional shapes had to be developed. Finally, technicians were trained, and an assembly line constructed for producing and measuring over 10 magnets per week.

Some have asked why this work wasn't farmed out to industry. Subsequently, that path has been successfully pursued at HERA and the LHC. I believe the answer is clear. The development of the Tevatron magnets was exploring a whole new technology. The magnet R&D program required rapid turnaround of model magnets and the ability to change the fabrication parameters and tooling. The close integration between the factory and the magnet R&D program provided this ability. On the other side, the invention of fast, highly-reproducible techniques for magnet production was a challenge that couldn't be separated from the magnet performance. Nowadays complete magnet fabrication specifications are available, and it is possible to benefit greatly from the expertise in mass production techniques that is available in industry. Before these specifications were available, involving industry in the process would have been expensive, time-consuming, and wasteful.

11.3 THE MAGNET

An adequate supply of conductor was the first priority. The theory of processing niobium titanium alloy into stable superconducting strand was understood, and the cabling technology to produce Rutherford cable had been studied in England. The problem was to manufacture high quality cable and strand in industry. The first step involved purchasing from industry a large quantity of highly purified niobium titanium alloy and having it fabricated into rods that were about 3 millimeters in diameter. This material was then portioned out to various companies for the manufacture of superconducting strand. The process was, and still is, proprietary and led to competition to produce strand with the highest possible current density. The physical properties of the strand were measured in the magnet test facility at Fermilab, and the companies that produced the best strand were rewarded with larger orders.

The wire was then cabled at New England Electric Company into 23 strand cable for use in the magnet program. Unlike the case of magnet fabrication, industry played a vital and crucial role in the development of the conductor. As the conductor became more available, the model magnet program increased in intensity. The mainstay of this program was the series of 1 foot-long magnets that could be quickly manufactured and that were used to learn about all of the quirks of SC magnets. They were not long enough for field quality studies, but all of the mechanical problems associated with the magnet and its fabrication were successfully studied. See Fig. 11.2.

The main challenge was to restrain the large magnetic force on the conductor so that it would be held accurately in position to less than 0.001". The positioning was critical because these were the first accelerator magnets in which the field shape was completely determined by the current distribution in contrast to iron magnets in which the field shape is governed by the shape of the iron in the pole faces. Frictional heat was an additional disastrous effect of conductor motion. The enthalpy of the conductor was so small at 4 K that even mJ/cm^3 of heat generation could force it past the superconducting transition temperature. Subsequent joule heating would then force the whole magnet to go into the normal state, and the coils would melt if the current was not reduced to zero in one or two seconds.

The solution to controlling the conductor motion was to apply a high static pressure to the coil when the stainless steel collars were applied. If the static

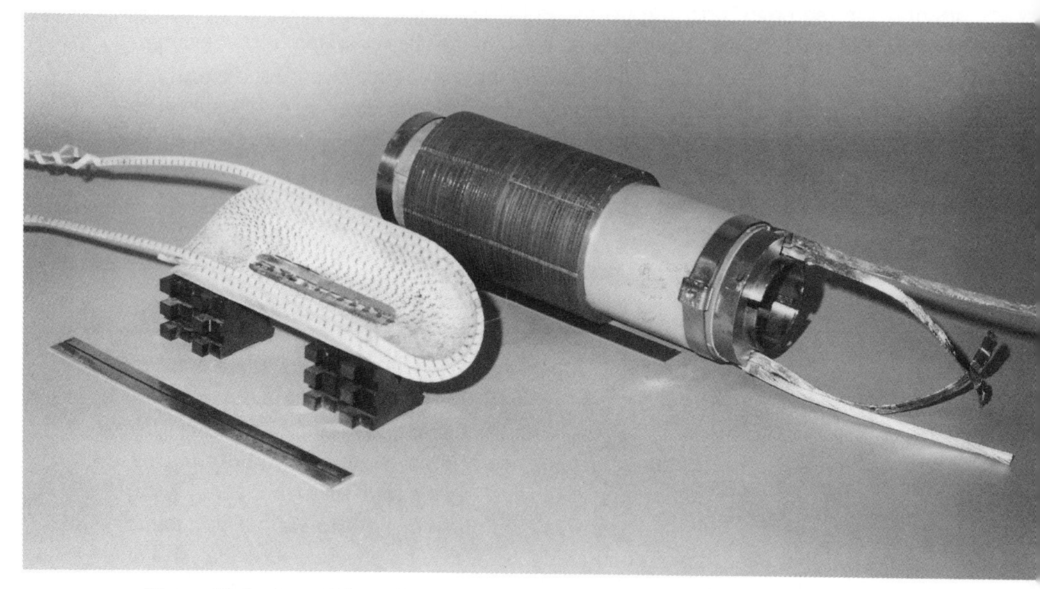

Figure 11.2: A partially collared model magnet from 1976. Shown also is a half-set of coils similar to the ones being collared.

pressure remained larger than any magnetic pressure, the coil shape would be determined by the shape of the stainless steel collars. The coil had a thermal contraction greater than the stainless steel collars causing the pressure to decrease as the temperature was lowered to 4 K. The 1 foot magnet program allowed us to study and ultimately solve this problem and many others.

11.4 MAGNETIC MEASUREMENT

The ultimate test of a magnet is the accuracy and reproducibility of its magnetic field. A successful accelerator requires controlling the field of the dipoles and quadrupoles to about one part in 10^4. Therefore, we built a large test facility. Using the magnetic field measurements, coil geometry was monitored and used to correct small errors in the fabrication. This closed a feedback loop around the factory.

There is a disadvantage to delaying magnetic measurements until the magnet is completed because considerable time may have elapsed between the construction of the coil and the completion of the magnet and cryostat. If errors are being made in the factory, they must be corrected quickly before more magnets are made. Magnetic measurement of the coil immediately after fabrication can detect these errors. Thus, instrumentation for measuring the field from the coil at low current and room temperature was developed. By measuring the field to a few parts in 10^4, it is possible to identify, at low cost, coils with geometrical defects. The source of the error can be rapidly corrected before more coils are fabricated. This technique is mandatory for monitoring completed magnets intended for modern machines where it is not practical to cool down and completely measure many thousand individual magnets.

11.5 CRYOGENICS

Fermilab broke new ground with the cryogenics system for the Tevatron. The cryostat brought liquid helium in contact with the beam tube, and a leak too small to detect with the conventional leak detector using gaseous helium, or a crack that opened up due to thermal stress as the magnet cooled down, could spell disaster for the machine vacuum when bathed with liquid helium. Not only would the magnet have to be completely disassembled, but such leaks could be extremely hard to locate. There were hundreds of welds on each cryostat that offered potential sources for such leaks. Careful quality control and training of welders finally solved this problem.

The large-scale distribution of liquid helium around a 6 kilometer ring also posed a challenge as did the liquid helium plant itself. A major concern was the length of time that would be necessary to warm up a sector of the ring and cool it down again if it was necessary to replace a magnet. As a result, the only cold mass was the coil assembly itself. The iron flux return was warm. This feature of the Tevatron is different from the more modern magnets that now use a cold iron yoke. The advantage of cooling the yoke is that it results in a simpler cryostat and the ability to use the cold iron to aid supporting the magnetic forces on the coil. Advances in refrigeration technology and the reliability demonstrated by superconducting magnets has put to rest the concern about long downtime for magnet replacement. The LHC is even being designed to operate below the lambda point for liquid helium at a temperature of about 1.8 K. Superfluid helium is a superb cooling fluid and large cryogenic plants can now be designed to operate reliability at this reduced temperature.

11.6 THE FUTURE

In 1975 we had a handful of successful 1-foot model magnets. By 1984 the Tevatron had a very successful fixed-target run, and in 1986 it started operation as a $\bar{p}p$ collider. It took only a little more than 10 years to go from the initial conception to an operating machine carrying out a physics program. The Tevatron pioneered the technology as can easily be seen by a casual comparison of the cross section with any of the modern SC accelerator magnets. See Fig. 11.3. The machine is now at its zenith, providing for the next 10 years the highest center-of-mass energy available. What can we learn from this history, and what is the next step? I would like to discuss a few of the high points of the collider physics experiments and then come back to these questions.

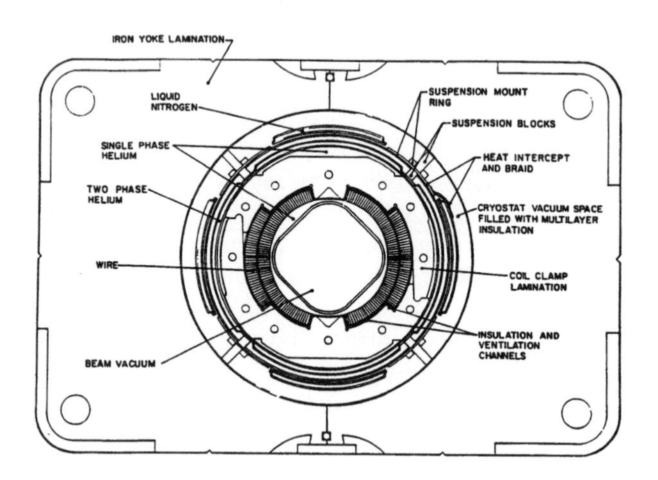

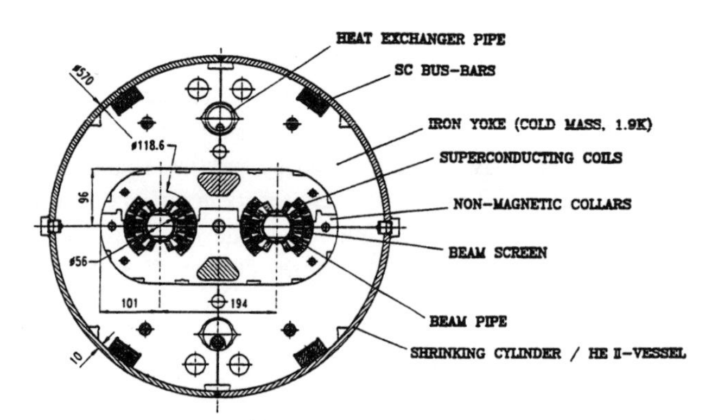

Figure 11.3: a) Cross Section of the Tevatron magnet. The iron return yoke is warm and surrounds the cryostat. The coil was immersed in LHe, and the cryostat also served as a closed system to deliver and return the liquid from a refrigerator to a string of 16 magnets.

b) Early version of the LHC two in one magnet. The collar supporting the coils is derived from the Tevatron, but the iron return yoke is enclosed within the cryostat and will supply superfluid LHe.

11.7 SOME PHYSICS RESULTS

It is not possible to review here all of the physics from the Tevatron. However, I would like to discuss some recent results that bear on the future. The first is the rather simple experiment of measuring the inclusive jet cross section as a function of jet p_T. The CDF curve is shown in Fig. 11.4 [1]. D0 has similar results, and the two experiments agree within the experimental errors.

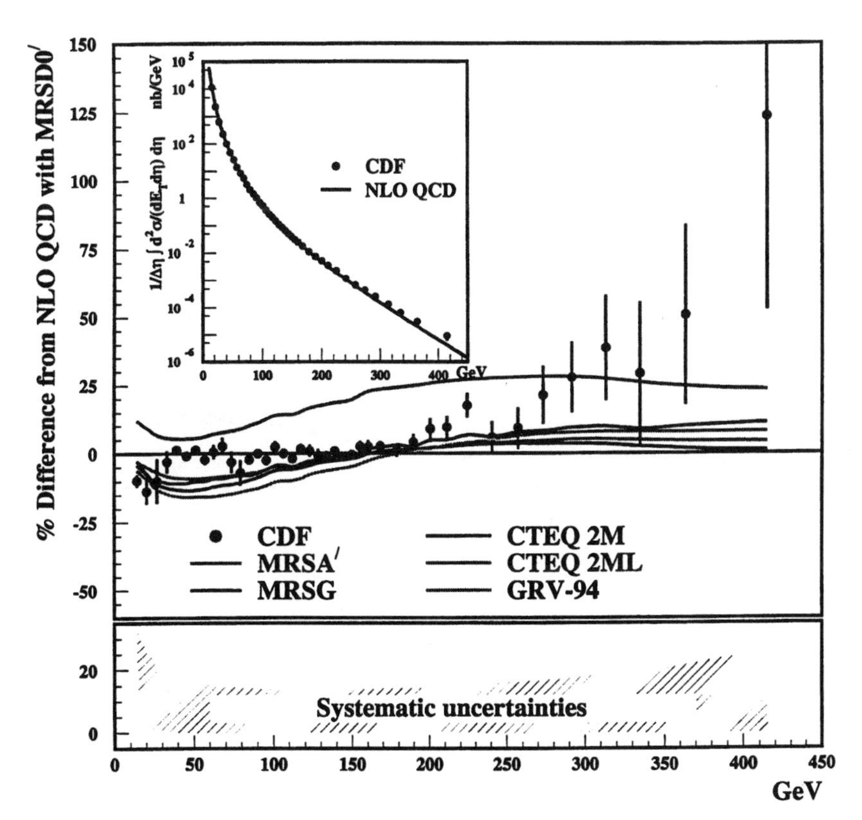

Figure 11.4: The inclusive jet spectrum versus P_t of the jet. The points are the CDF data and the line is a NLO QCD prediction. The ratio between the prediction for various PDFs and the data is shown at the bottom. The systematic error band is the shaded region at the very bottom.

The curve shows parton-parton scattering and is a basic test of QCD. At the high momentum transfers, it is also sensitive to parton substructure. The curve spans 9 orders of magnitude and shows remarkable agreement with the predictions of next leading order QCD, or NLO QCD. However, the basic $1/s$ variation in the cross section accounts for about three orders of magnitude. The rest is accounted for by the details of the parton distribution functions, PDF, which come from a compilation of the results of many experiments. The bottom of Fig. 11.4 shows the ratio of the measurement to the prediction using a given set of PDF. A factor of 2 disagreement is apparent. Does this indicate that that QCD is wrong? No, it appears that new PDF's can be constructed to encompass all of the experimental results. In addition, the angular distribution of dijet data has been studied and some results from D0 are shown in Fig. 11.5 [2]. This is an experiment that is complementary to the inclusive P_t jet spectrum, and it shows no deviation from the predictions of QCD. However, it is clear that precision tests of QCD are going to be exceedingly difficult.

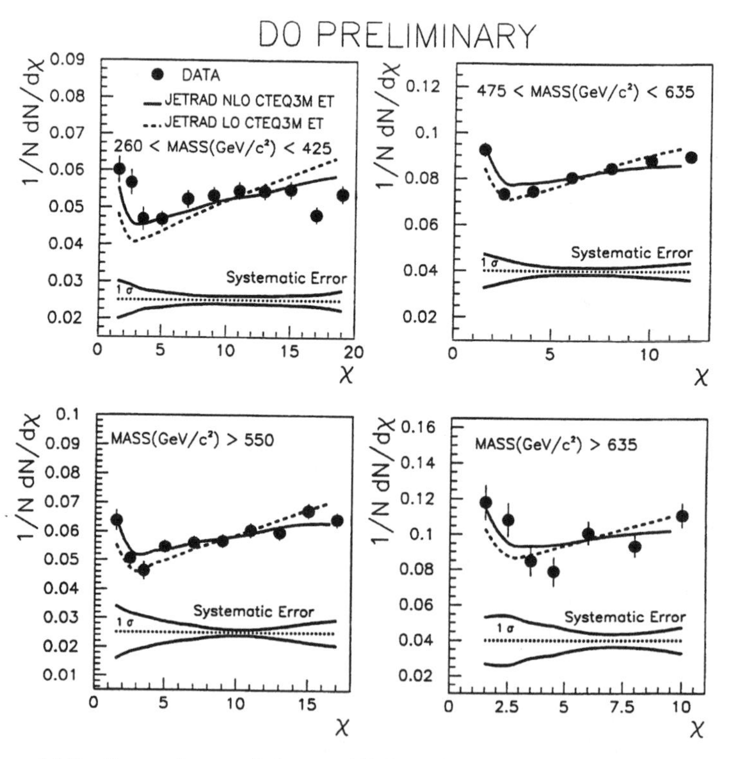

Figure 11.5: Comparisons of data to NLO and LO predictions of QCD using JETRAD with CTEQ3M and a renormalization scale of E_T. The error bars are statistical. Shown at the bottom of each plot is the plus and minus 1σ systematic error band.

I displayed this result for the amusement for our friends 50 years from now. Presumably, by then, precision tests of QCD will exist. But it is hard to believe that QCD will ever be tested to the precision that we have enjoyed with QED. Perhaps, as we go to higher momentum transfers, large deviations will occur that start to reveal a substructure of the underlying physics. The present results set a compositeness limit, Λ, greater than 1.5 TeV.

The next results pertain to b physics. Early in the planning stages of the Tevatron, many expressed the sentiment that hadron-hadron collisions were very dirty and would be difficult to analyze. One famous physicist likened it to throwing two garbage cans at each other. Few expected that b physics would be an important part of the program (CP violation even more unlikely!).

The silicon vertex detector at CDF was the first such instrument at a hadron collider, and it has opened up the whole field of b physics. Although proposed from the very first, its incredible success has surprised us all. The key has been to select events with a J/ψ decay to $\mu^+\mu^-$ and then use the SVX to tag events that show the J/ψ coming from a secondary vertex. Since the b lifetime is only about 1.6 ps, $c\tau$ is of the order of 500 μ which is easily measured by the SVX. The other necessary requirement is a very accurate charge particle tracking system to enable the various mass states to be reconstructed with high resolution.

A comparison of CDF and LEP measurements of some lifetimes is shown in Fig. 11.6 [3]. At LEP the cross section is low, but the b's from Z^0 decay have a high momentum that results in a large decay path, and the environment is very clean. In contrast, at the Tevatron the p_T of the b is small and the background is large, which makes the decay length short and harder to measure. The large production cross section offsets these difficulties.

In spite of the difficulties, it has been possible to search for and find new states. Fig. 11.7 [4] shows a mass peak from the elusive decay of the Λ_b to $\psi + \Lambda$. The mass is precisely measured to be 5621 $\pm$ 4 $\pm$ 3 MeV, showing the excellent precision obtainable at a hadron collider.

Finally, I would like to show an example of a measurement that uses the CDF detector in an unexpected manner. The states χ_{c1} and χ_{c2} decay to the J/ψ with the emission of a low energy gamma ray. If the gamma converts to an e^+e^- pair in the material of the detector, the energy of the gamma ray can be precisely determined by using the tracking detector to measure the energy and point of origin of the pair. Fig. 11.8 shows an x-ray of the inner detector using converted gamma rays, and Fig. 11.9 [5] shows the peaks from the $\chi \rightarrow J/\psi + \gamma$. This is an example of physics obtained through new and innovative techniques in data analysis.

The electroweak sector presents an opportunity for further tests of the Standard Model. From the lepton spectrum of the W decay both CDF and D0 have presented precision measurements for the W mass number. The accuracy of the numbers will be somewhat improved in the future. An interesting race will develop between the numbers available from the Tevatron experiments and those measured at LEP using W^+W^- pair production. We will return to this subject shortly.

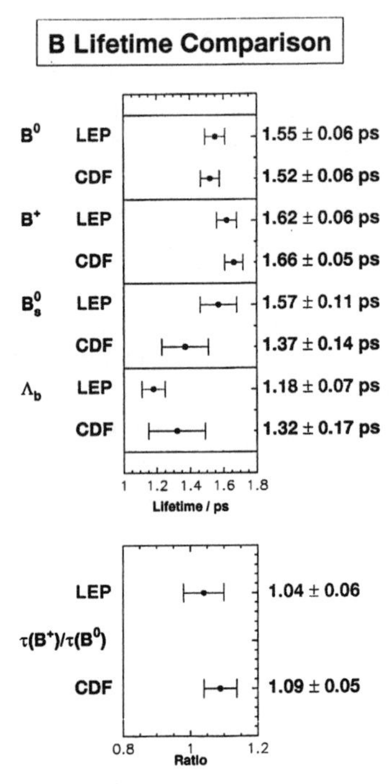

Figure 11.6: Comparison of various B lifetime measurements at CDF and LEP.

In addition to the mass of the W and Z, the trilinear coupling of the bosons is a sensitive test of the electroweak theory. CDF has measured some of these cross sections, and the results are shown in Fig. 11.10. A result from D0 on the anomalous moment of the Z is shown in Fig. 11.11 [6]. It is clear that there are no large deviations from the Standard Model, but precision tests will have to await more data.

The top mass has been measured by both experiments at the Tevatron. This measurement has posed a real challenge due to the small cross section and the complicated decay, but after the initial discovery, both experiments have greatly refined their measurements. The channel used is the lepton + 4 jet channel. The W from one top undergoes leptonic decay, and that from the other decays hadronically. This results in four jets, two of which are b jets, plus a lepton and a neutrino. The cleanest identification of the b jet comes from the observation of a

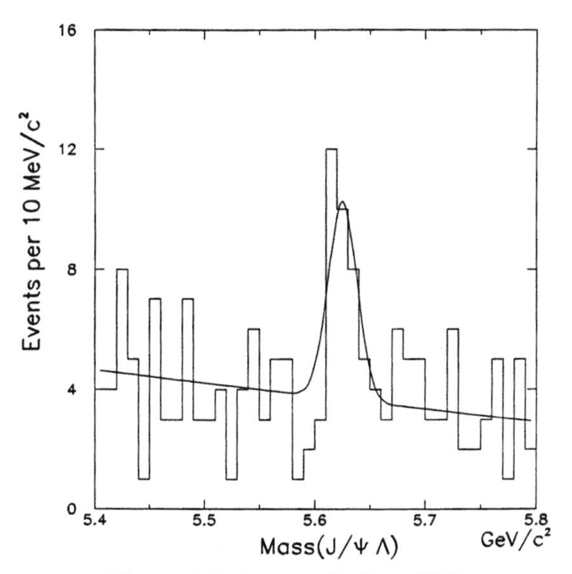

Figure 11.7: Λ_b mass plot from CDF.

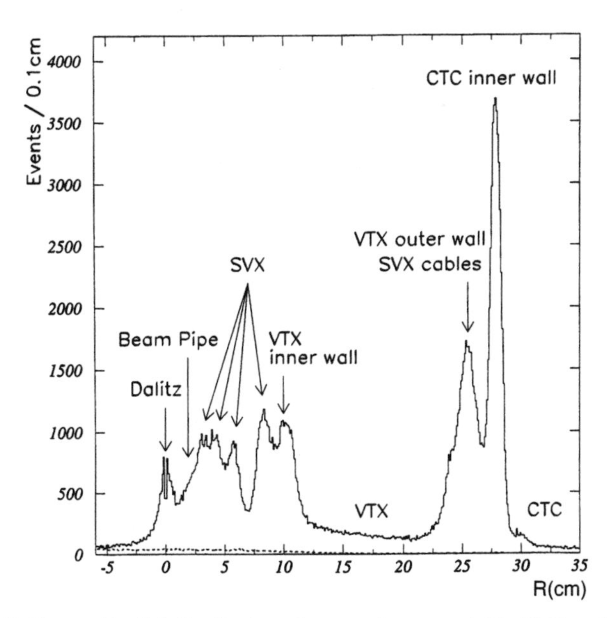

Figure 11.8: X-ray of radial distribution of conversion material in CDF as determined by using the pairs from gamma rays that convert in the material in the detector.

secondary vertex using the Silicon Vertex Detector in CDF. However, both experiments use the association of a lepton with a jet to identify the presence of a b quark.

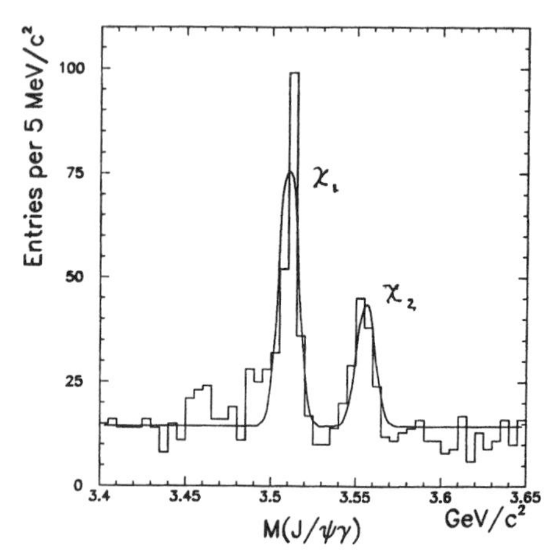

Figure 11.9: χ_1 and χ_2 peaks reconstructed from J/ψ + Gammas where the gamma energy and direction is derived from the converted pairs.

In addition to the two b jets, the candidate sample should show a W mass peak for the two non-b jets. Indeed, this is a nice test of the ability of the calorimetry to reconstruct a mass from jet data, and quantities of new data will furnish a major test of the accuracy of mass reconstruction through the use of hadron calorimetry. An event from CDF with all the decay products identified is shown in Fig. 11.12 [7].

A prediction for the mass of the Higgs is shown in Fig. 11.13 [8] using the best available mass values for the top and W. There is some hint that the Higgs may be rather light, although the experimental error still encompasses masses up to 1 TeV. It is interesting that the burden is on the W mass measurement, rather than that of the elusive top.

Finally we come to the search for exotic particles. There are many channels that have been used to seek some hint of new physics, and this is summarized in Table 11.1 [9]. So far, there is no hard evidence for any deviation from the Standard Model, although there does exist a very small set of intriguing "zoo" events. Whether these are from new physics or are just statistical fluctuations will have to await further data.

Table 11.1. Results of CDF New Particle Searches

Searches	Current CDF limit (GeV) (mostly Preliminary) Excluded region at 95% C.L.	data set (pb^{-1})
W' $\rightarrow$ eν (SM)	< 652	1a (20)
W' $\rightarrow$ WZ	< 560 (ref. model excluded)	1a+1b (110)
Z' $\rightarrow \ell\ell$ (SM)	< 690	1a+1b (110)
Z_ψ,Z_η,Z_χ,Z_I	< 580, 610, 585, 555	1a+1b (110)
Z_{LR}, Z_{ALRM}	< 620, 590	1a+1b (110)
Axigluon $\rightarrow$ dijet	200 < M < 930	1a+1b (103)
Technirho $\rightarrow$ dijet	250 < M < 500	1a+1b (103)
topgluon Γ = .1M	200 < M < 550	1a (20)
topgluon Γ = .5M	200 < M < 370	1a (20)
Leptoquark(2nd gen.)	< 180(scalar, β = 1)	1a+1b (70)
Leptoquark(3rd gen.)	< 99(scalar)	1a+1b (110)
Leptoquark(3rd gen.)	< 170, 225(vector,κ = 0,1)	1a+1b (110)
Pati-Salam LQ(B_s-eμ)	< 12100	1b (88)
Pati-Salam LQ(B_d-eμ)	< 18300	1b (88)
Composit, Scale (qqee)	< 3400(−),2400(+)	1a+1b (110)
Composit, Scale (qq$\mu\mu$)	< 3500(−),2900(+)	1a+1b (110)
$q^\star$(W+jet, γ+jet)	< 540	1a (20)
$q^\star \rightarrow$ dijet	200 < M < 750	1a+1b (103)
massive ch. stable pt.	< 190(color tripl. q)	1b (48)
gluino(MSSM)	< 180(all $\tilde{q}$ mass)	1b (80)
gluino(MSSM)	< 230($M_{\tilde{q}} = M_{\tilde{g}}$)	1b (80)
gaugino(MSSM)	< 68 ($\tilde{\chi}_1^\pm$, $\tilde{\chi}_2^0$)	1a+1b (110)
H$\pm$	< 150*	1a+1b (100)
H^0 ($p\bar{p} \rightarrow WH^0$, $H^0 \rightarrow$ bb)	> 15pb**	1a+1b (110)

* tanβ > 100,M_{top} = 175 GeV, σ_{top} = 5pb

** for 90 < M_{H° < 130 GeV, (in W $\rightarrow$ dijet channel)

CDF Electroweak Physics

$W\gamma, Z\gamma, WW, WZ, ZZ$ Production

PRL 75, 1017 (1995) : WW, WZ (lepton + jet channel)
PRL 74, 1941 (1995) : $Z\gamma$ Production
PRL 74, 1936 (1995) : $W\gamma$ Production

- **Leptonic Channel**

Diboson cross sections from CDF (preliminary)

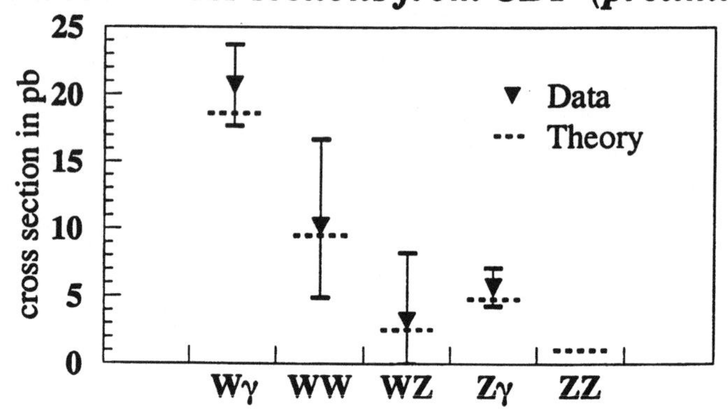

	L (pb^{-1})	Data (events)	Backgrounds (events)	σ_{exp} (pb)	σ_{theory} (pb)
$W\gamma$	67	109	26.4 ± 3.6	20.7 ± 3.0	18.6
WW	108	5	1.21 ± 0.30	$10.2^{+6.5}_{-5.3}$	9.5
WZ	110	1	$0.3^{+0.8}_{-0.3}$	$3.2^{+5.0}_{-3.2}$	2.5
$Z\gamma$	67	31	1.4 ± 0.4	5.7 ± 1.4	4.8
ZZ	120	1	?	?	1.0

Note that WW search requires *No Jets*

Figure 11.10: Trilinear production cross section limits from CDF.

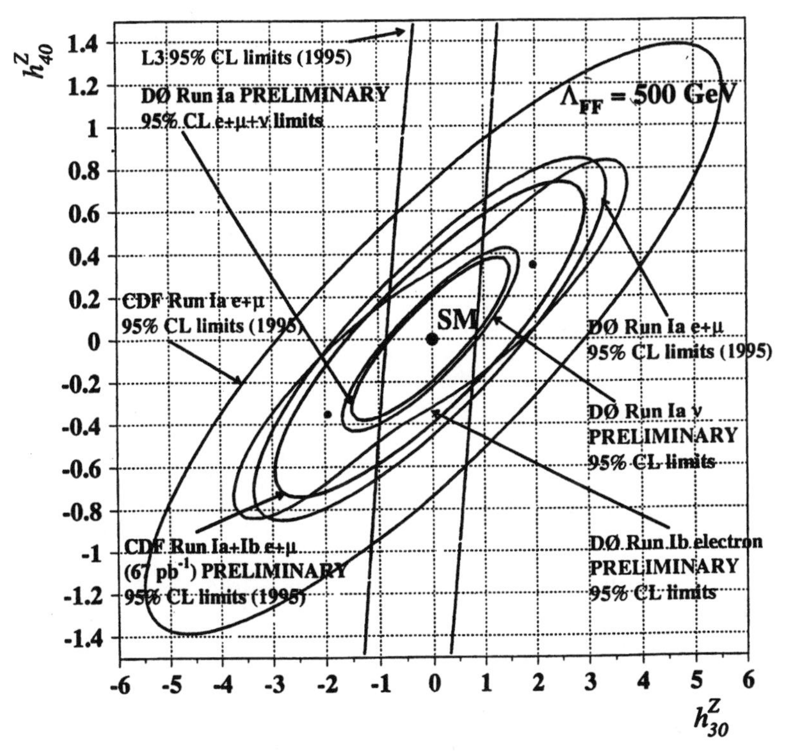

Figure 11.11: Limits on anomalous CP-conserving $ZZ\gamma$ couplings from $Z(ee)\gamma$, $Z(\nu\nu)\gamma$ production and from previous measurements. Dashed lines show unitarity contours for the form-factor scale Λ = 500 GeV.

11.8 THE NEAR FUTURE

The near future at Fermilab is rather well-defined for the Tevatron Collider Program. The luminosity of the machine will increase by a factor of 10 to 2×10^{32} as the new Main Injector comes on line. Both detectors are undergoing major upgrades. D0 is installing a silicon vertex detector and a superconducting solenoid, plus a tracking chamber that will allow the precision measurement of momentum. CDF is rebuilding a tracking system to accommodate the higher luminosity expected and is improving its muon system. A major new development will be the installation of instrumentation to trigger on events with the secondary vertex. The goal is to trigger on events containing a b, and if successful, will open up the possibility of studying CP violation in the b system.

The program which starts in 1999 will be very rich. Greatly improved mea-

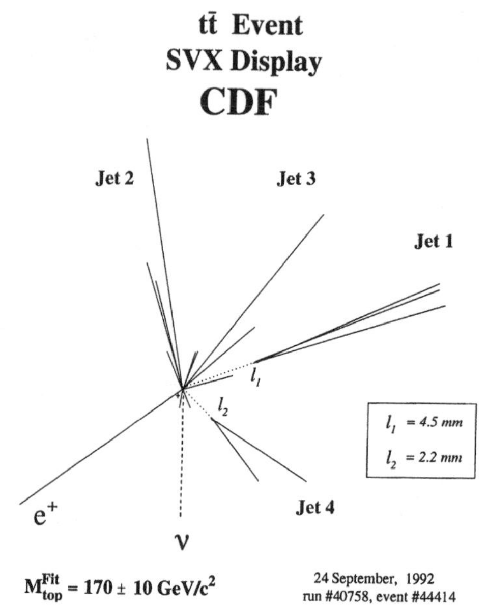

Figure 11.12: Top candidate showing the decay products. The mass of the two non b-jets is equal to the W-mass.

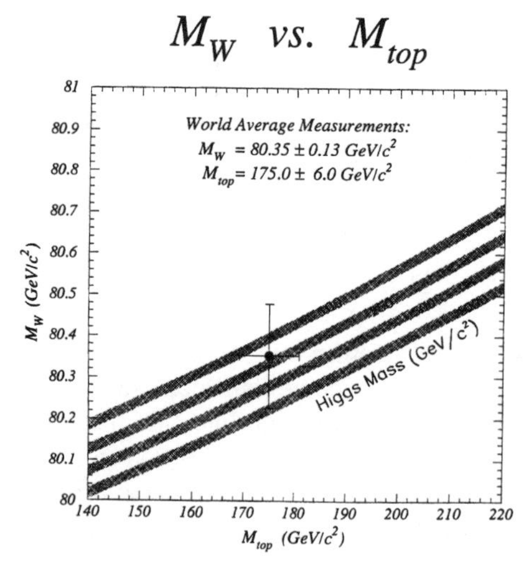

Figure 11.13: Plot showing the relation between the W mass, top mass, and the predicted Higgs mass. Lines, from top, correspond to Higgs masses of 100, 250, 500, 1000.

surements will be made of the W mass and of the trilinear boson coupling. The top quark mass will be measured to an accuracy of 1 to 2 GeV, and the various decay channels including their spin-imposed correlations will be studied in detail. Studies in b physics including CP violation will be a prominent part of the program. And finally, searches for exotic particles and for violation of the Standard Model will be pushed to new limits. This work will keep the Tevatron and its detectors well occupied until the LHC turns on.

11.9 THE FAR FUTURE

This has been a brief overlook at the Tevatron from its inception into the future when it will be outclassed by the LHC in both energy and luminosity. The community is now asking which type of machine to construct next. For the first time this is not a decision that will be made by any single laboratory, or even a single geographical region. It will be necessary to seek a world-wide consensus as to what machine to build. There will be a difficult transition from past procedures in which the future of an independent laboratory was largely under its direct control.

There are a number of options for the next step. They consist of a really large hadron collider or RLHC; an electron linear collider such as the JLC, NLC, or Tesla; and, finally, a newly considered entry, the muon collider. The choice will revolve around the question of what physics will be possible and how much the machine will cost.

The question of the physics reach is most difficult. We have dreams that SUSY will open up in the region immediately above present energies. However, the question of the exact energy scale is critical for the next machine. The Tevatron will ultimately answer the question whether SUSY sets in at a low mass scale. Otherwise, we must wait for the LHC which could be more than 10 years in the future. It may well be that we must learn to do things in series instead of the parallel approach that we have had in the past and, if this is the case, we have 10 years ahead of us for machine R&D.

In any case, the primary goal of machine design must be to reduce the cost per TeV. That is certainly the recognized challenge for the linear colliders. At Fermilab there is also an effort directed at trying to reduce the cost of a hadron collider by as much as a factor of 10 per TeV compared to the estimates made for the SSC. The R&D is directed at using superconducting coils in an iron yoke to make a very cheap, mass-produced magnet with a field of only 2 T. This approach recognizes the fact that the cost of the tunnel is only a small fraction of the cost of the magnet in modern machines like the LHC and the canceled SSC.

The muon collider is also an interesting possibility for a next generation machine. Although the idea came from Budker in 1964, it has been developed only recently. This development is documented in the Snowmass 1996 study [10].

Much more work, both theoretical and experimental, remains to be done before we can start to build such a collider. However, if the "next step" is delayed for long, it is possible the R&D could be completed, and the unique features of the machine developed into a very attractive project. It is for this reason that we should vigorously pursue the necessary studies.

A short description follows, but for complete details one should study the paper by R. Palmer that is included later in this compilation. Fig. 11.14 [10] shows a schematic drawing of a muon collider, and Table 11.2 gives the characteristics of a 2 TeV $\times$ 2 TeV collider studied in Ref. [10]. It starts with a 15 Hertz rapid cycling proton synchrotron whose energy is between 10 and 30 GeV and whose intensity is 2×10^{13} protons per pulse. The protons are directed onto a target that is embedded in a 20 T solenoidal field. The pions produced are captured by the solenoidal field and directed into a 200 meter long decay channel with an axial 5 T field. The muons are captured but occupy a large phase space. They must be collected into bunches, and the density phase space must be increased by a factor of about 10^6. The key to this cooling process is to introduce a momentum-dependent energy loss. Straggling and coulomb scattering introduce statistical noise. Nevertheless, it is possible to arrange matters so that muon cooling of the required amount actually takes place.

After forming the muons into bunches at a low energy (100 to 200 MeV), one must accelerate them to the energy required for collisions. This is done using a series of recirculating linacs to keep the acceleration time short. Finally, the two bunches of 2×10^{13} muons at 2 TeV are injected into a collider ring. If the field is about 8 Tesla, they will circulate for about 1,000 revolutions before decaying. A very low beta insertion produces a luminosity of 10^{35} per second.

Table 11.2. Parameters of collider rings.

		4 TeV	.5 Tev	Demo.
Beam energy	TeV	2	.25	.25
Beam γ		19,000	2,400	2,400
Repetition rate	Hz	15	15	2.5
Muons per bunch	10^{12}	2	4	4
Bunches of each sign		2	1	1
Normalized *rms* emittance ϵ^N	$10^{-6}\pi$ mrad	50	90	90
Bending Field	T	8.5	8.5	7.5
Circumference	km	7	1.2	1.5
Average ring mag. field B	T	6	5	4
Effective turns before decay		900	800	750
$\beta^\star$ at intersection	mm	3	8	8
rms beam size at I.P.	μm	2.8	17	17
Luminosity	$cm^{-2}s^{-1}$	10^{35}	5×10^{33}	6×10^{32}

The 4 TeV machine outlined above would fit on an existing site such as Fermilab. See Fig. 11.15. The development would be sequential for several reasons. The first is that much experimental work will be necessary to develop the muon source and collider technology. Thus, we envisage a demonstration machine or a first prototype of low energy. The cost of the machine can be divided into three parts: the source, the accelerator section, and the collider ring. The source is a fixed cost necessary for any collider, but the cost of the accelerator section and collider will scale with energy.

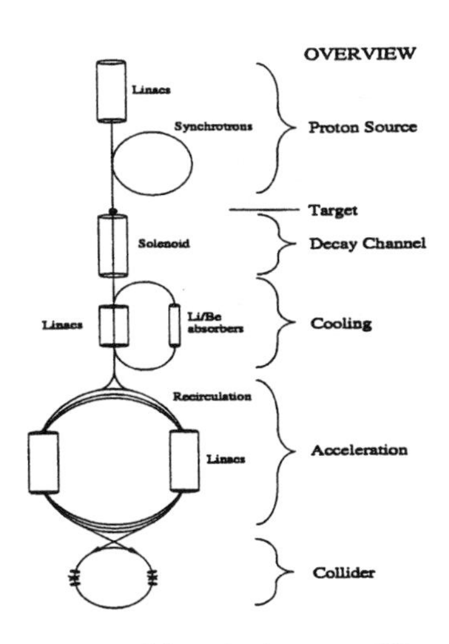

Figure 11.14: Schematic diagram of a muon collider. This does not represent the actual layout that would be used for such a machine, but shows the sequence of stages through which the muons pass.

We can imagine the development of the machine in concert with a physics program over a number of years. The early stages would involve an intense proton source that would make intense beams of pbars, K's, neutrinos, and muons. The high intensity would permit a look at rare decay modes of the kaon and muon which would provide us with additional handles on GUT scale physics. This would be complementary to the proton decay experiment. As we reach the limit on the "energy knob," we may have to seek much more exotic ways to test our theories. As an analogy, remember that neutral currents could have been discovered by careful study of the spectra of atoms!

The first collider has a number of possible target energies, one of which would explore the $t\bar{t}$ threshold. However, if SUSY should occur in the 200 to 400 GeV range, this collider would clearly be targeted at those energies. A unique property

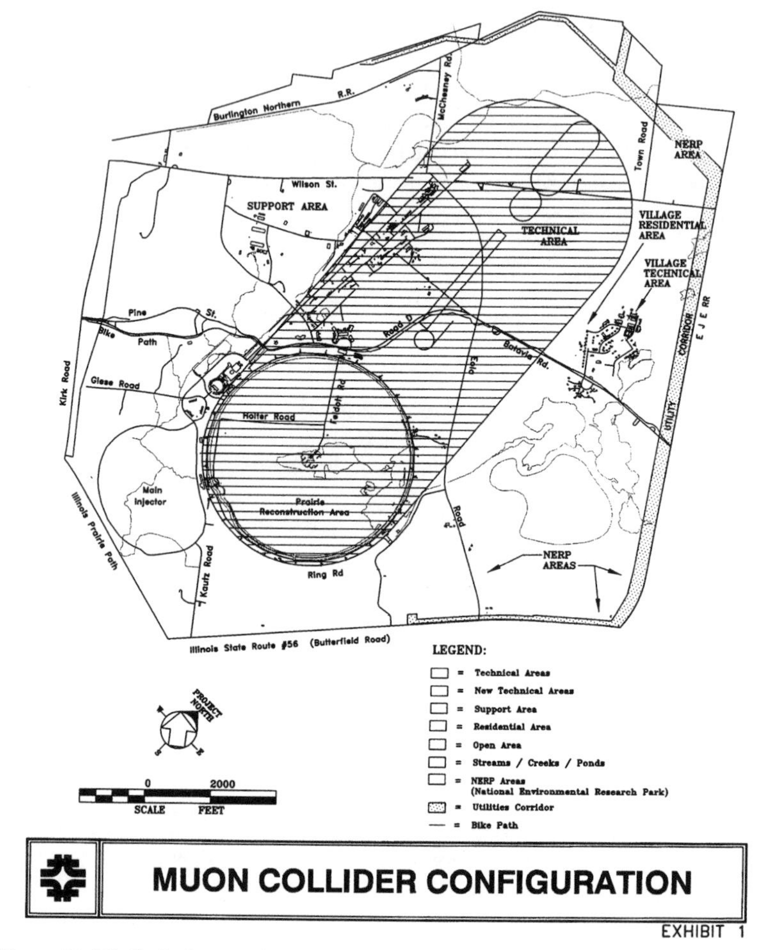

Figure 11.15: Scale layout showing that even a machine as large as 2 TeV × 2 TeV would fit on the Fermilab site.

of the muon collider is the fact that the coupling of the muons to the Higgs goes as the lepton mass squared. Thus, the s-channel production of the Higgs family becomes available for study. This physics has been extensively documented in Ref. [10]. It may require beams with very small energy spread if the Higgs mass is low and its width is narrow. The muon collider does have the possibility of very precisely defined beams, since there is only very weak synchrotron radiation from the relatively massive muons in the collider ring. However, preparing such beams for injection into the collider presents a real challenge for the designers.

In time, developments of the technology would lead to increased luminosity and to the possibility of polarized beams, albeit at the expense of reduced luminosity. Indeed, according to the theoretical studies, polarized beams play an important role in lepton collider physics. Unlike the electron collider, both muon beams can be polarized but only partially, and the luminosity decreases as the polarization increases.

11.10 CONCLUSION

For the moment we have achieved a goal of explaining almost everything that we can measure with our existing machines. Fifty years from now this era will probably be regarded by Princeton graduate students as the dawn of the Theory of Everything! A lot will depend on our wisdom in picking the next machine. I have tried to use the Tevatron as an example of how a new technology can be put in place in a period of ten years, and have sketched some of the resulting discoveries. I chose examples to show both expected as well as unexpected results. We can anticipate the same from the LHC as it matures.

Our history of predicting the future is dismal! We now face a new and difficult problem. The facilities we would like are too expensive for a single lab to build and operate on its own, and the direction in which we move requires our collective wisdom. More than the life of any single laboratory is at stake, and we must find a way to optimize the use of our resources. I have no doubt that some of the bright, young students that are in the present audience will have made these tough decisions that I have tried to outline and will be giving lectures at the 300th Birthday Party about the New Physics of Everything and how confused things were 50 years ago!

ACKNOWLEDGEMENTS

The author is grateful for much help and criticism from Ms. J. Jackson and for extensive help with preparing the manuscript from Ms. C. Picciolo.

REFERENCES

[1] "High p_T Jet Physics at the Tevatron Collider," L. Buckley-Geer for the CDF Collaboration, pub. Proceedings 1996 Divisional Meeting of the Division of Particles and Fields, American Physical Society, Minneapolis, MN, Aug. 10-15, 1996.

[2] "Measurement of Dijet Angular Distributions at CDF," F. Abe et al., The CDF Collaboration, Phys. Rev. Lett. **77** 5336 (1996).

[3] "Dijet Angular Distributions at D0," M. K. Fatyga for the D0 Collaboration, pub. Proceedings 1996 Divisional Meeting of the Division of Particles and Fields, American Physical Society, Minneapolis, MN, Aug. 10-15, 1996.

[4] "Observation of $\Lambda_{\rm b}^0 \rightarrow J/\psi\Lambda$ at the Fermilab Proton-Antiproton Collider," F. Abe et al., The CDF Collaboration, submitted to Phys. Rev. D August 28, 1996. FERMILAB-PUB-96/270-E.

[5] "Quarkonia Production at CDF," V. Papadimitriou for the CDF Collaboration, Published Proceedings Quarkonium Physics Workshop, University of Illinois at Chicago, Chicago, IL, June 1996. FERMILAB-CONF-96/402-E.

[6] "Recent D0 Results on Z_γ Production," G. Landsberg for the D0 Collaboration, pub. Proceedings 1996 Divisional Meeting of the Division of Particles and Fields, American Physical Society, Minneapolis, MN, Aug. 10-15, 1996.

[7] "Observation of Top Quark Production in $\bar{p}p$ Collisions with the CDF Detector at Fermilab," F. Abe et al., The CDF Collaboration, Phys. Rev. Lett. **74** 2626 (1995).

[8] "Recent Results on Top Quark Physics at CDF," M. Gallinaro, for the CDF Collaboration, pub. Proceedings QFTHEP 1996, St. Petersburg, Russia, Sept. 1996.

[9] "New Particle Searches at CDF," K. Maeshima for the CDF Collaboration, pub. Proceedings 28th International Conference on High Energy Physics (ICHEP'96), Warsaw, Poland, July 25-31, 1996.

[10] "$\mu^+\mu^-$ Collider: A Feasibility Study," edit. J. C. Gallardo, submitted to Snowmass 1996 Summer Study. FERMILAB-CONF- 96/092-E.

11.11 DISCUSSION

Session Chair: Leon Lederman
Rapporteur: Ioannis Kominis

TING: I am very impressed that you were able to measure the width of the W to about 34 MeV which is, in fact, even better than what we can do at LEP. Now, you also mentioned that you eventually you will able to do very clean b physics. How does this compare with the plans for b colliders?

TOLLESTRUP: The only way to answer you is to tell you that CDF expects $\Delta\sin(2\beta) = 0.09$ and I think that this could even decrease by a factor of 2. At present we have the largest sample in the world of $J/\Psi \rightarrow Ks$ on which to practice. There is also a very exciting new detector development that allows the detector to be triggered on a secondary vertex. If the detector can be triggered on events with two tracks that come from a single detached vertex, then, an enormous new era for b physics is opened. If this effort is successful, it may be possible to measure $\sin(2\alpha)$ as well. It's a new technology and it has to be developed.

SULAK: Regarding the synchrotron that Bill has pushed, can you say few words about the scaling arguments, as to why it is justified; I assume it is purely economic to use a low field iron magnet.The second question has to do with the muon collider, namely how you would ever make a detector that could survive within a γ-ray flux that comes along with the muons.

TOLLESTRUP: There are two proposals regarding the really LHC: The first, uses a high enough field so that the synchrotron radiation collapses the beam, and you get all the advantages of synchrotron radiation beam damping which is exploited in circular electron machines. The other option is to cut the cost/TeV, and to do that, you need a very simple magnet. The magnet Foster would like to use is constructed with a high-T_c superconductor, so that a large helium plant is not necessary. The cryostat is a a rather simple jacket, and the iron is extruded which can be done quiet accurately and inexpensively. Originally they were considering automatic boring machines which is becoming a highly developed technology. At Snowmass they talked to the geology people who told them that a larger tunnel may actually be less expensive. Regarding the second question: Since Snowmass we have reduced the background by about a factor of 10. There is about 15 Mwatt of electromagnetic radiation from muon decay electrons in the collider ring and this is not a trivial problem! The most difficult aspect comes from electrons that shower upstream of the interaction region. Bethe-Heitler muon pairs are generated at the level of about one per mil. These muons have an enormous range which make it difficult to shield the hadron calorimeter in the detector. Their main deleterious effect comes from deep inelastic scattering in the material of the hadron calorimeter which can result in localized clumps of energy being deposited in the calorimeter. Ways to cope with this problem are being investigated. Other radiation such as neutron and γ radiation is comparable to that projected for the LHC detectors.

CHAPTER 12

HIGH ENERGY COLLIDERS

R.B. PALMER AND J.C. GALLARDO
Center for Accelerator Physics
Brookhaven National Laboratory, Upton, NY

ABSTRACT

We consider the high energy physics advantages, disadvantages and luminosity requirements of hadron (pp, p$\bar{\text{p}}$), lepton (e^+e^-, $\mu^+\mu^-$) and photon-photon colliders. Technical problems in obtaining increased energy in each type of machine are presented. The relative sizes of the machines are also discussed.

12.1 INTRODUCTION

Particle colliders are only the latest evolution of a long history of devices used to study the violent collisions of particles on one another. Earlier versions used accelerated beams impinging on fixed targets. Fig. 12.1 shows the equivalent beam energy of such machines, plotted versus the year of their introduction. The early data given was taken from the original plot by Livingston [1]. For hadron, i.e., proton or proton-antiproton, machines (Fig. 12.1a), it shows an increase from around 10^5 eV with a rectifier generator in 1930, to 10^{15} eV at the Tevatron (at Fermilab near Chicago) in 1988. This represents an increase of more than a factor of about 33 per decade (the Livingston line, shown as the dashed line) over 6 decades. By 2005 we expect to have the Large Hadron Collider (at CERN, Switzerland) with an equivalent beam energy of 10^{17} eV, which will almost exactly continue this trend. The SSC, had we built it on schedule, would, by this extrapolation, have been a decade too early!

The rise in energy of electron machines shown (Fig. 12.1b) is slightly less dramatic; but, as we shall discuss below, the relative effective physics energy of lepton machines is greater than for hadron machines, and thus the effective energy gains for the two types of machine are comparable.

These astounding gains in energy ($\times 10^{12}$) have been partly bought by greater expenditure: increasing from a few thousand dollars for the rectifier, to a few billion dollars for the LHC ($\times 10^{6}$). The other factor ($\times 10^{6}$) has come from new ideas. Linear e^+e^-, $\gamma - \gamma$, and $\mu^+\mu^-$colliders are new ideas that we hope will continue this trend, but it will not be easy.

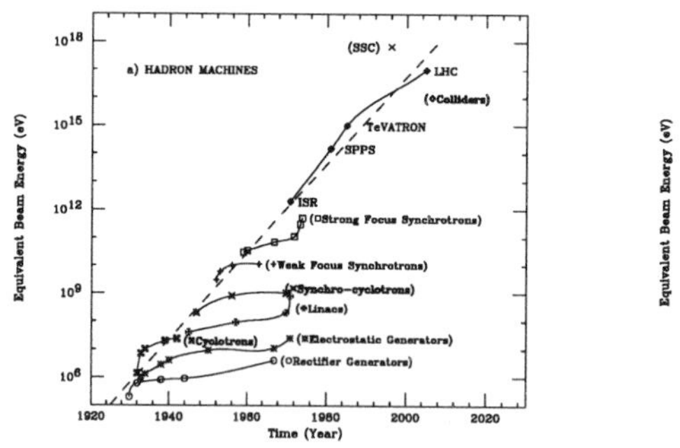

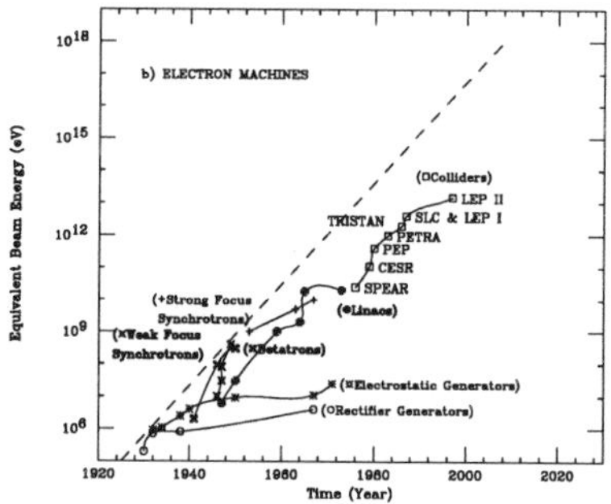

Figure 12.1: The Livingston Plots: Equivalent beam energy of colliders versus the year of their introduction; (a) for Hadron Machines and (b) for Lepton Machines.

12.2 PHYSICS CONSIDERATIONS

12.2.1 GENERAL

Hadron-hadron colliders (pp or p$\bar{\text{p}}$) generate interactions between the many constituents of the hadrons (gluons, quarks, and antiquarks); the initial states are not defined, and most interactions occur at relatively low energy, generating a very large background of uninteresting events. The rate of the highest energy events is a little higher for antiproton-proton machines because the antiproton contains valence antiquarks that can annihilate the quarks in the proton. But this is a small effect for colliders above a few TeV, when the interactions are dominated by interactions between quarks and antiquarks in their seas, and between the gluons. In either case, the individual parton-parton interaction energies (the energies used for physics) are a relatively small fraction of the total center of mass energy. This is a disadvantage when compared with lepton machines. An advantage, however, is that all final states are accessible. In addition, as we saw in Fig. 12.1, hadron machines have been available with higher energies than lepton devices, and, as a result, most initial discoveries in elementary particle physics have been made with hadron machines.

In contrast, lepton-antilepton colliders generate interactions between the fundamental point-like constituents in their beams, reactions are relatively simple to understand, the full machine energies are available for physics, and there is negligible background of low energy events. If the center of mass energy is set equal to the mass of a suitable state of interest, then there can be a large cross section in the s-channel, in which a single state is generated by the interaction. In this case, the mass and quantum numbers of the state are constrained by the initial beams. If the energy spread of the beams is sufficiently narrow, then precision determination of masses and widths are possible.

A gamma-gamma collider, like the lepton-antilepton machines, would have all the machine energy available for physics, and would have well-defined initial states; but these states would be different from those with the lepton machines, and thus be complementary to them.

For most purposes (technical considerations aside) e^+e^- and $\mu^+\mu^-$ colliders would be equivalent. But in the particular case of s-channel Higgs boson production, the cross section, being proportional to the mass squared, is more than 40,000 times greater for muons than electrons. When technical considerations are included, the situation is more complicated. Muon beams are harder to polarize and muon colliders will have much higher backgrounds from decay products of the muons. On the other hand, muon collider interactions will require less radiative correction and will have less energy spread from beamstrahlung.

Each type of collider has its own advantages and disadvantages for high energy physics: they are complementary.

12.2.2 Required Luminosity for Lepton Colliders

In lepton machines the full center of mass of the leptons is available for the final state of interest, and a "physics energy" $E_{\rm phy}$ can be defined that is equal to the total center of mass energy:

$$E_{\rm phy} = E_{\rm CM}. \tag{12.1}$$

Since fundamental cross sections fall as the square of the center of mass energies involved, for a given rate of events, the luminosity of a collider must rise as the square of its energy. A reasonable target luminosity is one that would give 10,000 events per unit of R per year (the cross section for lepton pair production is one R, the total cross section is about 20 R, and is somewhat energy-dependent as new channels open up):

$$\mathcal{L}_{\rm req.} \approx 10^{34}\,({\rm cm^{-2}s^{-1}})\left(\frac{E_{\rm phy}}{1\,({\rm TeV})}\right)^2. \tag{12.2}$$

12.2.3 THE EFFECTIVE PHYSICS ENERGIES OF HADRON COLLIDERS

Hadrons, being composite, have their energy divided between their various constituents. A typical collision of constituents will thus have significantly less energy than that of the initial hadrons. Studies done in Snowmass 82 and 96 suggest that, for a range of studies, and given the required luminosity (as defined in Eq. 12.2), then the hadron machine's effective "physics" energy is between about 1/3 and 1/10 of its total. We will take a value of 1/7:

$$E_{\mathrm{phy}}(\pounds = \pounds_{\mathrm{req.}}) \approx \frac{E_{\mathrm{CM}}}{7}.$$

The same studies have also concluded that a factor of 10 in luminosity is worth about a factor of 2 in effective physics energy, this being approximately equivalent to:

$$E_{\mathrm{phy}}(\pounds) = E_{\mathrm{phy}}(\pounds = \pounds_{\mathrm{req.}}) \left(\frac{\pounds}{\pounds_{\mathrm{req}}}\right)^{0.3}.$$

From which, with Eq. 12.2, one obtains:

$$E_{\mathrm{phy}} \approx \left(\frac{E_{\mathrm{CM}}}{7(\mathrm{TeV})}\right)^{0.6} \left(\frac{\pounds}{10^{34}(\mathrm{cm}^{-2}\mathrm{s}^{-1})}\right)^{0.2} (\,\mathrm{TeV}). \tag{12.3}$$

Table 12.1 gives some examples of this approximate "physics" energy.

Table 12.1: Effective Physics Energy of Some Hadron Machines

Machine	CM Energy TeV	Luminosity $\mathrm{cm}^{-2}\mathrm{s}^{-1}$	Physics Energy TeV
ISR	.056	10^{32}	0.02
Tevatron	1.8	7×10^{31}	0.16
LHC	14	10^{34}	1.5
VLHC	60	10^{34}	3.6

It must be emphasized that this effective physics energy is not a well-defined quantity. It should depend on the physics being studied. The initial discovery of a new quark, like the top, can be made with a significantly lower "physics" energy than that given here. And the capabilities of different types of machines have intrinsic differences. The above analysis is useful only in making very broad comparisons between machine types.

12.3 HADRON-HADRON MACHINES

12.3.1 LUMINOSITY

An antiproton-proton collider requires only one ring, compared with the two needed for a proton-proton machine (the antiproton has the opposite charge to the proton and can thus rotate in the same magnet ring in the opposite direction; protons going in opposite directions require two rings with bending fields of the opposite sign), but the luminosity of an antiproton-proton collider is limited by the constraints in antiproton production. A luminosity of at least 10^{32} cm^{-2}s^{-1} is expected at the antiproton-proton Tevatron; and a luminosity of 10^{33} cm^{-2}s^{-1} may be achievable, but the LHC, a proton-proton machine, is planned to have a luminosity of 10^{34} cm^{-2}s^{-1}: an order of magnitude higher. Since the required luminosity rises with energy, proton-proton machines seem to be favored for future hadron colliders.

The LHC and other future proton-proton machines might [2] be upgradable to 10^{35} cm^{-2}s^{-1}, but radiation damage to a detector would then be a severe problem. The 60 TeV Really Large Hadron Colliders (RLHC: high and low field versions) discussed at Snowmass are being designed as proton-proton machines with luminosities of 10^{34} cm^{-2}s^{-1}, and it seems reasonable to assume that this is the highest practical value.

12.3.2 SIZE AND COST

The size of hadron-hadron machines is limited by the field of the magnets used in their arcs. A cost minimum is obtained when a balance is achieved between costs that are linear in length, and those that rise with magnetic field. The optimum field will depend on the technologies used both for the the linear components (tunnel, access, distribution, survey, position monitors, mountings, magnet ends, etc.) and those of the magnets themselves, including the type of superconductor used.

The first hadron collider, the 60 GeV ISR at CERN, used conventional iron pole magnets at a field less than 2 T. The only current hadron collider, the 2 TeV Tevatron, at FNAL, uses NbTi superconducting magnets at approximately 4 K giving a bending field of about 4.5 T. The 14 TeV Large Hadron Collider (LHC), under construction at CERN, plans to use the same material at 1.8 K yielding bending fields of about 8.5 T.

Future colliders may use new materials allowing even higher magnetic fields. Model magnets have been made with Nb_3Sn, and studies are underway on the use of high T_c superconductor. $Bi_2Sr_2Ca_1Cu_2O_8$ (BSCCO) material is currently available in useful lengths as powder-in-Ag tube processed tape. It has a higher critical temperature and field than conventional superconductors, but, even at 4 K, its current density is less than Nb_3Sn at all fields below 15 T. It is thus unsuitable for most accelerator magnets. In contrast, $YBa_2Cu_3O_7$ (YBCO) material has a current density above that for Nb_3Sn (4 K), at all fields and temperatures below

20 K. But this material must be deposited on specially treated metallic substrates and is not yet available in lengths greater than 1 m. It is reasonable to assume, however, that it will be available in useful lengths in the not-too-distant future for hadron colliders.

A parametric study was undertaken to learn what the use of such materials might do for the cost of colliders. 2-in-1 cosine theta superconducting magnet cross sections (in which the two magnet coils are circular in cross section, have a cosine theta current distribution, and are both enclosed in a single iron yoke) were calculated using fixed criteria for margin, packing fraction, quench protection, support, and field return. Material costs were taken to be linear in the weights of superconductor, copper stabilizer, aluminum collars, iron yoke, and stainless steel support tube. The cryogenic costs were taken to be inversely proportional to the operating temperature and linear in the outer surface area of the cold mass. Tunnel, access, vacuum, alignment, focusing, and diagnostic costs were taken to be linear with tunnel length. The relative values of the cost dependencies were scaled from LHC estimates.

Results are shown in Fig. 12.2. Costs were calculated assuming NbTi at (a) 4 K, and (b) 1.8 K, Nb_3 Sn at (c) 4 K and YBCO High T_c at 20 K (d) and (e). NbTi and Nb_3 Sn costs per unit weight were taken to be the same; YBCO was taken to be either equal to NbTi (in (d)), or 4 times NbTi (in (e)). It is seen that the optimum field moves from about 6 T for NbTi at 4 K to about 12 T for YBCO at 20 K; while the total cost falls by almost a factor of 2.

One may note that the optimized cost per unit length remains approximately constant. This might have been expected: at the cost minimum, the cost of linear and field dependent terms are matched, and the total remains about twice that of the linear terms.

The above study assumes this particular type of magnet and may not be indicative of the optimization for radically different designs. A group at FNAL [3] is considering an iron-dominated, alternating gradient, continuous, single turn collider magnet design (Low field RLHC). Its field would be only 2 T and circumference very large (350 km for 60 TeV), but with its simplicity and with tunneling innovations, it is hoped to make its cost lower than the smaller high field designs. There are, however, greater problems in achieving high luminosity with such a machine than with the higher field designs.

12.4 CIRCULAR e^+e^- MACHINES

12.4.1 LUMINOSITY

The luminosities of most circular electron-positron colliders have been between 10^{31} and 10^{32} cm^{-2}s^{-1}; CESR is fast approaching 10^{33} cm^{-2}s^{-1}, and machines are now being constructed with even higher values. Thus, at least in principle, luminosity does not seem to be a limitation (although it may be noted that the

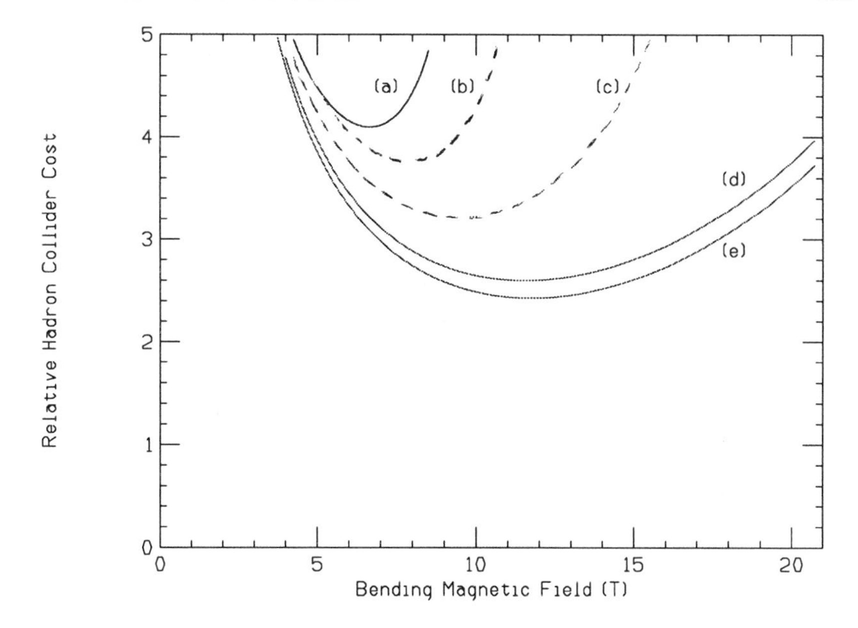

Figure 12.2: Relative costs of a collider as a function of its bending magnetic field, for different superconductors and operating temperatures.

0.2 TeV electron-positron collider LEP has a luminosity below the requirement of Eq. 12.2).

12.4.2 SIZE AND COST

At energies below 100 MeV, using a reasonable bending field, the size and cost of a circular electron machine is approximately proportional to its energy. But at higher energies, if the bending field B is maintained, the energy lost $\Delta V_{\rm turn}$ to synchrotron radiation rises rapidly,

$$\Delta V_{\rm turn} \propto \frac{E^4}{R\, m^4} \propto \frac{E^3\, B}{m^4}, \tag{12.4}$$

and soon becomes excessive (R is the radius of the ring). A cost minimum is then obtained when the cost of the ring is balanced by the cost of the rf needed to replace the synchrotron energy loss. If the ring cost is proportional to its circumference, and the rf is proportional to its voltage, then the size and cost of an optimized machine rises as the square of its energy.

The highest energy circular e^+e^- collider is the LEP at CERN which has a circumference of 27 km, and will achieve a maximum center of mass energy of about 0.2 TeV. Using the predicted scaling, a 0.5 TeV circular collider would have to have a 170 km circumference, and would be very expensive.

12.5 e^+e^- LINEAR COLLIDERS

For energies much above that of LEP (0.2 TeV) it is probably impractical to build a circular electron collider. The only possibility, then, is to build two electron linacs facing one another. Interactions occur at the center, and the electrons pass through the collision region only once. The size of such colliders is now dominated by the length of the two linacs and is inversely proportional to the average accelerating gradient in those structures. In current proposals [4] using conventional rf, these lengths are far greater than the circumferences of hadron machines of the same beam energy, but as noted in section 12.3, the effective physics energy of a lepton machine is higher than that of a hadron machine with the same beam energy, thus offsetting some of this disadvantage.

12.5.1 LUMINOSITY

The luminosity $\pounds$ of a linear collider can be written:

$$\pounds = \frac{1}{4\pi E} \frac{N}{\sigma_x} \frac{P_{\rm beam}}{\sigma_y} n_{\rm collisions}, \tag{12.5}$$

where σ_x and σ_y are average beam spot sizes including any pinch effects, and we take σ_x to be much greater than σ_y. E is the beam energy, $P_{\rm beam}$ is the total beam power, and, in this case, $n_{\rm collisions} = 1$. This can be expressed [8] as,

$$\pounds \approx \frac{1}{4\pi E} \frac{n_\gamma}{2r_o\alpha\, U(\Upsilon)} \frac{P_{\rm beam}}{\sigma_y}, \tag{12.6}$$

where the quantum correction $U(\Upsilon)$ is given by

$$U(\Upsilon) \approx \sqrt{\frac{1}{1+\Upsilon^{2/3}}} \tag{12.7}$$

with

$$\Upsilon \approx \frac{2F_2 r_o^2}{\alpha} \frac{N\,\gamma}{\sigma_z\,\sigma_x}, \tag{12.8}$$

$F_2 \approx 0.43$, r_o is the classical electromagnetic radius, α is the fine-structure constant, and σ_z is the *rms* bunch length. The quantum correction Υ is close to unity for all proposed machines with energy less than 2 TeV, and this term is often omitted [5]. Even in a 5 TeV design [6], an Υ of 21 gives a suppression factor of only 3. n_γ is the number of photons emitted by one electron as it passes through the other bunch. If n_γ is significantly greater than one, then problems are encountered with backgrounds of electron pairs and mini-jets, or with unacceptable beamstrahlung energy loss. Thus n_γ can be taken as a rough criterion of these effects and constrained to a fixed value. We then find

$$\pounds \propto \frac{1}{E} \frac{P_{\text{beam}}}{\sigma_y \, U(\Upsilon)},$$

which may be compared to the required luminosity that increases as the square of energy, giving the requirement:

$$\frac{P_{\text{beam}}}{\sigma_y \, U(\Upsilon)} \propto E^3. \tag{12.9}$$

It is this requirement that makes it hard to design very high energy linear colliders. High beam power demands high efficiencies and heavy wall power consumption. A small σ_y requires tight tolerances, low beam emittances, and strong final focus. And a small value of $U(\Upsilon)$ is hard to obtain because of its weak dependence on Υ ($\propto \Upsilon^{-1/3}$).

12.5.2 CONVENTIONAL RF

The gradients for structures have limits that are frequency-dependent, but the real limit on accelerating gradients in these designs comes from a trade-off between the cost of rf power against the cost of length. The use of high frequencies reduces the stored energy in the cavities, reducing the rf costs and allowing higher accelerating gradients: the optimized gradients being roughly proportional to the frequency up to a limit of approximately 250 MeV/m at a frequency of the order of 100 GHz. One might thus conclude that higher frequencies should always be preferred. There are, however, counterbalancing considerations from the requirements of luminosity.

Figure 12.3, using parameters from current 0.5 TeV linear collider proposals [4], plots some relevant parameters against the rf frequency. One sees that as the frequencies rise,

- the required alignment tolerances get tighter;
- the resolution of beam position monitors must also be better; and
- despite these better alignments, the calculated emittance growth during acceleration is worse; and
- the wall-power to beam-power efficiencies are also less.

Thus while length and cost considerations may favor high frequencies, luminosity considerations would prefer lower frequencies.

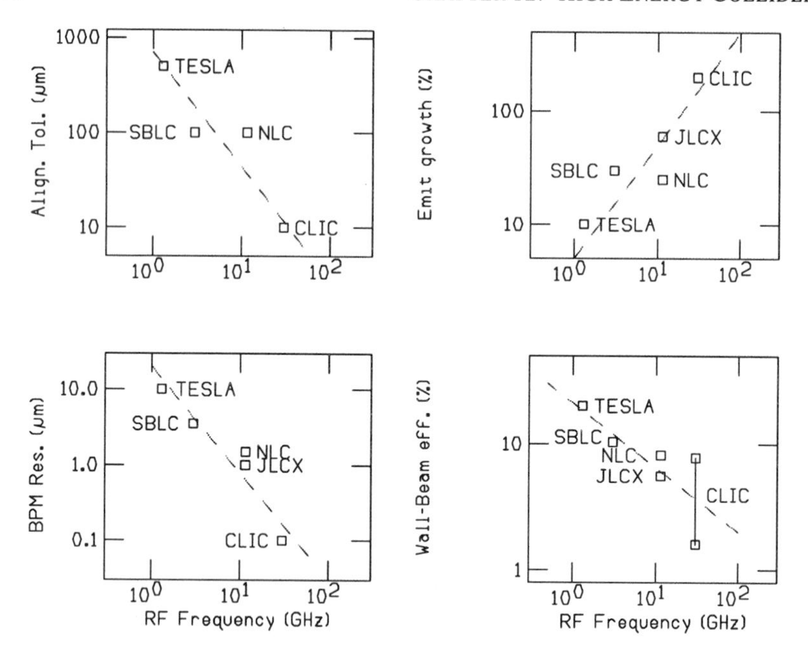

Figure 12.3: Dependence of some sensitive parameters of 0.5 TeV proposed linear colliders as a function of their rf frequencies.

12.5.3 SUPERCONDUCTING RF

If, however, the rf costs can be reduced, for instance when superconducting cavities are used, then there will be no trade-off between rf power cost and length. Higher gradients would lower the length and thus the cost. Unfortunately, the gradients achievable in currently operating niobium superconducting cavities is lower than that planned in the higher frequency conventional rf colliders. Theoretically, the limit is about 40 MV/m, but practically, 25 MV/m is as high as seems possible. Nb_3Sn and high T_c materials may allow higher field gradients in the future.

The removal of the requirements for very high peak rf power allows the choice of longer wavelengths (the TESLA collaboration is proposing 23 cm at 1.3 GHz), thus greatly relieving the emittance requirements and tolerances for a given luminosity.

At the current 25 MeV per meter gradients, the length and cost of a superconducting machine is probably higher than for the conventional rf designs. With greater luminosity more certain, its proponents can argue that it is worth the greater price. If, using new superconductors, higher gradients become possible, thus reducing lengths and costs, the advantages of a superconducting solution might become overwhelming.

12.5.4 AT HIGHER ENERGIES

At higher energies (as expected from Eq. 12.9), obtaining the required luminosity gets harder. Fig. 12.4 shows the dependency of some example machine parameters with energy. SLC is taken as the example at 0.1 TeV, NLC parameters at 0.5 and 1 TeV, and 5 and 10 TeV examples are taken from a review paper by one of the authors [6]. One sees that:

- the assumed beam power rises approximately as E;
- the vertical spot sizes fall approximately as E^{-2};
- the vertical normalized emittances fall even faster: $E^{-2.5}$; and
- the momentum spread due to beamstrahlung has been allowed to rise.

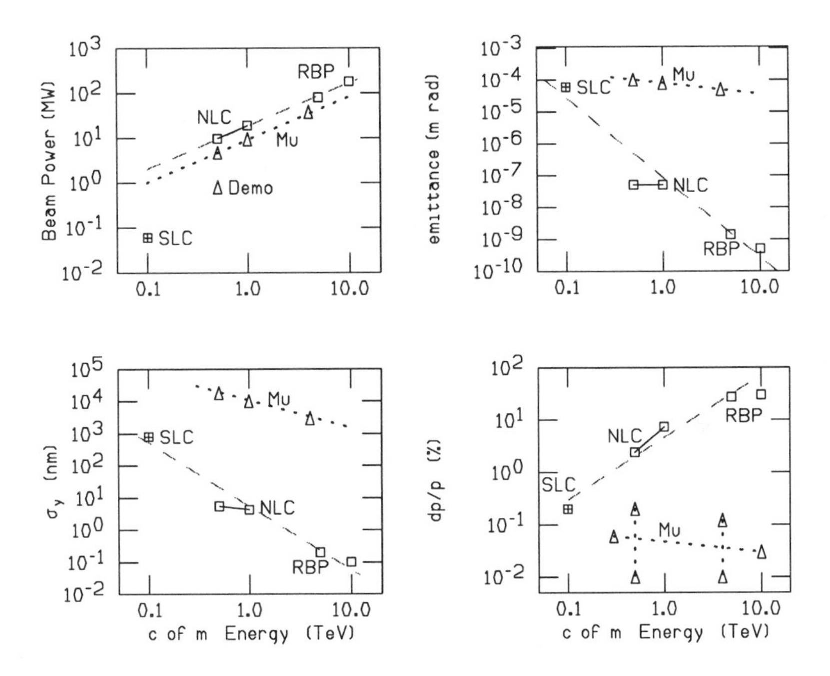

Figure 12.4: Dependence of some sensitive parameters on linear collider energy, with comparison of same parameters for $\mu^+\mu^-$ colliders.

These trends are independent of the acceleration method, frequency, etc., and indicate that as the energy and required luminosity rise, so the required beam powers, efficiencies, emittances, and tolerances will all get harder to achieve. The use of higher frequencies or exotic technologies that would allow the gradient to rise will, in general, make the achievement of the required luminosity even more difficult. It may well prove impractical to construct linear electron-positron colliders, with adequate luminosity, at energies above a few TeV.

12.6 $\gamma - \gamma$ COLLIDERS

A gamma-gamma collider [9] would use opposing electron linacs, as in a linear electron collider, but just prior to the collision point laser beams would be Compton backscattered off the electrons to generate photon beams that would collide at the IP instead of the electrons. If suitable geometries are used, the mean photon-photon energy could be 80% or more of that of the electrons, with a luminosity about 1/10th.

If the electron beams, after they have backscattered the photons, are deflected, then backgrounds from beamstrahlung can be eliminated. The constraint on N/σ_x in Eq. 12.5 is thus removed, and one might hope that higher luminosities would now be possible by raising N and lowering σ_x. Unfortunately, to do this, one needs sources of bunches with larger numbers of electrons and smaller emittances, and one must find ways to accelerate and focus such beams without excessive emittance growth. Conventional damping rings will have difficulty doing this [10]. Exotic electron sources would be needed, and methods using lasers to generate [11] or cool [12] the electrons and positrons are under consideration.

Clearly, gamma-gamma collisions can and should be made available at any future electron-positron linear collider to add physics capability; whether they can give higher luminosity for a given beam power is less clear.

12.7 $\mu^+\mu^-$ COLLIDERS

12.7.1 ADVANTAGES AND DISADVANTAGES

The possibility of muon colliders was introduced by Skrinsky et al. [13] and Neuffer [14] and has been aggressively developed over the past two years in a series of meetings and workshops [15, 16, 17, 18, 19].

The main advantages of muons, as opposed to electrons, for a lepton collider are:

- The synchrotron radiation that forces high energy electron colliders to be linear is (see Eq. 12.4) inversely proportional to the fourth power of mass. It is negligible in muon colliders. Thus a muon collider can be circular. In practice this means it can be smaller. The linacs for the SLAC proposal for a 0.5 TeV Next Linear Collider would be 20 km long. The ring for a muon collider of the same energy would be only about 1.3 km circumference.

- The luminosity of a muon collider is given by the same formula (Eq. 12.5) as given above for an electron positron collider, but there are two significant changes: 1) The classical radius r_o is now that of the muon and is 200 times smaller; and 2) the number of collisions a bunch can make $n_{\text{collisions}}$ is no longer 1, but is now limited only by the muon lifetime and becomes related to the average bending field in the muon collider ring, with

$$n_{\text{collisions}} \approx 150\, B_{\text{ave}}$$

With an average field of 6 Tesla, $n_{\text{collisions}} \approx 900$. These two effects give muons luminosity advantage of more than 10^5 *in principle.* As a result, for the same luminosity, the required beam power, spot sizes, emittances, and energy spread are far less in $\mu^+\mu^-$ colliders than in e^+e^- machines of the same energy. The comparison is made in Fig. 12.4.

- The suppression of synchrotron radiation induced by the opposite bunch (beamstrahlung) allows the use of beams with lower momentum spread, and QED radiation is reduced.
- s-channel Higgs production is enhanced by a factor of $(m_\mu/m_e)^2 \approx 40000$. This, combined with the lower momentum spreads, would allow more precise determination of Higgs masses, widths, and branching ratios.

But there are problems with the use of muons:

- Muons can be obtained from the decay of pions made by higher energy protons impinging on a target. But in order to obtain enough muons, a high intensity proton source is required with very efficient capture of the pions, and muons from their decay.
- The selection of fully polarized muons is inconsistent with the requirements for efficient collection. Polarizations only up to 50 % are practical, and some loss of luminosity is inevitable (e^+e^- machines can polarize the e^-'s up to $\approx$ 85 %).
- Because the muons are made with very large emittance, they must be cooled, and this must be done very rapidly because of their short lifetime. Conventional synchrotron, electron, or stochastic cooling is too slow. Ionization cooling [20] is the only clear possibility, but does not cool to very low emittances.
- Because of the short lifetime of muons, conventional synchrotron acceleration would be too slow. Recirculating accelerators or pulsed synchrotrons must be used.
- Because they decay while stored in the collider, muons radiate the ring and detector with decay electrons. Shielding is essential and backgrounds will be high.

12.7.2 DESIGN STUDIES

A collaboration, lead by BNL, FNAL and LBNL, with contributions from 18 institutions has been studying a 4 TeV, high luminosity machine and presented a Feasibility Study [19] to the 1996 Snowmass Workshop. The basic parameters of this machine are shown schematically in Fig. 12.5 and given in Table 12.2. Fig. 12.6 shows a possible layout of such a machine.

Table 12.2 also gives the parameters of a 0.5 TeV demonstration machine based on the AGS as an injector. It is assumed that a demonstration version based on upgrades of the FERMILAB, or CERN machines would also be possible.

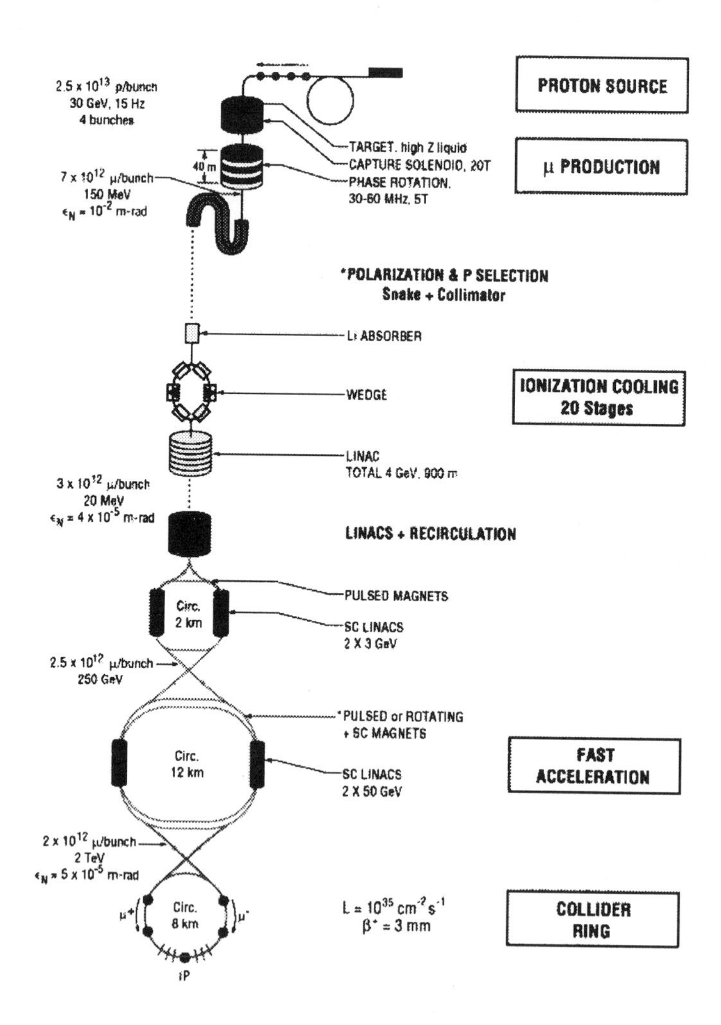

Figure 12.5: Overview of a 4 TeV Muon Collider

Table 12.2: Parameters of Collider Rings

cm Energy	TeV	4	.5
Beam energy	TeV	2	.25
Beam γ		19,000	2,400
Repetition rate	Hz	15	2.5
Proton driver energy	GeV	30	24
Protons per pulse		10^{14}	10^{14}
Muons per bunch		$2\ 10^{12}$	$4\ 10^{12}$
Bunches of each sign		2	1
Beam power	MW	38	.7
Norm. *rms* emit. ϵ_n	π mm mrad	50	90
Bending Fields	T	9	9
Circumference	Km	8	1.3
Ave. Bending Fields	T	6	5
Effective turns		900	800
β^* at intersection	mm	3	8
rms bunch length	mm	3	8
rms I.P. beam size	μm	2.8	17
Chromaticity		2000-4000	40-80
$\beta_{\max}$	km	200-400	10-20
Luminosity	$cm^{-2}s^{-1}$	10^{35}	10^{33}

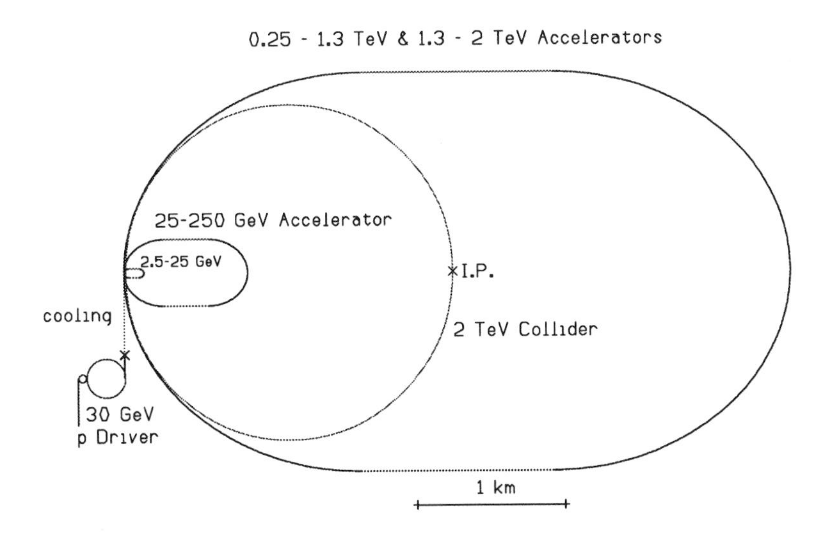

Figure 12.6: Layout of the collider and accelerator rings.

The main components of the 4 TeV collider would be:

- A proton source with KAON [21] like parameters (30 GeV, 10^{14} protons per pulse, at 15 Hz).
- A liquid metal target surrounded by a 20 T hybrid solenoid to make and capture pions.
- A 5 T solenoidal channel to allow the pions to decay into muons, with rf cavities to decelerate the fast ones that come first, while accelerating the slower ones that come later. Muons from pions in the 100-500 MeV range emerge in a 6 m long bunch at 150 $\pm$ 30 MeV.
- A solenoidal snake and collimator to select the momentum, and thus the polarization, of the muons.
- A sequence of 20 ionization cooling stages, each consisting of: a) energy loss material in a strong focusing environment for transverse cooling; b) linac reacceleration; and c) lithium wedges in dispersive environments for cooling in momentum space.
- A linac and/or recirculating linac pre-accelerator, followed by a sequence of pulsed field synchrotron accelerators using superconducting linacs for rf.
- An isochronous collider ring with locally corrected low beta (β=3 mm) insertion.

12.7.3 STATUS AND REQUIRED R AND D

Muon colliders are promising, but they are in a far less developed state than hadron or e^+e^- machines. No muon collider has ever been built. Much theoretical and experimental work will be needed before we will even know if they are possible. In particular, theoretical work is needed on the cooling sequence, on the collider ring, and on estimations of background in the detectors. The highest priority experimental work is:

- Demonstration of ionization cooling;
- Demonstration of liquid targets, solenoid pion capture, and the use of rf near such a source;
- Construction of model pulsed magnets for the accelerator and large aperture superconducting quadrupoles for the intersection region of the collider.

12.8 COMPARISON OF MACHINES

In Fig. 12.7, the effective physics energies (as defined by Eq. 12.3) of representative machines are plotted against their total tunnel lengths. We note:

- Hadron Colliders: It is seen that the energies of machines rise with their size, and that this rise is faster than linear ($E_{\text{eff}} \propto L^{1.3}$). This extra rise is a reflection of the increases in bending magnetic field used as new technologies and materials have become available.

- Circular Electron-Positron Colliders: As expected from the cost optimization discussed in Sect. 12.4, the energies of these machines rise approximately as the square root of their size.

- Linear Electron-Positron Colliders: The SLAC Linear Collider is the only existing machine of this type. One example of a proposed machine (the NLC) is plotted. The line drawn has the same slope as for the hadron machines and implies a similar rise in accelerating gradient as technologies advance.

- Muon-Muon Colliders: Only the 4 TeV collider, discussed above, and the

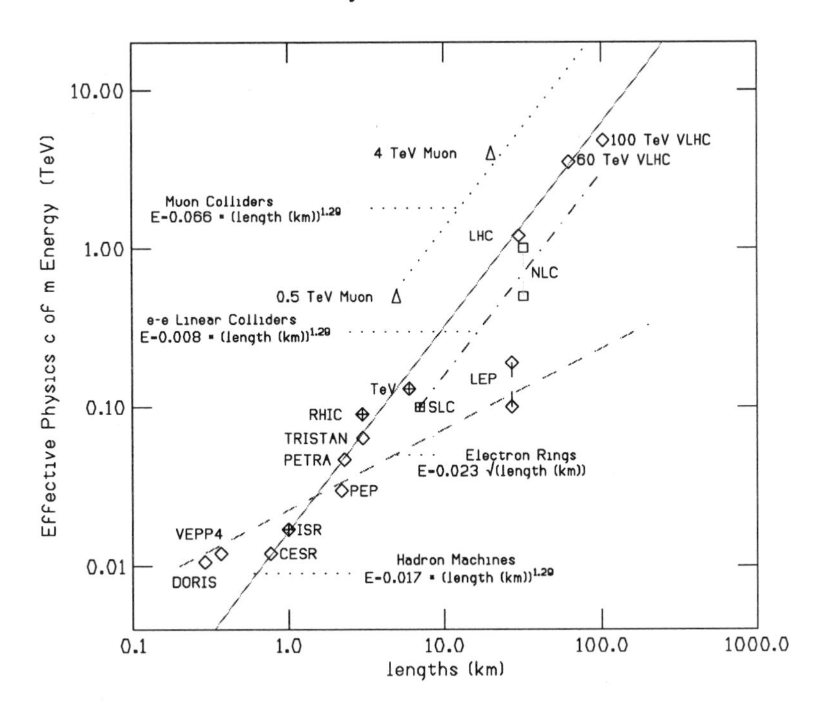

Figure 12.7: Effective physics energies of colliders as a function of their total length.

0.5 TeV *demonstration machine* have been plotted. The line drawn has the same slope as for the hadron machines.

It is noted that the muon collider offers the greatest energy per unit length. This is also apparent in Fig. 12.8, in which the footprints of a number of proposed machines are given on the same scale.

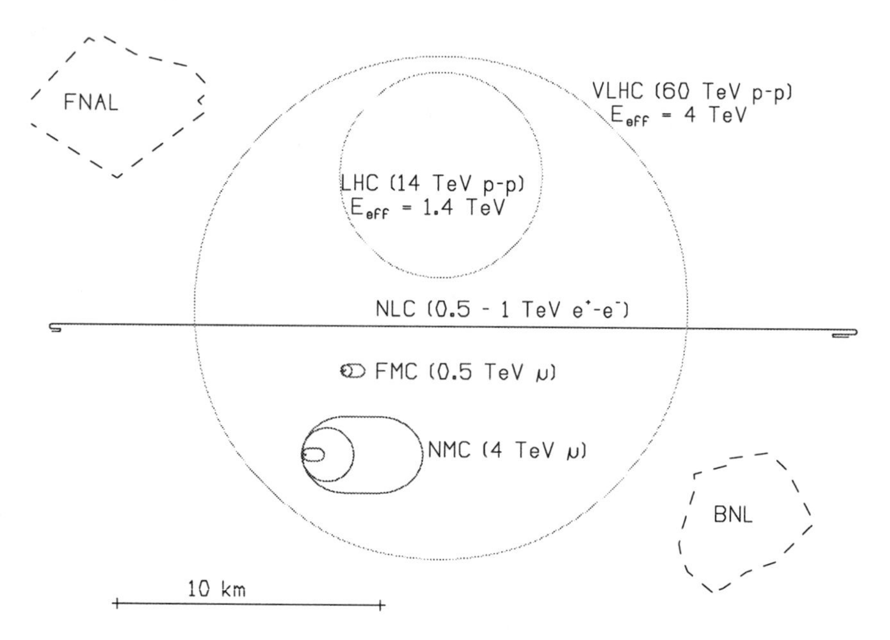

Figure 12.8: Approximate sizes of some possible future colliders.

12.9 CONCLUSIONS

Our conclusions, with the caveat that they are indeed only our opinions, are:

- The LHC is a well optimized and appropriate next step towards high *effective physics* energy.
- A Very Large Hadron Collider, with energy greater than the SSC (e.g., 60 TeV CM) and cost somewhat less than the SSC, may well be possible with the use of high T_c superconductors that may become available.
- A "Next Linear Collider" is the only clean way to complement the LHC with a lepton machine, and the only way to do so soon. But it appears that

even a 0.5 TeV collider may be more expensive than the LHC, has significantly less *effective physics energy*, and will be technically challenging. Obtaining the design luminosity may not be easy.

- Extrapolating conventional rf e^+e^- linear colliders to energies above 1 or 2 TeV will be very difficult. Raising the rf frequency can reduce length and probably cost for a given energy, but obtaining luminosity increasing as the square of energy, as required, may not be feasible.

- Laser-driven accelerators are becoming more realistic and can be expected to have a significantly lower cost per TeV. But the ratio of luminosity to wall power is likely to be significantly worse than for conventional rf-driven machines. Colliders using such technologies are thus unlikely to achieve the very high luminosities needed for physics research at higher energies.

- A higher gradient superconducting Linac collider using Nb_3Sn or high T_c materials, if it became technically possible, could be the most economical way to attain the required luminosities in a higher energy e^+e^- collider.

- Gamma-gamma collisions can and should be obtained at any future electron-positron linear collider. They would add physics capability to such a machine, but, despite their freedom from the beamstrahlung constraint, may not achieve higher luminosity.

- A Muon Collider, being circular, could be far smaller than a conventional rf e^+e^- linear collider of the same energy. Very preliminary estimates suggest that it would also be significantly less expensive. The ratio of luminosity to wall power for such machines above 2 TeV may be better than that for electron positron machines, and extrapolation to a center of mass energy of 4 TeV does not seem unreasonable. If research and development can show that it is practical, then a 0.5 TeV muon collider could be a useful complement to e^+e^- colliders, and, at higher energies (e.g., 4 TeV), could be a viable alternative.

ACKNOWLEDGEMENTS

We acknowledge important contributions from many colleagues, especially those that contributed to the feasibility study submitted to the Snowmass Workshop 96 Proceedings [19] from which much of the material and some text for this report, has been taken: In particular we acknowledge the contributions of the editors of each one of the chapters of the $\mu^+\mu^-$ Collider: A Feasibility Study: V. Barger, J. Norem, R. Noble, H. Kirk, R. Fernow, D. Neuffer, J. Wurtele, D. Lissauer, M. Murtagh, S. Geer, N. Mokhov, and D. Cline. This research was supported by the U.S. Department of Energy under Contract No. DE-ACO2-76-CH00016 and DE-AC03-76SF00515.

REFERENCES

[1] M.S. Livingston, *High Energy Accelerators* (Interscience, New York, 1954).

[2] A.W. Chao, R.B. Palmer, L. Evans, J, Gareyte, R.H. Siemann, in *Hadron Colliders (SSC/LHC)*, proceedings of the 1990 Summer Study on High Energy Physics, Snowmass, 1990, p. 667.

[3] S. Holmes for the RLHC Group, *Summary Report*, presentation at the Snowmass Workshop 96, to be published.

[4] *International Linear Collider Technical Review Committee Report*, SLAC-R-95-471, 1995.

[5] See for example, H. Murayama and M. Peskin, *Physics Opportunities of* e^+e^- *Linear Colliders*, SLAC-PUB-7149/LBNL-38808/UCB-PTH-96/18, June 1996; to appear in Annual Review of Nuclear and Particle Physics.

[6] R.B. Palmer, *Prospects for High Energy* E^+E^- *Linear Colliders*, Annu. Rev. Nucl. Part. Sci. **40** 529-92 (1990).

[7] N. Akasaka, *Dark current simulation in high gradient accelerating structure* EPAC96 Proceedings, Barcelona, Spain, 1996, (Institute of Physics Publishing), p. 483 Sitges.

[8] P. Chen and K. Yokoya, Phys. Rev. D **38** 987 (1988); P. Chen, SLAC-PUB-4823 (1987); K. Yokoya and P. Chen, lecture at the US-CERN Accelerator School, Hilton Head, South Carolina, 1990 (unpublished); Report No. KEK 91-2,1991.

[9] V. Telnov, Nucl. Instrum. and Methods, **A294** 72 (1990; *A Second Interaction Region for Gamma-Gamma, Gamma-Electron and Electron-Electron Collisions for NLC*, edited by K-J Kim, LBNL-38985, LLNL-UCRL-ID 124182, SLAC-PUB-95-7192.

[10] R.B. Palmer, *Accelerator parameters for* $\gamma - \gamma$ *colliders*; Nucl. Instrum. and Methods, **A355** 150-153 (1995).

[11] P. Chen and R. Palmer, *Coherent Pair Creation as a Positron Source for Linear Colliders*, edited by J. Wurtele (AIP Press, Conference Proceedings, 1993), p. 279.

[12] V. Telnov, *Laser Cooling of Electron Beams for linear colliders*; NSF-ITP-96-142 and SLAC-PUB 7337

[13] E. A. Perevedentsev and A.N. Skrinsky in *Proceedings of the 12th International Conference on High Energy Accelerators, 1983*, edited by F.T. Cole and R. Donaldson, p. 485; A.N. Skrinsky and V.V. Parkhomchuk, Sov. J. Nucl. Phys. **12** 3 (1981); *Early Concepts for* $\mu^+\mu^-$ *Colliders and High Energy* μ *Storage Rings*, proceedings of the Physics Potential & Development of $\mu^+\mu^-$ Colliders, 2nd Workshop, Sausalito, CA, edited by D. Cline (AIP Press, Woodbury, New York, 1995).

[14] D. Neuffer, IEEE Trans. **NS-28** 2034 (1981).

[15] *Proceedings of the Mini-Workshop on $\mu^+\mu^-$ Colliders: Particle Physics and Design, Napa CA*, Nucl Instrum. Methods, **A350** (1994); *Proceedings of the Muon Collider Workshop*, February 1993, Los Alamos National Laboratory Report LA- UR-93-866 (1993) and Physics Potential & Development of $\mu^+\mu^-$ Colliders 2nd Workshop, Sausalito, CA, edited by D. Cline (AIP Press, Woodbury, New York, 1995).

[16] Transparencies at the *2 + 2 TeV $\mu^+\mu^-$ Collider Collaboration Meeting*, Feb 6-8, 1995, BNL, compiled by Juan C. Gallardo; transparencies at the *2 + 2 TeV $\mu^+\mu^-$ Collider Collaboration Meeting*, July 11-13, 1995, FERMILAB, compiled by Robert Noble; *Proceedings of the 9th Advanced ICFA Beam Dynamics Workshop*, edited by J.C. Gallardo (AIP Press, Conference Proceedings 372 1996).

[17] D.V. Neuffer and R.B. Palmer, *Proceedings European Particle Accelerator Conference, London, 1994*; M. Tigner, in Advanced Accelerator Concepts, Port Jefferson, NY 1992, AIP Conf. Proc. **279** 1 (1993).

[18] R.B. Palmer et al., *Monte Carlo Simulations of Muon Production*, proceedings of the Physics Potential & Development of $\mu^+\mu^-$ Colliders 2nd Workshop, Sausalito, CA, edited by D. Cline, (AIP Press, Woodbury, New York 1995) p. 108; R.B. Palmer et al., *Muon Collider Design*, proceedings of the Symposium on Physics Potential & Development of $\mu^+\mu^-$ Colliders, Nucl. Phys B (Proc. Suppl.) 51A (1996).

[19] *$\mu^+\mu^-$ collider, A Feasibility Study*, BNL-52503, Fermilab-Conf-96/092, LBNL-38946, submitted to the Proceedings of the Snowmass96 Workshop.

[20] A.N. Skrinsky and V.V. Parkhomchuk, *Methods of cooling beams of charged particles*, Sov. J. Part. Nucl. **12** 223 (1981); D. Neuffer, *Principles and Applications of Muon Cooling*, Part. Acc. **14** 75 (1983); R.C. Fernow and J.C. Gallardo, *Muon Transverse Ionization Cooling: Stochastic Approach*, Phys. Rev. E **52** 1039 (1995).

[21] *KAON Factory Study*, Accelerator Design Study Report, TRIUMF, Vancouver, B.C. Canada.

12.10 DISCUSSION

Session Chair: Richard Taylor
Rapporteur: Kentaro Nagamine

TAYLOR: I was afraid we might run over, but all is well because we spent hardly any time on what is wrong with muon colliders compared to the length of time we spent on what was wrong with linear colliders. (laugh)

HENRY: So what's wrong with the muon colliders? (laugh)

PALMER: I can tell you which parts of the muon collider keep me awake at night. It changes, of course, from week to week. Enormous progress has been made even in last few months. The collider lattice was a problem, but doesn't worry me any more. We also had a serious difficulty in the transverse cooling lattices. When we first tried tracking particles through, some muons never came out. They were hitting resonances. Now we understand that problem and have tracked through transverse cooling sections that work.

But we have not done energy cooling yet. We know theoretically how to do it, but we haven't got a realistic lattice and tracked muons through it. Having been burnt once, I will have sleepless nights until we get past that hurdle.

The collider ring may have instability problems that are not fully understood. We think that it will have to have BNS damping applied by using RF quadrupoles, but we haven't worked that out.

We haven't done many things that need to be done, but I do not yet see any insuperable problems. I do not sleep that badly. (laugh)

VOGT: Have you considered using surface muons which have been considered at KEK as an alternative muon source?

PALMER: Yes, but we need bunches with very large numbers of muons in order to get luminosity. It seems to be difficult to get them from surface muons. And there is a more basic problem: we need both charges, I do not think this is possible with surface muons.

WITTEN: What fraction of muons decay before entering the ring?

PALMER: With the parameters we've considered, about three-fourths are lost. Half decay during the cooling sequence, and half of the remainder decay during acceleration.

MANN: It would be interesting to hear about the shielding problems that arise in the muon-muon collider.

PALMER: I think I know what you're trying to get at. (laugh) The radiation from decay electrons in the ring itself can be shielded relatively easily. Dumping 2 TeV muons is more difficult because it takes 2 km of concrete to stop them, but that, too, is OK. The problem you may be hinting at, which I didn't mention because we are not yet sure about the calculation, is radiation from neutrinos. Muons decaying in a straight section of the collider ring produce a neutrino beam with opening angle of $1/\gamma$ that, for a 4 TeV collider, is only a meter or two wide

35 km away. The neutrino cross section is small, but rising linearly with energy, and there are 20 mega-watts of power in that beam. The resulting radiation level appears to be close to the legal limit. You can't shield it and it always breaks ground somewhere because unfortunately the earth is round. (laugh) It rises as the fourth power of the energy and is only inverse with the machine depth. Thus, even if a 4 TeV $\mu^+\mu^-$ collider is just OK, a 10 TeV collider is probably impractical.

LEDERMAN: Going back to the beginning of your talk on the Livingston Plot, you said that there is a 10^6 rise of cost rise for a 10^{12} rise in energy so we are 10^6 cleverer.(laugh) Did you include inflation in those numbers?

PALMER: Yes, I did, but I may not have done it right. Down the bottom I had 100 KeV and I said to myself what I could buy that now for a few thousand dollars. This is not fair because in 1930 you could not buy one and it must have taken quite a bit of labor to build one. I did not try and estimate that cost.

CHAPTER 13

VISTAS IN THEORETICAL PHYSICS

EDWARD WITTEN
School of Natural Sciences
Institute for Advanced Study, Princeton, NJ

Back in the late 1960s—halfway back to Princeton's 200th anniversary conference!—there was something called the "dual resonance model" of strong interactions. It was based on a kind of "duality"—whence the name—between two things that are usually quite different, namely "t-channel" and "s-channel" scattering.

Scattering in the s-channel is resonance production, and usually dominates at specific energies and any scattering angle, while t-channel scattering is the relativistic version of scattering from a potential, which usually dominates at high energy and small angle. In conventional quantum field theory, they are governed by the Feynman diagrams of Figs. 13.1(a) and 13.1(b). With only finitely many particles exchanged, it is impossible for the s channel and t channel diagrams to be equal. But, inspired by the apparently limitless number of hadron resonances, physicists wondered whether there could be a theory of infinitely many particles in which the sum of s-channel exchanges would equal the sum of t-channel exchanges, as suggested in Fig. 13.1(c).

At first, this seemed well-nigh impossible, but it turned out not only to be possible but to have a simple explanation. For this, think of a hadron not in terms of a point particle but in terms of a little string with quarks at the ends, as in Fig. 13.2. (The quarks carry charges but in this model are massless.) If hadrons have such a microscopic description, there will naturally be infinitely many of them associated with different modes of vibration of the string.

Once one takes that step, one can generalize ordinary Feynman diagrams to string diagrams. One simply replaces the lines in a Feynman diagram—which represent the propagation of a point particle in space-time—by ribbons (or, in the case of closed strings, tubes) that represent the propagation of a string. Duality between s-channel and t-channel scattering now becomes the simple statement, illustrated in Fig. 13.3, that once the lines are replaced by ribbons that join smoothly, the s-channel and t-channel diagrams are the same topologically.

Such was the impact of this insight that the name "dual resonance model" went into eclipse (freeing up the word "duality" for later uses!) and the subject was renamed "string theory." String theory failed as a theory of hadrons, or at least QCD (the $SU(3)$ gauge theory with quarks) proved to be much more powerful, while the string theory (being too soft in the deep Euclidean region) was inadequate in its known form.

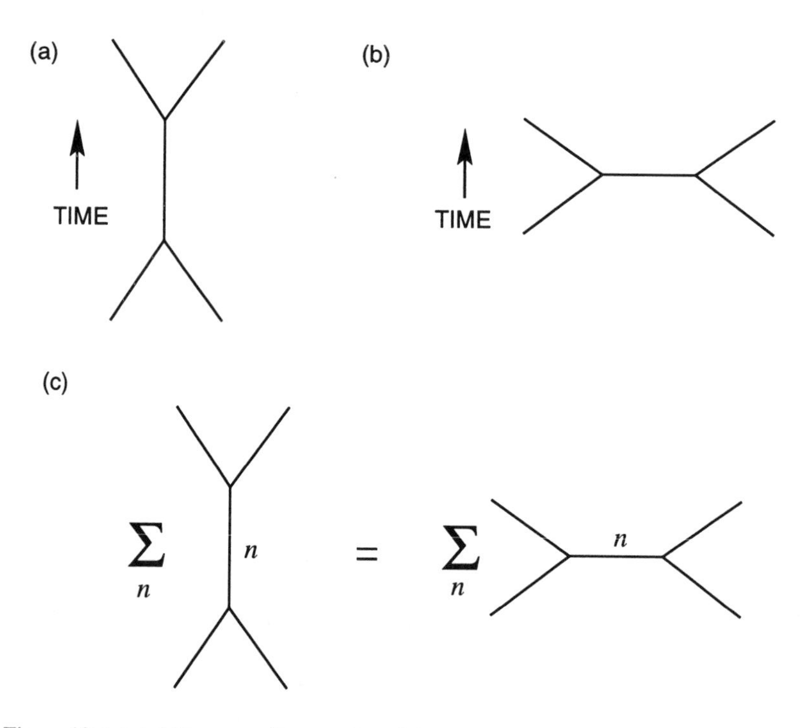

Figure 13.1: (a) A Feynman diagram describing s-channel scattering. (b) A Feynman diagram describing t-channel scattering. (c) The "duality" hypothesis, which asserts that s-channel and t-channel scattering are equal, if one sums over exchanges of infinitely many types of particle.

But the qualitative picture of a meson as an open string with quarks at the end is quite reminiscent of QCD. In the string picture, if one stretches a string so that it breaks, a new $q - \bar{q}$ pair appears and one gets two strings, each of which has a $q - \bar{q}$ pair at its ends. This corresponds to the familiar—but still rather mysterious—phenomenon of quark confinement in QCD. It is hard to believe that this relation is just an accident. Following important work by 't Hooft [1974], many physicists have believed that if we could understand the large N limit of

QCD (generalized to a theory with $SU(N)$ gauge group) we would understand "why" the string picture is in some ways so close to the truth. The hope is that QCD for large N has a description as a string model with $1/N$ as the coupling constant. I do not believe that such a description is just around the corner (in fact, I think progress is more likely to come, not directly, but as a by-product of other things I will talk about later), but it is a reasonable goal for the $300^{mathrmth}$ anniversary conference.

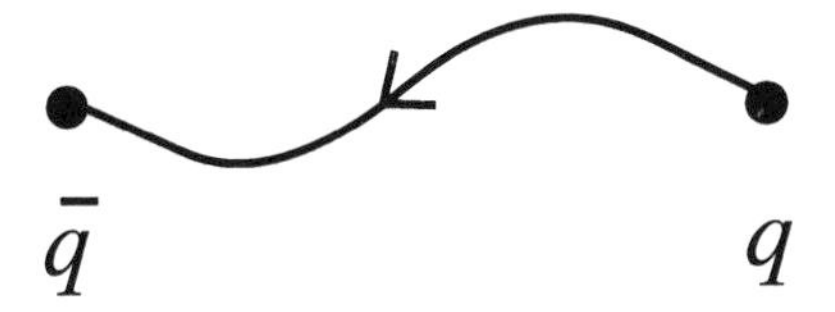

Figure 13.2: A meson understood as an oriented string with a quark at one end and an antiquark at the other.

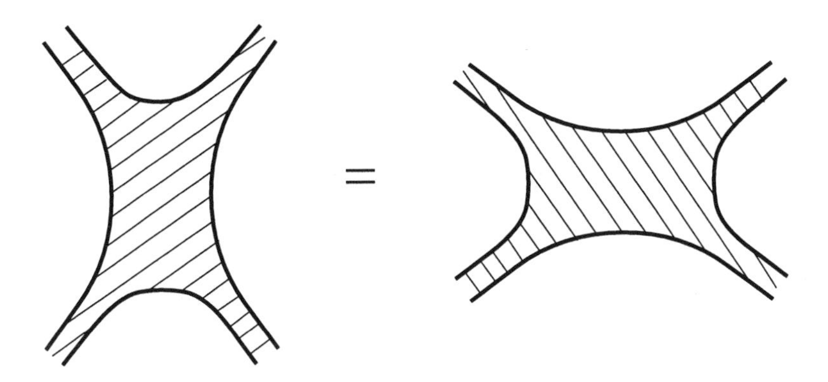

Figure 13.3: In going from field theory to string theory, the lines of Figs. 13.1(a) and 13.1(b) are replaced by ribbons that join smoothly. The duality hypothesis of Fig. 13.1(c) now becomes a simple topological statement.

13.1 PREDICTIONS AND POST-DICTIONS IN THE CLASSIC PERIOD

To achieve this kind of "duality" along the lines suggested in Fig. 13.3, one must face the problem of quantizing the vibrations of a relativistic string. Studies of this problem in the 1970s gave many surprising discoveries, including the following:

1. The ground state of the open string behaves like a Yang-Mills gauge boson.

A theory with open strings behaves at long distances like non-abelian gauge theory plus small corrections.

2. Open strings do not make sense without closed strings, and the quantization of the closed string gives *a massless graviton*. As a result, it turns out that *any* relativistic quantum string theory looks at long distances like *a quantum theory incorporating general relativity*, in dramatic contrast to the situation in standard quantum field theory where incorporating gravity is apparently impossible.

3. Incorporating fermions in the theory required the invention of an altogether new symmetry, *supersymmetry*, in which, roughly, the coordinates of space-time are enriched with the inclusion of anticommuting quantum variables. (To be more precise, quantizing the string gave directly the *world-sheet* supersymmetry of Ramond, which was adapted to *space-time* supersymmetry by Wess and Zumino; the relation between them was discovered later by Gliozzi, Scherk, and Olive. Supersymmetry was also conceived independently in the USSR by Gol'fand and Likhtman.)

The characteristic pre- and post-dictions of string theory, as understood in the classical period, are thus

- Yang-Mills gauge invariance
- general relativity
- supersymmetry.

Of these, the first two are, of course, pillars of physics, and the search for the third is one of the main goals of present and future accelerators.

13.2 THE MYSTERY OF THE COSMOLOGICAL CONSTANT

By the early 1980s, it was clear that one could use string theory to make consistent theories of quantum gravity interacting with gauge bosons and charged fermions (i.e., a pseudo-standard model), but there was a general flaw: the weak interactions had to conserve parity. This was overcome in 1984 with the Green-Schwarz anomaly cancellation. This discovery soon led to the $E_8 \times E_8$ heterotic string, and models of particle physics (plus quantum gravity) that were strikingly elegant, and rather close to the real thing.

In fact, by the mid-1980s, it seemed relatively clear that the main obstacle to making these models more realistic was the lack of understanding of the cosmological constant. The vanishing (or extreme smallness, though I personally doubt it) of the cosmological constant is a mystery mainly about *quantum* gravity, since, classically, one can simply declare the cosmological constant to be zero. Moreover, the problem is sharpest in a theory that has no adjustable dimensionless parameter by which the cosmological constant could be adjusted to zero; string theory has this characteristic. Indeed, only in string theory is the problem completely well-posed.

The problem, however, needs to be stated even more precisely: Explain why there is a *stable vacuum*, with *vanishing* energy density, *after* supersymmetry breaking. Spontaneous supersymmetry breaking *per se* is not the problem, since there are natural mechanisms for this which, moreover, can naturally give exponentially small supersymmetry breaking; that is what we hope to observe at TeV energies in the real world. The question is why supersymmetry breaking results in a stable vacuum of zero energy. I believe that if we understood this, we would be better placed to build models of supersymmetry breaking that would be closer to the truth, and the phenomenology would improve (just as it did in 1984 when the obstruction to parity violation in weak interactions was removed). Incidentally, I expect this to happen before the large N limit of QCD is understood, though I would be happy to be proved wrong.

Hopefully, with better models of supersymmetry breaking, one would be able to make some qualitative predictions about the "soft supersymmetry breaking" terms that should be uncovered if supersymmetry is, in fact, relevant to physics at the next generation of experiments. There is not now a compelling scenario concerning the nature of supersymmetry breaking, with or without string theory; therefore, we would learn much, above and beyond any confirmation of existing theories, if supersymmetry is discovered and it becomes possible to probe supersymmetry-breaking experimentally.

In practice, with our present unsatisfactory models of supersymmetry breaking, we never find a stable vacuum in which there is a cosmological constant. One always finds, instead, an unstable vacuum and runaway behavior whenever there appears to have been a cosmological constant. This is probably not a coincidence, though we do not know for sure to what extent it reflects a physical principle and to what extent it reflects limitations on what calculations we can do.

13.3 THE BIG QUESTIONS

I have spoken a lot about string theory without saying what it is. In a way, that is the life of a string theorist. The theory has its roots in the "dual" amplitude, discovered by Veneziano in 1968, which first exhibited how s-channel and t-channel scattering could be equivalent. It has developed since then, by a long process of tinkering, without much *a priori* understanding of what the theory really is.

The "big questions" in the subject are questions like these:

• What is the theory whose perturbation expansion physicists have grappled with since 1968?

• What are the basic ingredients or building blocks by which this theory should be constructed?

• What is the analog of the uncertainty principle of quantum mechanics, or the equivalence principle of general relativity?

From what we already know, I think the answers to those questions must be staggeringly beautiful, and involve changes in physical concept and technique at least on the scale of what happened with quantum mechanics and general relativity.

If the answers are found, I think it will probably become clear that the theory is right; it will be simplified (at least in part) and taught to undergraduates,[1] and all this will be a major theme at the 300th anniversary meeting.

In practice, since 1968, progress has come mainly not by asking the "biggest" questions, but through a more modest process of tinkering and developing the known formalism. Here are some examples, some of which I already mentioned:

• Asking how to include fermions and stumbling on a trail that led to supersymmetry.

• Asking how the weak interactions could violate parity, leading to anomaly cancellation, the heterotic string, and a much better insight about how the standard model might arise from string theory.

• Generalizing the "old" perturbative string theory to allow for more general conformal field theories, leading to a framework in which—as seen in many concrete calculations—the familiar concept of "space-time" is only an idealization.

• And most recently, extrapolating from the traditional regime of weakly coupled perturbative string theory—where we lived peaceful and sheltered lives for many years—to the regime of strong coupling, where quantum effects are big and surprises lurk.

I will say a few words about the last point. First of all, making such an extrapolation normally sounds hopeless, as the strong coupling region is usually out of reach. But we learned in the last few years, originally in field theory,

[1]The time lag between when a substantial part of the big questions are answered and when the theory begins to be taught to undergraduates will probably be similar to the number of years that elapsed from 1905 until Princeton University first taught special relativity to freshmen.

that such extrapolations can, indeed, be convincingly made in the supersymmetric case. It takes practice, a sharp eye, and an odd bag of tricks.

What we have learned from this enterprise is, notably, that *there is only one theory*. In the last generation, the quest for superunification of the forces of nature focused mainly on five or six different theories. There were five string theories—Type I, Type IIA, Type IIB, heterotic $E_8 \times E_8$, and heterotic $SO(32)$—and a wild card, eleven-dimensional supergravity. These theories were discovered in different ways in the period from about 1970 to 1984, and by now they have been intensively studied. We now understand that these five or six different theories are different limiting cases of one theory, sometimes tentatively called M theory.

We have learned, moreover, that—in the context of string theory—strings are only, at most, the "first among equals." Other objects, such as the "D-branes," enter on an equivalent footing at a more fundamental level. The D-branes are extended objects of various dimension—points, membranes, etc.—whose role has really been understood only since Polchinski's work of last year. For weak coupling, the D-branes are very heavy and "non-perturbative." For strong coupling they are "dual" to strings; they are exchanged with strings by non-perturbative symmetries. One indication that these objects are fundamental is that understanding them has made it possible for the first time to count—at least in some special situations—the quantum states of a black hole. Since the work of Bekenstein and Hawking over twenty years ago, we have known that a black hole is characterized by an *entropy* S depending on its mass, angular momentum, and charge. If black hole entropy is like any other entropy in physics, the existence of a non-zero S should mean that there is a large quantum degeneracy of the black hole; the number of states should be roughly e^S.

Counting the quantum states of a black hole is a problem that necessarily involves both *gravity* and *quantum mechanics*, so it makes sense only in string theory. The interpretation of certain black holes as collections of D-branes has, indeed, made it possible to count states, and to check that the degeneracy is the exponential of the Bekenstein-Hawking entropy.

To do this, you need to know something about D-brane interactions. These are of quite an interesting nature. Roughly speaking, a single D-brane has a position $\vec{X}$ in space, with components $X_1, X_2, \ldots, X_d$ (d being the number of spatial dimensions). But in the case of N identical D-branes, the X_i become $N \times N$ matrices that generally do not commute. These matrices can be interpreted as defining positions in the classical sense only when they commute and can be simultaneously diagonalized.

This, then, might be the germ of a new kind of noncommutativity in physics—a modern heir of the Heisenberg formula $[p, x] = -i\hbar$. But how close to the truth is it likely to be?

13.4 OUTLOOK

In string theory, we have peeled layers off the onion for 28 years now, and it is still hard to guess how many more layers and surprises there are. In past years, when asked how long the process would take, that has always been my answer. It may well still be the correct answer.

However, I mentioned the black hole state counting problem not merely for its intrinsic beauty and fame. The fact that earlier we could not count the quantum states—or elementary components—of a black hole, and now we can, may be a hint that we are close to closure, that we are now at least close to knowing the basic ingredients in which the theory should be described.

If so, the noncommutative "position" matrices of a D-brane may be a good clue about the answers to the "big" questions. A proposal in that direction has recently been made (by Banks, Fischler, Shenker, and Susskind). If their proposal, or something like it, is right, we will probably be doing a completely new kind of physics, with non-commutativity built in at a much more fundamental level, long before the 300th anniversary of Princeton University.

13.5 DISCUSSION

Session Chair: David Gross
Rapporteur: Rajesh Gopakumar

SHENKER: This is an exhilarating period in string theory because things are changing very rapidly. What we are saying now may be obsolete in fifty weeks, let alone fifty years. One point I'd like to re-emphasize: supersymmetry has played a central role in recent developments. It may even be that properties like locality, which one had thought to be fundamental, might actually emerge as a consequence of extensive supersymmetric cancellations between nonlocal degrees of freedom.

VERLINDE: I had two questions. Firstly, in trying to deepen our understanding of the conceptual issues and questions of string theory, are there any guiding principles? Or do you think progress will come from "tinkering" with the formalism? Secondly, what is it that we need in order to make contact with the four dimensional world?

WITTEN: For the second question, I must once again emphasize the distinction between a less fundamental and more fundamental level of understanding. The distinction between the solutions of a theory and the theory itself. Once we understand the latter, it will be easier to understand the former. As for the first question, progress in string theory has historically come from trying to understand pressing concrete problems. At the moment, the most significant such problem is that of the cosmological constant. Conceptual jumps may be needed such as, for example, the recent matrix model proposal.

SAMIOS: What if we don't find Supersymmetry in the TeV machines?

WITTEN: Well, then nature would not have been kind enough to make Supersymmetry relevant to electroweak symmetry breaking. Strictly speaking, it is not necessary in string theory that SUSY breaking be seen at the TeV scale. But just as nature has provided the binary pulsar for humans to detect gravitational waves and verify general relativity, it is hopeful that she will be similarly kind in this case. In any case, progress in string theory will be faster if we find supersymmetry at the TeV scale.

KHURI: I'd like to hark back to an early prediction of yours; at the Shelter Island conference of 1983 (before the superstring revolution), you expressed an intention to talk about strings if John Schwarz did not. This, I remember, caused surprise among the organisers since you were not working, at least openly, on strings until then. Why were you betting on string theory?

WITTEN: Because neither then, nor now, were there any competing answers nearly as credible as string theory.

CHAPTER 14

THE FUTURE OF PARTICLE PHYSICS AS A NATURAL SCIENCE

FRANK WILCZEK

School of Natural Sciences
Institute for Advanced Study, Princeton, NJ

ABSTRACT

In the first part of the talk, I give a low-resolution overview of the current state of particle physics—the triumph of the Standard Model and its discontents. I review and re-endorse the remarkably direct and (to me) compelling argument that existing data, properly interpreted, point toward a unified theory of fundamental particle interactions and toward low-energy supersymmetry as the near-term future of high energy physics as a natural science. I then attempt, as requested, some more 'visionary'—i.e., even lower resolution—comments about the farther future. In that spirit, I emphasize the continuing importance of condensed matter physics as a source of inspiration and potential application, in particular for expansion of symmetry concepts, and of cosmology as a source of problems, applications, and perhaps ultimately, limitations.

14.1 TRIUMPH OF THE STANDARD MODEL

The core of the Standard Model [1, 2, 3] of particle physics is easily displayed in a single figure, here Fig. 14.1. There are gauge groups $SU(3) \times SU(2) \times U(1)$ for the strong, weak, and electromagnetic interactions. The gauge bosons associated with these groups are minimally coupled to quarks and leptons according to the scheme depicted in the figure. The non-abelian gauge bosons within each of the $SU(3)$ and $SU(2)$ factors also couple, in a canonical minimal form, to one another. The $SU(2) \times U(1)$ group is spontaneously broken to the $U(1)$ of electromagnetism. This breaking is parameterized in a simple and (so far) phenomenologically adequate way by including an $SU(3) \times SU(2) \times U(1)$ $(1, 2, -\frac{1}{2})$ scalar 'Higgs' field which condenses, that is, has a non-vanishing expectation value in the ground state. Condensation occurs at weak coupling if the bare (mass)^2 associated with the Higgs doublet is negative.

The fermions fall into five separate multiplets under $SU(3) \times SU(2) \times U(1)$, as depicted in the figure. The color $SU(3)$ group acts horizontally; the weak $SU(2)$ vertically, and the hypercharges (equal to the average electric charge) are as indicated. Note that left- and right-handed fermions of a single type generally transform differently. This reflects parity violation. It also implies that fermion masses, which of course connect the left- and right-handed components, only arise upon spontaneous $SU(2) \times U(1)$ breaking.

Only one fermion family has been depicted in Fig. 14.1; of course in reality there are three repetitions of this scheme. Also not represented are all the complications associated with the masses and Cabibbo-like mixing angles among the fermions. These masses and mixing angles are naturally accommodated as parameters within the Standard Model, but I think it is fair to say that they are not much related to its core ideas—more on this below.

$$\begin{array}{ccccc} \mathrm{SU(3)} & \times & \mathrm{SU(2)} & \times & \mathrm{U(1)} \\ 8\ \text{gluons} & & W^{\pm}, Z & & \gamma \\ & & \uparrow & \text{—} & \uparrow \\ & & & & \text{mixed} \end{array}$$

$$\overset{\mathrm{SU(3)}}{\longleftrightarrow}$$

$$\mathrm{SU(2)} \updownarrow \begin{pmatrix} u_L^r & u_L^w & u_L^b \\ d_L^r & d_L^w & d_L^b \end{pmatrix} \frac{1}{6} \qquad \begin{matrix} (u_R^r & u_R^w & u_R^b)\,\frac{2}{3} \\ (d_R^r & d_R^w & d_R^b) - \frac{1}{3} \end{matrix}$$

$$\begin{pmatrix} \nu_L \\ e_L \end{pmatrix} - \frac{1}{2} \qquad e_R - 1$$

Figure 14.1: The core of the Standard Model: the gauge groups and the quantum numbers of quarks and leptons. There are three gauge groups, and five separate fermion multiplets (one of which, e_R, is a singlet). Implicit in this figure are the universal gauge couplings—exchanges of vector bosons—responsible for the classic phenomenology of the strong, weak, and electromagnetic interactions. The triadic replication of quark and leptons, and the Higgs field whose couplings and condensation are responsible for $SU(2) \times U(1)$ breaking and for fermion masses and mixings, are not indicated.

With all these implicit understandings and discrete choices, the core of the Standard Model is specified by three numbers—the universal strengths of the strong, weak, and electromagnetic interactions. The electromagnetic sector, QED, has been established as an extraordinarily accurate and fruitful theory for several decades now. Let me now briefly describe the current status of the remainder of the Standard Model.

Some recent stringent tests of the electroweak sector of the Standard Model are summarized in Fig. 14.2. In general, each entry represents a very different experimental arrangement, and is meant to test a different fundamental aspect of the theory, as described in the caption. There is precisely one parameter (the mixing angle) available within the theory, to describe all these measurements. As you can see, the comparisons are generally at the level of a per cent accuracy or so. Overall, the agreement appears remarkably good, especially to anyone familiar with the history of weak interactions.

Some recent stringent tests of the strong sector of the Standard Model are summarized in Fig. 14.3 [5]. Again a wide variety of very different measurements are represented, as indicated in the caption. A central feature of the theory (QCD) is that the value of the coupling, as measured in different physical processes, will depend in a calculable way upon the characteristic energy scale of the process. The coupling was predicted—and evidently is now convincingly measured—to decrease as the inverse logarithm of the energy scale: asymptotic freedom. Again, all the experimental results must be fit with just one parameter—the coupling at any single scale, usually chosen as M_Z. As you can see, the agreement between theory and experiment is remarkably good. The accuracy of the comparisons is at the 1-2 % level.

Let me emphasize that these figures barely begin to do justice to the evidence for the Standard Model. Several of the results in them summarize quite a large number of independent measurements, any one of which might have falsified the theory. For example, the single point labeled 'DIS' in Fig. 14.3 describes literally hundreds of measurements in deep inelastic scattering with different projectiles and targets and at various energies and angles, which must—if the theory is correct—all fit into a tightly constrained pattern.

I last reviewed this situation on a related occasion several months ago. At that time, there were reported discrepancies between experimental observations and the Standard Model prediction of the branching ratio R_b of the Z into b quarks, and also with the Standard Model (QCD) prediction of inclusive jet production at high transverse energy. In the meantime these discrepancies have come to seem much less significant: for R_b, mostly because of the inclusion of new data; for the jet production, because of a better appreciation of the uncertainties in existing structure function parametrizations. Thus once (or rather twice) again, the Standard Model has survived the challenges that inevitably accompany stiff scrutiny. Another small but long-standing and annoying anomaly, the slightly high value of the strong coupling $\alpha_s(M_Z)$ inferred from the width of the Z has also disappeared—largely, I am told, because the effect of passing trains perturbing the beam energy and thus causing a spurious 'widening' has now been accounted for!

The central theoretical principles of the Standard Model have been in place for

	Measurement with Total Error	Systematic Error	Standard Model	Pull
$\alpha(m_Z^2)^{-1}$	128.896 ± 0.090	0.083	128.907	−0.1
a) LEP				
line-shape and lepton asymmetries:				
m_Z [GeV]	91.1863 ± 0.0020	0.0015	91.1861	0.1
Γ_Z [GeV]	2.4946 ± 0.0027	0.0017	2.4960	−0.5
σ_h^0 [nb]	41.508 ± 0.056	0.055	41.465	0.8
R_ℓ	20.778 ± 0.029	0.024	20.757	0.7
$A_{FB}^{0,\ell}$	0.0174 ± 0.0010	0.007	0.0159	1.4
+ correlation matrix				
τ polarisation:				
$\mathcal{A}_\tau$	0.1401 ± 0.0067	0.0045	0.1458	−0.9
$\mathcal{A}_e$	0.1382 ± 0.0076	0.0021	0.1458	−1.0
b and c quark results:				
R_b^0	0.2179 ± 0.0012	0.0009	0.2158	1.8
R_c^0	0.1715 ± 0.0056	0.0042	0.1723	−0.1
$A_{FB}^{0,b}$	0.0979 ± 0.0023	0.0010	0.1022	−1.8
$A_{FB}^{0,c}$	0.0733 ± 0.0049	0.0026	0.0730	0.1
+ correlation matrix				
$q\overline{q}$ charge asymmetry:				
$\sin^2\theta_{eff}^{lept}$ $(\langle Q_{FB}\rangle)$	0.2320 ± 0.0010	0.0008	0.23167	0.3
b) SLD				
$\sin^2\theta_{eff}^{lept}$ (A_{LR})	0.23061 ± 0.00047	0.00014	0.23167	−2.2
R_b^0	0.2149 ± 0.0038	0.0021	0.2158	−0.2
$\mathcal{A}_b$	0.863 ± 0.049	0.032	0.935	−1.4
$\mathcal{A}_c$	0.625 ± 0.084	0.041	0.667	−0.5
c) $p\overline{p}$ and νN				
m_W [GeV] $(p\overline{p})$	80.356 ± 0.125	0.110	80.353	0.0
$1 - m_W^2/m_Z^2$ (νN)	0.2244 ± 0.0042	0.0036	0.2235	0.2
m_t [GeV] $(p\overline{p})$	175 ± 6	4.5	172	0.5

Table 1:

Figure 14.2: A recent compilation of precision tests of electroweak theory, from [4], to which you are referred for details. Despite some 'interesting' details, clearly the evidence for electroweak $SU(2) \times U(1)$, with the simplest doublet-mediated symmetry breaking pattern, is overwhelming.

nearly twenty-five years. Over this interval the quality of the relevant experimental data has become incomparably better, yet no essential modifications of these venerable principles has been required. Let us now praise the Standard Model:

- The Standard Model is here to stay, and describes *a lot*.

  Since there is quite direct evidence for each of its fundamental ingredients

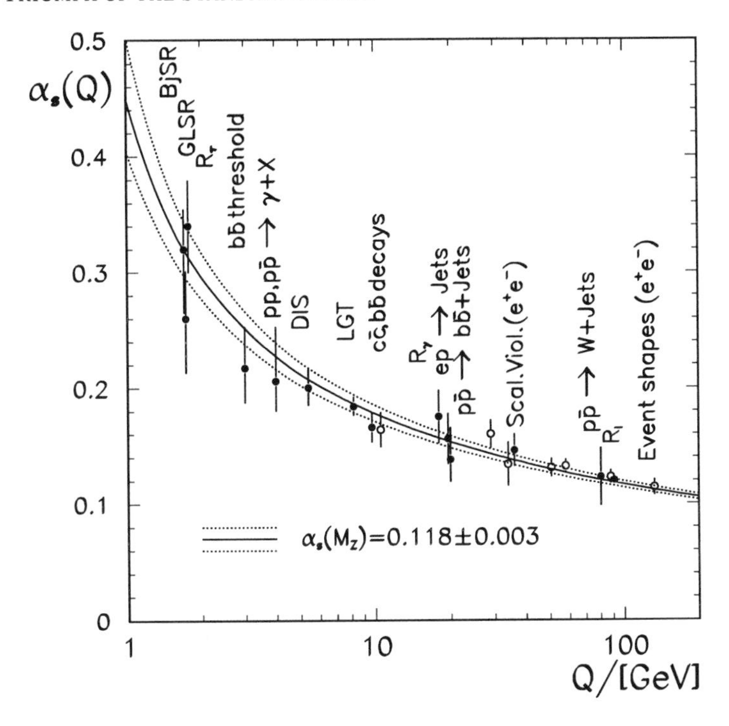

Figure 14.3: A recent compilation of tests of QCD and asymptotic freedom, from [5], to which you are referred for details. Results are presented in the form of determinations of the effective coupling $\alpha_s(Q)$ as a function of the characteristic typical energy-momentum scale involved in the process being measured. Clearly the evidence for QCD in general, and for the decrease of effective coupling with increasing energy-momentum scale (asymptotic freedom) in particular, is overwhelming.

(i.e., its interaction vertices), and since the Standard Model provides an extremely economical packaging of these ingredients, I think it is a safe conjecture that it will be used, for the foreseeable future, as the working description of the phenomena within its domain. And this domain includes a very wide range of phenomena—indeed not only what Dirac called "all of chemistry and most of physics,"[1] but also the original problems of radioactivity and nuclear interactions which inspired the birth of particle physics in the 1930s, and much that was unanticipated.

- The Standard Model is a *principled* theory.

Indeed, its structure embodies a few basic principles: special relativity, locality, and quantum mechanics, which lead one to quantum field theories, local sym-

[1] Dirac was referring, here, to quantum electrodynamics.

metry (and, for the electroweak sector, its spontaneous breakdown), and renormalizability (minimal coupling). The last of these principles, renormalizability, may appear rather technical and perhaps less compelling than the others; we shall shortly have occasion to re-examine it in a larger perspective. In any case, the fact that the Standard Model is principled in this sense is profoundly significant: it means that its predictions are precise and unambiguous, and generally cannot be modified 'a little bit' except in very limited, specific ways. This feature makes the experimental success especially meaningful, since it becomes hard to imagine that the theory could be approximately right without in some sense being exactly right.

• The Standard Model *can be extrapolated.*

Specifically because of the asymptotic freedom property, one can extrapolate using the Standard Model from the observed domain of experience to much larger energies and shorter distances. Indeed, the theory becomes simpler—the fundamental interactions are all effectively weak—in these limits. The whole field of very early Universe cosmology depends on this fact, as do the impressive semi-quantitative indications for unification and supersymmetry I shall be emphasizing momentarily.

14.2 DEFICIENCIES OF THE STANDARD MODEL

Just because it is so comprehensive and successful, we should judge the Standard Model by demanding criteria. It is clearly an important part of the Truth; the interesting question becomes: How big a part? Critical scrutiny reveals several important shortcomings of the Standard Model:

• The Standard Model contains scattered multiplets with peculiar hypercharge assignments.

While little doubt can remain that the Standard Model is essentially correct, a glance at Fig. 14.1 is enough to reveal that it is not a complete or final theory. The fermions fall apart into five lopsided pieces with peculiar hypercharge assignments; this pattern needs to be explained. Also the separate gauge theories, which are quite mathematically and conceptually similar, are fairly begging to be unified.

• The Standard Model supports the possibility of strong P and T violation [6].

There is a near-perfect match between the necessary 'accidental' symmetries of the Standard Model, dictated by its basic principles as enumerated above, and the observed symmetries of the world. The glaring exception is that there is an allowed—gauge invariant, renormalizable—interaction which, if present, would induce significant violation of the discrete symmetries P and T in the strong interaction. This is the notorious θ term. θ is an angle which *a priori* one might expect to be of order unity, but in fact is constrained by experimental limits on the neutron electric dipole moment to be $\theta \lesssim 10^{-8}$. This problem can be addressed by postulating a sort of quasi-symmetry (Peccei-Quinn [7] symmetry), which roughly speaking corresponds to promoting θ to a dynamical variable—a quantum field. The quanta of this field [8], *axions*, provide an interesting dark matter candidate [9]. Other possibilities for explaining the absence of strong P and T violation have been proposed, but they require towers of hypotheses which seem to me quite fragile.

In no way, of course, should the absence of strong P and T violation be taken as evidence against QCD itself. For practical purposes, one can simply take θ as a parameter to be fixed experimentally. One finds it to be very small—and is done with it!

• The Standard Model does not address family problems.

There are several distinct 'family problems,' ranging from the extremely qualitative (digital—why *three* families?) to the semi-qualitative (some distinctive patterns—why does like couple to like, with small mixing angles?) to the straightforward but most challenging goal of doing full justice to experience by computing (analog) experimental numbers with controlled, small fractional errors:

Why are there three repeat families? Rabi's famous question regarding the muon—"Who ordered *that*?"—still has no convincing answer.

How does one explain the very small electron mass? The dimensionless coupling associated with the electron mass, that is its strength of Yukawa coupling to the Higgs field, is about $g_e \sim 2 \times 10^{-6}$. It is almost as small as the limits on the θ angle (suggesting, perhaps, that strong P and T violation is just around the corner?). This question can of course be generalized—all the fermion masses, with one exception, are sufficiently small to beg qualitative explanation. The exception, of course, is the t quark. It is a fascinating and important possibility that roughly the observed value of t quark mass at low energies might result after running from a wide range of fundamental couplings at a high scale [10]. If so, one would have a satisfactory qualitative explanation of the value of this parameter.

Why do the weak currents couple approximately in the order of masses? That is, light with light, heavy with heavy, and intermediate with intermediate. Why are the mixings what they are—small, but not miniscule? The same, for the CP violating phase in the weak currents (parameterized invariantly by Jarlskog's J) [11]? —and, by the way, are we sure that $\theta \ll J$? And so on ...

• The Standard Model does not allow non-vanishing neutrino masses.

This is the only entry on my list that has a primarily *experimental* motivation. At present there are three quite different experimental hints for non-vanishing neutrino masses: the solar neutrino problem [12], the atmospheric neutrino problem [13], and the Los Alamos oscillation experiment [14]. The Standard Model in its conventional form does not allow non-zero neutrino masses. However I would like to mention that only a very minimal extension of the theory is necessary to accommodate such masses. One can add a complete $SU(3) \times SU(2) \times U(1)$ singlet fermion N_R to the model. N_R can be given, consistent with all symmetries and with the requirement of renormalizability, a Majorana mass M. Note especially that such a mass does not violate $SU(2) \times U(1)$. Likewise, N_R can have a Yukawa coupling to the ordinary lepton doublets through the Higgs field. Then condensation of the Higgs field activates the "see-saw" mechanism [15] to give a small mass for the observed neutrinos; with $M \lesssim 10^{15}$ Gev a range of experimentally and perhaps cosmologically interesting neutrino masses can be accommodated.

• Gravity is not included in the Standard Model.

This really represents (at least) two logically separate problems.

First there is the ultraviolet problem, the notorious non-renormalizability of quantum gravity. This provides an appropriate context, in which to introduce the modern perspective toward the whole concept of renormalizability.

Suppose that one were to be naïve, and simply add the Einstein Lagrangian for general relativity to the Standard Model, of course coupling the matter fields appropriately (minimally). Following Feynman and many others, one could then derive, formally, the perturbation series for any physical process. One would find, however, that the integrals over closed loops generally diverge at the high-energy (ultraviolet) end. Indeed the graviton coupling has, in units where $\hbar = c = 1$, dimensions of M_{Planck}^{-1}.[2] Here $M_{\text{Planck}} \approx 10^{19}$ Gev is a measure of the stiffness of space-time. Thus higher and higher order terms will, on dimensional grounds, introduce higher and higher factors of the ultraviolet cutoff to compensate. However if we determine (by notional experiments) the couplings at a given scale $\Lambda << M_{\text{Planck}}$ and calculate corrections by only including energy-momenta between the scale $P < \Lambda$ of interest and Λ, the successive terms in perturbation theory will be accompanied by positive powers of $\frac{\Lambda}{M_{\text{Planck}}}$ and will be small. Thus we can, for example, consistently set all non-minimal couplings to zero at any chosen energy-momentum scale well below the Planck scale. They will then be negligibly small for all practically accessible scales. For different choices of the scale they will be different, but since they are negligible in any case that hardly matters. This procedure is, in practice, the one we always adopt—and the Standard Model peacefully coexists with gravity, so long as we refuse to consider $P \gtrsim M_{\text{Planck}}$.

However, from this perspective a second problem looms larger than ever. The energy (and negative pressure) density of matter-free space, the notorious cosmological term, occurs as the coefficient λ of the identity term in the action: $\delta\mathcal{L} = \lambda \int \sqrt{g}$. It has dimensions of (mass)^4, and on phenomenological grounds

[2] This occurs because the kinetic energy for the graviton arises from the Einstein action $\propto M_{\text{Planck}}^2 \int \sqrt{g} R$ so that in expanding about flat space one must take $g_{\mu\nu} = \eta_{\mu\nu} + \frac{1}{M_{\text{Planck}}} h_{\mu\nu}$ in order to obtain a properly normalized quadratic kinetic term for h.

we must suppose $\lambda \lesssim (10^{-12}\ \text{Gev})^4$. The question is: where does such a tiny scale come from? What is so special about the present state of the universe, that the value of the effective λ for it, which one might naïvely expect to reflect contributions from much higher scales, is so effectively zeroed?

There may also be problems with forming a fully consistent quantum theory even of low-energy processes involving black holes [16].

Finally I will mention a question that I think has a rather different status from the foregoing, although many of my colleagues would put it on the same list, and maybe near the top:

The Standard Model begs the question "*Why* does $SU(2) \times U(1)$ symmetry break?"

To me, this is an example of the sort of metaphysical question that could easily fail to have a meaningful answer. There is absolutely nothing wrong, logically, with the classic implementation of the Higgs sector as described earlier. Nevertheless, one might well hunger for a wider context in which to view the existence of the Higgs doublet and its negative (mass)^2—or a suitable alternative.

14.3 UNIFICATION: SYMMETRY

Each of the deficiencies of the Standard Model mentioned in the previous section has provoked an enormous literature, literally hundreds or thousands of papers. Obviously I cannot begin to do justice to all this work. Here I shall concentrate on the first question, that of deciphering the message of the scattered multiplets and peculiar hypercharges. Among our questions, this one has so far inspired the most concrete and compelling response—a story whose implications range far beyond the question which inspired it.

Given that the strong interactions are governed by transformations among three color charges—say RWB for red, white, and blue—while the weak interactions are governed by transformations between two others—say GP for green and purple—what could be more natural than to embed both theories into a larger theory of transformations among all five colors? This idea has the additional attraction that an extra U(1) symmetry commuting with the strong SU(3) and weak SU(2) symmetries automatically appears, which we can attempt to identify with the remaining gauge symmetry of the Standard Model, that is hypercharge. For while in the separate SU(3) and SU(2) theories we must throw out the two gauge bosons which couple respectively to the color combinations R+W+B and G+P, in the SU(5) theory we only project out R+W+B+G+P, while the orthogonal combination (R+W+B)-$\frac{3}{2}$(G+P) remains.

Georgi and Glashow [17] originated this line of thought, and showed how it could be used to bring some order to the quark and lepton sector, and in particular to supply a satisfying explanation of the weird hypercharge assignments in the Standard Model. As shown in Fig. 14.4, the five scattered SU(3)×SU(2)×U(1)

SU(5): 5 colors RWBGP

10: 2 different color labels (antisymmetric tensor)

$$\begin{array}{lccc} u_L: & RP, & WP, & BP \\ d_L: & RG, & WG, & BG \\ u^c_L: & RW, & WB, & BR \\ & (\bar{B}) & (\bar{R}) & (\bar{W}) \\ e^c_L: & GP & & \\ & () & & \end{array} \qquad \begin{pmatrix} 0 & u^c & u^c & u & d \\ & 0 & u^c & u & d \\ & & 0 & u & d \\ & * & & 0 & e \\ & & & & 0 \end{pmatrix}$$

5̲̄: 1 anticolor label

$$\begin{array}{lccc} d^c_L: & \bar{R}, & \bar{W}, & \bar{B} \\ e_L: & \bar{P} & & \\ \nu_L: & \bar{G} & & \end{array} \qquad (d^c \quad d^c \quad d^c \quad e \quad \nu)$$

$$\boxed{Y = -\tfrac{1}{3}\,(R+W+B) + \tfrac{1}{2}\,(G+P)}$$

Figure 14.4: Organization of the fermions in one family in $SU(5)$ multiplets. Only two multiplets are required. In passing from this form of displaying the gauge quantum numbers to the form familiar in the Standard Model, one uses the bleaching rules R+W+B = 0 and G+P = 0 for $SU(3)$ and $SU(2)$ color charges (in antisymmetric combinations). Hypercharge quantum numbers are identified using the formula in the box, which reflects that within the larger structure $SU(5)$ one only has the combined bleaching rule R+W+B+G+P = 0. The economy of this figure, compared to Fig 14.1, is evident.

multiplets get organized into just two representations of $SU(5)$. It is an extremely non-trivial fact that the known fermions fit so smoothly into $SU(5)$ multiplets.

In all the most promising unification schemes, what we ordinarily think of as matter and anti-matter appear on a common footing. Since the fundamental gauge transformations do not alter the chirality of fermions, in order to represent the most general transformation possibilities one should use fields of one chirality, say left, to represent the fermion degrees of freedom. To do this, for a given fermion, may require a charge conjugation operation. Also, of course, once we contemplate changing strong into weak colors it will be difficult to prevent quarks and leptons from appearing together in the same multiplets. Generically, then, one expects that in unified theories it will not be possible to make a global distinction between matter and anti-matter and that both baryon number B and lepton number L will be violated, as they definitely are in $SU(5)$ and its extensions.

As shown in Fig. 14.4, there is one group of ten left-handed fermions that have all possible combinations of one unit of each of two different colors, and another group of five left-handed fermions that each carry just one negative unit of some color. (These are the ten-dimensional antisymmetric tensor and the complex conjugate of the five-dimensional vector representation, commonly referred to as the "five-bar".) What is important for you to take away from this discussion is

SO(10): 5 bit register

$$(\pm\pm\pm\pm\pm)\ :\ \underline{\text{even}}\ \#\ \text{of}\ -$$

$$\begin{array}{rlcc} & (++-|+-) & 6 & (\mathrm{u_L}, \mathrm{d_L}) \\ 10: & (+--|++) & 3 & \mathrm{u_L^c} \\ & (+++|--) & 1 & \mathrm{e_L^c} \\ \bar{5}: & (+--|--) & \bar{3} & \mathrm{d_L^c} \\ & (---|+-) & \bar{2} & (\mathrm{e_L}, \nu_\mathrm{L}) \\ 1: & (+++|++) & 1 & \mathrm{N_R} \end{array}$$

Figure 14.5: Organization of the fermions in one family, together with a right-handed neutrino degree of freedom, into a single multiplet under $SO(10)$. The components of the irreducible spinor representation, which is used here, can be specified in a very attractive way by using the charges under the $SO(2) \otimes SO(2) \otimes SO(2) \otimes SO(2) \otimes SO(2)$ subgroup as labels. They then appear as arrays of $\pm$ signs, resembling binary registers. There is the rule that one must have an even number of - signs. Strong $SU(3)$ acts on the first three components, weak $SU(2)$ on the final two. The $SU(5)$ quantum numbers are displayed in the left-hand column, the number of entries with each sign-pattern just to the right, and finally the usual Standard Model designations on the far right.

not so much the precise details of the scheme, but the idea that *the structure of the Standard Model, with the particle assignments gleaned from decades of experimental effort and theoretical interpretation, is perfectly reproduced by a simple abstract set of rules for manipulating symmetrical symbols.* Thus, for example, the object RB in this figure has just the strong, electromagnetic, and weak interactions we expect of the complex conjugate of the right-handed up-quark, without our having to instruct the theory further. If you've never done it I heartily recommend to you the simple exercise of working out the hypercharges of the objects in Fig. 14.4 and checking against what you need in the Standard Model—after doing it, you'll find it's impossible ever to look at the standard model in quite the same way again.

Although it would be inappropriate to elaborate the necessary group theory here, I'll mention that there is a beautiful extension of $SU(5)$ to the slightly larger group $SO(10)$, which permits one to unite all the fermions of a family into a single multiplet [18]. In fact, the relevant representation for the fermions is a 16-dimensional spinor representation. Some of its features are depicted in Fig. 14.5. The 16th member of a family in $SO(10)$, beyond the 15 familiar degrees of freedom with a Standard Model family, has the quantum numbers of the right-handed neutrino N_R as mentioned above. This emphasizes again how easy and natural is the extension of the Standard Model to include neutrino masses using the see-saw mechanism.

14.4 UNIFICATION: DYNAMICS, AND A BIG HINT OF SUPERSYMMETRY [19]

14.4.1 THE CENTRAL RESULT

We have seen that simple unification schemes are successful at the level of *classification*; but new questions arise when we consider the dynamics which underlies them.

Part of the power of gauge symmetry is that it fully dictates the interactions of the gauge bosons, once an overall coupling constant is specified. Thus if SU(5) or some higher symmetry were exact, then the fundamental strengths of the different color-changing interactions would have to be equal, as would the (properly normalized) hypercharge coupling strength. In reality the coupling strengths of the gauge bosons in SU(3)×SU(2)×U(1) are observed not to be equal, but rather to follow the pattern $g_3 \gg g_2 > g_1$.

Fortunately, experience with QCD emphasizes that couplings "run." The physical mechanism of this effect is that in quantum field theory the vacuum must be regarded as a polarizable medium, since virtual particle-anti-particle pairs can screen charge. Thus one might expect that effective charges measured at shorter distances, or equivalently at larger energy-momentum or mass scales, could be different from what they appear at longer distances. If one had only screening, then the effective couplings would grow at shorter distances as one penetrates deeper inside the screening cloud. However it is a famous fact [3] that due to paramagnetic spin-spin attraction of like charge vector gluons [20], these particles tend to *antiscreen* color charge, thus giving rise to the opposite effect—asymptotic freedom—that the effective coupling tends to shrink at short distances. This effect is the basis of all perturbative QCD phenomenology, which is a vast and vastly successful enterprise, as we saw in Fig. 14.3.

For our present purpose of understanding the disparity of the observed couplings, it is just what the doctor ordered. As was first pointed out by Georgi, Quinn, and Weinberg [21], if a gauge symmetry such as SU(5) is spontaneously broken at some very short distance, then we should not expect that the effective couplings probed at much larger distances (such as are actually measured at practical accelerators) will be equal. Rather, they will all have been affected to a greater or lesser extent by vacuum screening and anti-screening, starting from a common value at the unification scale but then diverging from one another at accessible accelerator scales. The pattern $g_3 \gg g_2 > g_1$ is just what one should expect, since the antiscreening or asymptotic freedom effect is more pronounced for larger gauge groups, which have more types of virtual gluons.

The marvelous thing is that the running of the couplings gives us a truly quantitative handle on the ideas of unification, for the following reason. To fix the relevant aspects of unification, one basically needs only to fix two parameters: the scale at which the couplings unite, which is essentially the scale at which

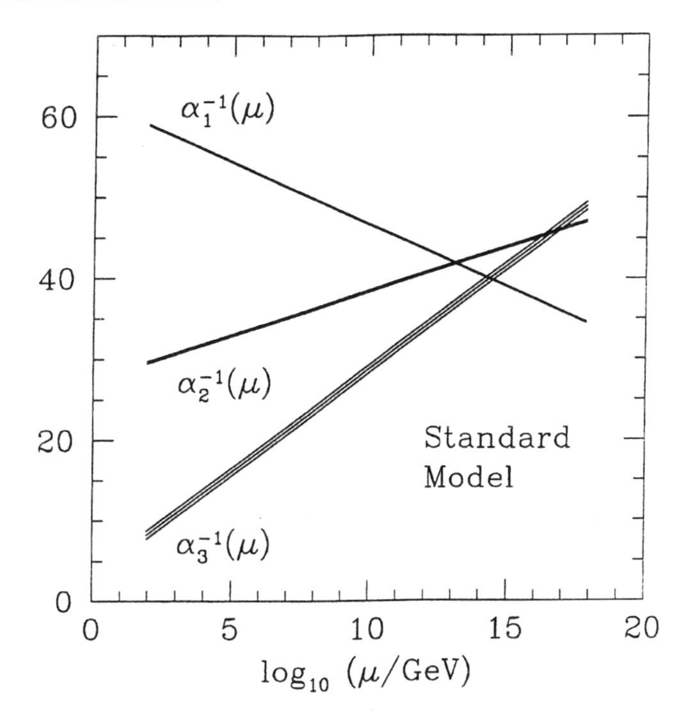

Figure 14.6: Evolution of Standard Model effective (inverse) couplings toward small space-time distances, or large energy-momentum scales. Notice that the physical behavior assumed for this figure is the direct continuation of Fig. 14.3, and has the same conceptual basis. The error bars on the experimental values at low energies are reflected in the thickness of the lines. Note the logarithmic scale. The qualitative aspect of these results is extremely encouraging for unification and for extrapolation of the principles of quantum field theory, but there is a definite small discrepancy with recent precision experiments.

the unified symmetry breaks; and their value when they unite. Given these, one calculates three outputs: the three *a priori* independent couplings for the gauge groups SU(3)×SU(2)×U(1) of the Standard Model. Thus the framework is eminently falsifiable. The miraculous thing is how close it comes to working (Fig. 14.6).

The unification of couplings occurs at a very large mass scale, $M_{\mathrm{un.}} \sim 10^{15}$ Gev. In the simplest version, this is the magnitude of the scalar field vacuum expectation value that spontaneously breaks SU(5) down to the standard model symmetry SU(3)×SU(2)×U(1), and is analogous to the scale $v \approx 250$ Gev for electroweak symmetry breaking. The largeness of this large scale mass scale is important in several ways:

- It explains why the exchange of gauge bosons that are in SU(5) but not in

SU(3)×SU(2)×U(1), which re-shuffles strong into weak colors and generically violates the conservation of baryon number, does not lead to a catastrophically quick decay of nucleons. The rate of decay goes as the inverse fourth power of the mass of the exchanged gauge particle, so the baryon-number violating processes are predicted to be far slower than ordinary weak processes, as they had better be.

• $M_{\text{un.}}$ is significantly smaller than the Planck scale $M_{\text{Planck}} \sim 10^{19}$ Gev at which exchange of gravitons competes quantitatively with the other interactions, but not ridiculously so. This indicates that while the unification of couplings calculation itself is probably safe from gravitational corrections, the unavoidable logical next step in unification must be to bring gravity into the mix.

• Finally one must ask how the tiny ratio of symmetry-breaking mass scales $v/M_{\text{un.}} \sim 10^{-13}$ required arises dynamically, and whether it is stable. This is the so-called gauge hierarchy problem, which I shall discuss in a more concrete form momentarily.

The success of the GQW calculation in explaining the observed hierarchy $g_3 \gg g_2 > g_1$ of couplings and the approximate stability of the proton is quite striking. In performing it, we assumed that the known and confidently expected particles of the Standard Model exhaust the spectrum up to the unification scale, and that the rules of quantum field theory could be extrapolated without alteration up to this mass scale—thirteen orders of magnitude beyond the domain they were designed to describe. It is a triumph for minimalism, both existential and conceptual.

However, on further examination it is not quite good enough. Accurate modern measurements of the couplings show a small but definite discrepancy between the couplings, as appears in Fig. 14.6. And heroic, dedicated experiments to search for proton decay did not find it [22]; they currently exclude the minimal SU(5) prediction $\tau_p \sim 10^{31}$ yrs. by about two orders of magnitude.

Given the scope of the extrapolation involved, perhaps we should not have hoped for more. There are several perfectly plausible bits of physics that could upset the calculation, such as the existence of particles with masses much higher than the electroweak but much smaller than the unification scale. As virtual particles these would affect the running of the couplings, and yet one certainly cannot exclude their existence on direct experimental grounds. If we just add particles in some haphazard way things will only get worse: minimal SU(5) nearly works, so the generic perturbation from it will be deleterious. This is a major difficulty for so-called technicolor models, which postulate many new particles in complicated patterns. Even if some *ad hoc* prescription could be made to work, that would be a disappointing outcome from what appeared to be one of our most precious, elegantly straightforward clues regarding physics well beyond the Standard Model.

Fortunately, there is a theoretical idea which is attractive in many other ways, and seems to point a way out from this impasse. That is the idea of supersymmetry [23]. Supersymmetry is a symmetry that extends the Poincare symmetry of special relativity (there is also a general relativistic version). In a supersymmetric theory one has not only transformations among particle states with different energy-momentum but also between particle states of different *spin*. Thus spin 0 particles can be put in multiplets together with spin $\frac{1}{2}$ particles, or spin $\frac{1}{2}$ with spin 1, and so forth.

Supersymmetry is certainly not a symmetry in nature: for example, there is certainly no bosonic particle with the mass and charge of the electron. More generally if one defines the R-parity quantum number

$$R \equiv (-)^{3B+L+2S} ,$$

which should be accurate to the extent that baryon and lepton number are conserved, then one finds that all currently known particles are R even whereas their supersymmetric partners would be R odd. Nevertheless, there are many reasons to be interested in supersymmetry, and especially in the hypothesis that supersymmetry is effectively broken at a relatively low scale, say ≈ 1 TeV. Anticipating this for the moment, let us consider the consequences for running of the couplings.

The effect of low-energy supersymmetry on the running of the couplings was first considered long ago [24], well before the discrepancy described above was evident experimentally. One might have feared that such a huge expansion of the theory, which essentially doubles the spectrum, would utterly destroy the approximate success of the minimal SU(5) calculation. This is not true, however. To a first approximation, roughly speaking because it is a space-time as opposed to an internal symmetry, supersymmetry does not affect the group-theoretic structure of the unification of couplings calculation. The absolute rate at which the couplings run with momentum is affected, but not the relative rates. The main effect is that the supersymmetric partners of the color gluons, the gluinos, weaken the asymptotic freedom of the strong interaction. Thus they tend to make its effective coupling decrease and approach the others more slowly. Thus their merger requires a longer lever arm, and the scale at which the couplings meet increases by an order of magnitude or so, to about 10^{16} GeV. Also the common value of the effective couplings at unification is slightly larger than in conventional unification ($\frac{g^2_{\text{un.}}}{4\pi} \approx \frac{1}{25}$ *versus* $\frac{1}{40}$). This increase in unification scale significantly reduces the predicted rate for proton decay through exchange of the dangerous color-changing gauge bosons, so that it no longer conflicts with existing experimental limits.

Upon more careful examination there is another effect of low-energy supersymmetry on the running of the couplings, which although quantitatively small has become of prime interest. There is an important exception to the general rule that adding supersymmetric partners does not immediately (at the one loop level) affect the relative rates at which the couplings run. This rule works for particles that come in complete SU(5) multiplets, such as the quarks and leptons (which, since they do not upset the full SU(5) symmetry, have basically no effect) or for the supersymmetric partners of the gauge bosons, because they just renormalize the existing, dominant effect of the gauge bosons themselves. However, there is one peculiar additional contribution, from the supersymmetric partner of the Higgs doublet. It affects only the weak SU(2) and hypercharge U(1) couplings. (On phenomenological grounds the SU(5) color triplet partner of the Higgs doublet must be extremely massive, so its virtual exchange is not important below the unification scale. *Why* that should be so, is another aspect of the hierarchy problem.) Moreover, for slightly technical reasons even in the minimal supersymmetric model it is necessary to have two different Higgs doublets with opposite hypercharges.[3] The main effect of doubling the number of Higgs fields and including their supersymmetric partners is a sixfold enhancement of the asymmetric Higgs field contribution to the running of weak and hypercharge couplings. This causes a small, accurately calculable change in the calculation. From Fig. 14.7 you see that it is a most welcome one. Indeed, in the minimal implementation of supersymmetric unification, it puts the running of couplings calculation right back on the money [25].

Since the running of the couplings with scales depends only logarithmically on the mass scale, the unification of couplings calculation is not terribly sensitive to the precise scale at which supersymmetry is broken, say between 100 GeV and 10 TeV. (To avoid confusion later, note that here by "the scale at which supersymmetry is broken" I mean the typical mass splitting between Standard Model particles and their supersymmetric partners. The phrase is frequently used in a different sense, referring to the largest splitting between supersymmetric partners in the entire world-spectrum; this could be much larger, and indeed in popular models it almost invariably is. The ambiguous terminology is endemic in the literature; fortunately, the meaning is usually clear from the context.) There have been attempts to push the calculation further, in order to address this question of the supersymmetry breaking scale, but they are controversial. For example, comparable uncertainties arise from the splittings among the very large number of particles with masses of order the unification scale, whose theory is poorly developed and unreliable. Superstring theory suggests [26] many possible ways in which the simple calculation described here might go wrong;[4] if we take the favorable result of this calculation at face value, we must conclude that none of them happen.

In any case, if we are not too greedy the main points still shine through:

- If supersymmetry is to fulfill its destiny of elucidating the hierarchy problem in

[3]Perhaps the simplest, though not the most profound, way to appreciate the reason for this has to do with anomaly cancelation. The minimal spin-1/2 supersymmetric partner of the Higgs doublet is chiral and has non-vanishing hypercharge, introducing an anomaly. By including a partner for the anti-doublet, one cancels this anomaly.

[4]Indeed, in the simplest superstring-inspired models it is not entirely easy to accommodate the 'low' value of the unification scale compared to the Planck scale.

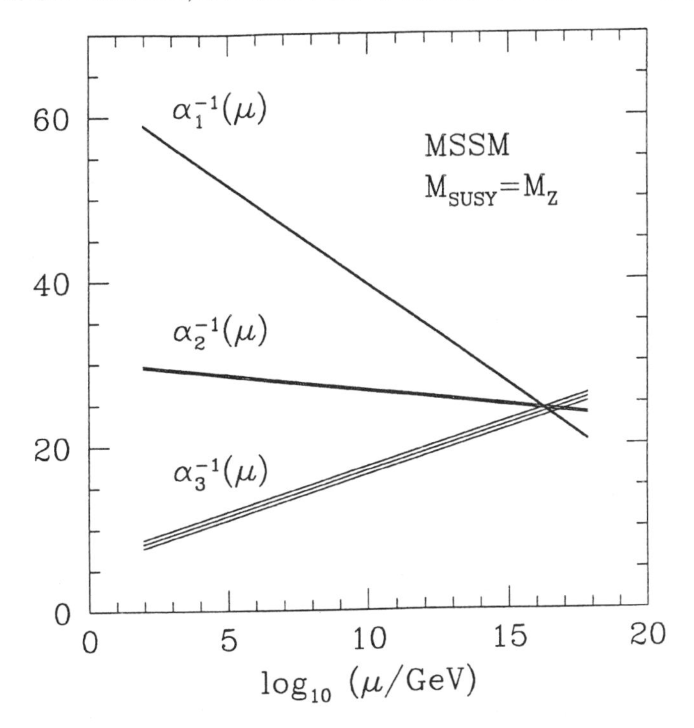

Figure 14.7: Evolution of the effective (inverse) couplings in the minimal extension of the Standard Model, to include supersymmetry. The concepts and qualitative behaviors are only slightly modified from Fig. 14.6 (a highly non-trivial fact!) but the quantitative result is changed, and comes into adequate agreement with experiment. I would like to emphasize that results along these lines were published well before the difference between Fig. 14.6 and Fig. 14.7 could be resolved experimentally, and in that sense one has already derived a successful *prediction* from supersymmetry.

a straightforward way, then the supersymmetric partners of the known particles cannot be much heavier than the SU(2)×U(1) electroweak breaking scale, i.e., they should not be beyond the expected reach of LHC.

• If we assume this to be the case then the meeting of the couplings takes place in the simplest minimal models of unification, to adequate accuracy, without further assumption. This is a most remarkable and non-trivial fact.

14.4.2 IMPLICATIONS

The preceding result, taken at face value, has extremely profound implications:

• Quantum field theory, and specifically its characteristic vacuum polarization effects leading to asymptotic freedom and running of the couplings, continue to work quantitatively up to energy scales many orders of magnitude beyond where they were discovered and established.

I would like to emphasize also some negative implications of this: there are things that might have, but do *not*, happen. It might have happened that the known particles are some complicated composites of more elementary objects, or that many additional strong couplings appeared at higher energies (technicolor), or that additional dimensions became dynamically active, or that particle physics simply dissolved into some amorphous mess. Unless Fig. 14.7 is a cruel joke on the part of Mother Nature, none of this happens, or at least the complications are in a strong, precise sense walled off from the Standard Model and the dynamical evolution of its couplings.

• Supersymmetry, in its virtual form, has already been discovered.

14.4.3 WHY SUPERSYMMETRY IS A GOOD THING

Thus has Nature spoken, in a promissory whisper. Many of us are seduced, because She is telling us something we want to hear:

• You will notice that we have made progress in uniting the gauge bosons with each other, and the various quarks and leptons with each other, but not the gauge bosons with the quarks and leptons. It takes supersymmetry—perhaps spontaneously broken—to make this feasible.

• Supersymmetry was invented in the context of string theory, and seems to be necessary for constructing consistent string theories containing gravity (critical string theories) that are at all realistic.

• Most important for present purposes, supersymmetry can help us to understand the vast disparity between weak and unified symmetry breaking scales mentioned above. This disparity is known as the gauge hierarchy problem. It actually raises several distinct problems, including the following. In calculating radiative corrections to the $(\text{mass})^2$ of the Higgs particle from diagrams of the type shown in Fig. 14.8 one finds an infinite, and also large, contribution. By this I mean that the divergence is quadratic in the ultraviolet cutoff. No ordinary symmetry will make its coefficient vanish. If we imagine that the unification scale provides the cutoff, we find that the radiative correction to the $(\text{mass})^2$ is much larger than the final value we want. (If the Higgs field were composite, with a soft form factor, this problem might be ameliorated. Following that road leads to technicolor, which as mentioned before seems to lead us far away from our best source of inspiration.) As a formal matter, one can simply cancel the radiative correction against a large bare contribution of the opposite sign, but in the absence of some deeper motivating principle this seems to be a horribly ugly procedure. Now in a supersymmetric theory for any set of virtual particles circulating in the loop, there will also be another graph with their supersymmetric partners circulating. If the partners were accurately degenerate, the contributions would cancel. Otherwise, the threatened quadratic divergence will be cut off only at virtual momenta such that the difference in $(\text{mass})^2$ between the virtual particle and its supersymmetric partner is relatively negligible. Thus we will be assured adequate cancelation if and only if supersymmetric partners are not too far split in mass—in the present context, if the splitting is not much greater than the weak scale. This is (a crude version of) the most important *quantitative* argument which suggests the relevance of "low-energy" supersymmetry.

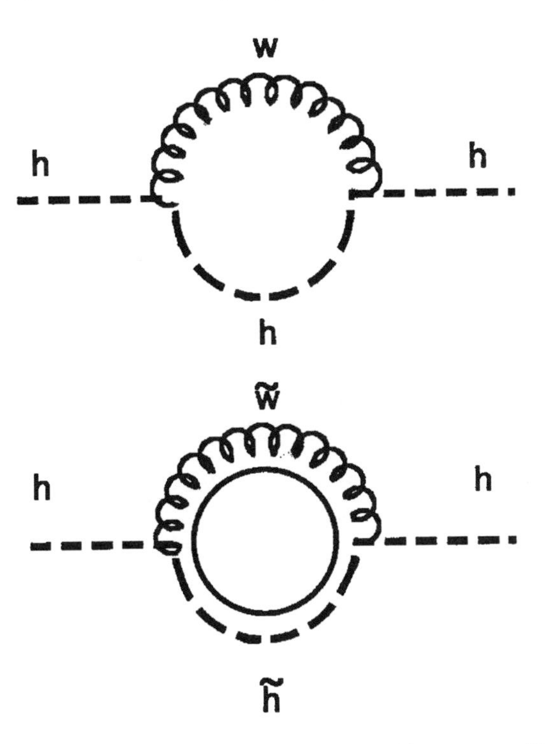

Figure 14.8: Contributions to the Higgs field self-energy. These graphs give contributions to the Higgs field self-energy which separately are formally quadratically divergent, but when both are included the divergence is removed. In models with broken supersymmetry a finite residual piece remains. If one is to obtain an adequately small finite contribution to the self-energy, the mass difference between Standard Model particles and their superpartners cannot be too great. This—and essentially only this—motivates the inclusion of virtual superpartner contributions in Fig. 14.7 beginning at relatively low scales.

• Supersymmetric field theories have many special features, which make them especially interesting, and perhaps promising, phenomenologically.

I cannot be very specific about this here, both because there are as yet no canonical models and because the subject is excessively technical, but let me just mention some appropriate concepts: radiative $SU(2) \times U(1)$ breaking associated with the heavy top quark; doublet-triplet splitting mechanisms; approximate flat directions for generating large mass hierarchies. Supersymmetric models also have additional mechanisms for neutral flavor-changing processes and CP violation, which are dangerously large generically, but in appropriate models can be suppressed down to a level which is interesting—but not *too* interesting—experimentally.

All this provides, in my opinion, a very good specific brief for optimism about the future of experimental particle physics exploring the high-energy frontier, and also—with somewhat less certainty—the frontier of small exotic flavor-changing and CP violating processes. We can already discern, at the limit of our vision, the shores of a strange new world not too far away, where we can realistically hope to land and explore.

14.5 THE FARTHER FUTURE: CONNECTIONS [27]

Up to this point I have discussed the near-to-medium future of particle physics from what might be called a traditional or internal perspective. According to that view, the field is defined as the search for the fundamental laws of nature, in a reductionist sense: laws that cannot be be explained in terms of anything simpler, the end-answers to repeated queries 'Why?' [28]. (I have also foregone any substantial discussion of string theory, partly because my colleague Edward Witten will be discussing it here, but also partly to emphasize, in the spirit of scientific conservatism—*hypothesis non fingo*—how far we can get without using it. Indeed, it remains a great challenge to develop string theory to the point that it becomes a functional part of natural science, in the sense of yielding characteristic, specific insight into concrete physical phenomena.)

I think it is vitally important, in doing justice to the significance of this reductionist activity, also to discuss its external connections—specifically, its broader role in extending our understanding and appreciation of physical phenomena, whether or not these involve clean application of the "fundamentals." That could easily turn into a long vague discourse, but I will try to be brief and usefully specific, realizing that this involves considerable risk of touting the wrong horses. Also, in line with my title I will not discuss connections to mathematics, philosophy, theology, science fiction [29], ...

14.5.1 MATTER

The exchange of ideas between particle and condensed matter physics has a long and glorious history [30]. It is almost uncanny how almost every one of the basic conceptual ingredients of the Standard Model is mirrored in some facet of condensed matter theory. Soon after Einstein inferred the field-particle connection for photons, he applied it to solids, introducing the phonon concept. The band theory of metals, and specifically the hole concept, was developed in parallel with Dirac's theory of electrons and positrons. In more recent times, the distillation of ideas about spontaneous symmetry breaking provoked by BCS theory ramified both into the effective theory of the strong interaction and the foundations of electroweak theory; and the concepts of running couplings and asymptotic scale invariance, which became prominent in the theory of second-order phase transitions, were crucial in elucidating the modern microscopic theory of the strong interaction.

There is a profound underlying reason why theoretical concepts developed for understanding physical phenomena on vastly different energy and distance scales, and separated by several layers of 'reduction,' find dual usage. It is because in each domain the same principles of *symmetry* and *locality* are basic. Since these basic principles will, I believe, continue to guide us, there is every prospect that a fruitful exchange of ideas will continue.

More specifically, in recent years investigations stimulated by the discovery of the quantum Hall effect have uncovered an amazing wealth of structures. One has learned to use gauge theories of a highly non-trivial kind to characterize the various states, now including even nonabelian theories (which have many strange aspects); to exploit conformal field theories in analyzing the behavior at boundaries; to predict and recognize baby skyrmions both theoretically and experimentally; to realize new forms of confinement. Unusual realizations of symmetry, holomorphic functions, and non-commuting spatial variables are quite prominent in the theory. It would not be appropriate to discuss any of these topics in detail now, but I would like briefly to mention the primitive observation that in many ways opens the subject—a subject that I commend to the attention of all high-energy theorists. In a strong magnetic field the particle Lagrangian naïvely simplifies as

$$\mathcal{L} = \frac{m}{2}(\dot{x}^2 + \dot{y}^2) + B\dot{x}y \to B\dot{x}y \, .$$

This limiting Lagrangian is rather peculiar: it leads to a vanishing Hamiltonian, and identifies By as the canonical momentum conjugate to x! Thus the spatial coordinates no longer commute; also, the original rotational symmetry between x and y has become a true canonical (not point) transformation. We are, of course, just describing the rather trivial quantum mechanics of the lowest Landau level in an exotic fashion; but it is striking how easily and naturally unusual realizations of symmetry arise here. It is also intriguing to contemplate what has happened from the opposite perspective: what was a non-commuting momentum variable, viewed from within the lowest Landau level, has been promoted to a commuting, and manifestly symmetrical, variable in the overlying theory.

Another, more down-to-earth source of connections between high energy and condensed matter physics is that high energy physics is supposed, after all, to describe actual matter under extreme conditions. We should therefore address the obvious, qualitative physical questions about this matter: what is it like in bulk, and does it undergo interesting phase transitions as a function of density and/or temperature? Several fascinating possibilities for hadronic matter have been suggested: a quark-gluon plasma, with restoration of chiral symmetry at high temperature, seems a safe bet; pion or especially kaon condensation, and strange matter, are possibilities. Some of these possibilities will be probed by projected relativistic heavy-ion collision experiments. Closer to home, it is quite disappointing that there is still no convincing first-principles explanation of such basic phenomena as the existence of a hard core and the saturation of ordinary nuclear forces. These questions provide very worthy challenges for future theory, and in my opinion are receiving too little attention. Serious attempts to address them will require new methods, probably with a significant numerical component, which (if found) would very likely have implications for many-body problems more generally. As the power of computers increases, our inability to calculate is ever more embarrassing.

Here is a specific question of a different, but related, sort: the classic Lanford-Dyson-Lieb discussion of 'stability of matter' breaks down for bosons. What does this mean for supersymmetric matter in bulk? What does it mean for the ground state of ordinary matter, if the world is approximately supersymmetric?

14.5.2 COSMOS

It is the earliest moments of the Big Bang, when extraordinary energies and densities were achieved, which provide the obvious arena for future high energy physics in the physical world. Opportunities and challenges are readily identified:

• It makes good sense to extrapolate toward the *very* early Universe.

As Fig. 14.3 shows us in hard data, the strong coupling runs toward a small effective value at high energy. Fig. 14.6, and especially Fig. 14.7, emphasize why the seductive assumption that in crucial respects particle physics remains simple and weakly coupled up to extraordinarily high energies is very hard to resist, because it leads to a strikingly successful account of the unification of couplings. Thus fundamental particle physics becomes profoundly simpler (though superficially more complex) toward the earliest moments of the Big Bang.

The scale for the running of inverse couplings, once they are large, is logarithmic. This feature naturally connects mass scales identifiable in particle physics experiments with exponentially (in the inverse couplings) larger scales. Numerically, as we have seen, one is lead in this way close to the Planck scale. Thus there is direct, though of course extremely limited, evidence that quantum field theory at weak coupling governs the interactions of the known particles up to energies (temperatures, densities) nearly as high as one could reasonably hope.

• Cosmic phase transitions happened.

Since QCD and asymptotic freedom are firmly established, we can say with complete confidence that the effective low-energy degrees of freedom in the strong interaction—hadrons with confined color and spontaneously broken chiral symmetry—are quite different from the fundamental colored quark and gluon degrees of freedom which manifest themselves at high temperature. The transition between them must be accompanied by a rapid crossover and perhaps by a phase transition. Indeed, the existence of a phase transition can be rigorously established in models closely related to real-world QCD, such as the variant of QCD where the $m_u = m_d = 0$ or the pure-glue theory. The nature of the transition, perhaps surprisingly, seems to depend sensitively upon the spectrum of light quarks.

The behavior of QCD at high temperatures is in principle, and to some extent in practice, calculable. The QCD crossover or transition that occurred during the Big Bang is even in some rough sense *reproducible*, and will be approximated in future relativistic heavy ion collisions.

Similarly, since the electroweak $SU(2) \times U(1) \rightarrow U(1)$ breaking pattern is firmly established, there is an excellent chance that another phase transition, associated with $SU(2) \times U(1)$ restoration at high temperature, occurs. It is not guaranteed that there is a strict phase transition, since there is no gauge-invariant order parameter for the Higgs phase, though at weak coupling one certainly expects a sharp transition. One of the most interesting projects for future accelerators, which has important implications for cosmology, will be to map out the relevant parameters so that we can characterize this crossover or transition.

These 'established' examples encourage one to speculate about the possibility of phase transitions associated with unification, or perhaps occurring in some hidden sector.

Cosmological phase transitions have many possible consequences for the history of the universe and for observational cosmology, including:

Defects of all sorts, including textures, strings, and monopoles that could persist even to the present day. Truly stable domain walls must be avoided or inflated away, but appropriately long-lived ones could be important in the prior evolution of the universe.

Inflation, if one gets trapped in a metastable condition by a barrier or by weakness of the relevant coupling that drives the transition.

Gravity waves, if the scale of the transition is high or if the transition is sufficiently violent.

Baryogenesis, under the same qualitative conditions.

It will be fascinating to discover which, if any, of these possibilities is realized by the electroweak transition. There is also a great opportunity, if one can establish compelling, sufficiently detailed models for unification, or a hidden sector, or any number of other possibilities, to examine their cosmological implications.

• We have specific, credible dark matter candidates, notably the lightest R-odd particle (LSP).

I have already discussed this.

• We have significant motivation for the possibility of additional very light fields.

I would like to conclude this discussion of 'applications' by alluding to a circle of ideas which, though it can be made to sound quite fantastic, is I think, actually deeply implicit in much current thinking about particle physics and cosmology.

The axion is perhaps the most well-motivated and studied exemplar of a family of related fields including familons, dilatons, and moduli fields that in one way or another embody the idea that what we ordinarily consider 'constants' might not be fundamental parameters fixed in the very formulation of the laws of nature, but rather can be considered usefully as dynamical entities. In a theory with only one fundamental, dimensional parameter, such as superstring theory appears to be, there is clearly a sense in which all dimensionless couplings are dynamical variables. They might, nevertheless, behave effectively as constants either if it costs a very large energy-density to excite them at all (e.g., massive fields) or if ordinary matter couples very weakly to long-wavelength fluctuations of the field (e.g., stiff light fields with derivative couplings).

In the latter case one might anticipate long-range forces mediated by the exchange of the field. I believe that experiments to look for such forces are among the most fundamental which can presently be attempted. They address in a concrete way the question: are the constants of nature uniquely determined to be what we observe by dynamical laws, or are they 'frozen accidents' imprinted at the Big Bang? Or are they presently relaxing towards some more favorable value?[5] Note that although a rapidly (on cosmological time scales) oscillating field represents non-relativistic matter, a sufficiently slowly varying field might appear as a contribution to the effective cosmological term.

If these ideas are along the right lines, it could be misguided to seek a unique Lorentz-invariant 'vacuum' state as a model for our present world. One might instead be required, at a fundamental level, to seek the boundary conditions for particle physics from cosmology (and *vice versa*).

At this point, our discussion of *applications* of particle physics, as traditionally understood, has modulated into a discussion of its possible ultimate *limita-*

[5]The standard picture of axion production in the Big Bang is essentially an example of this: the axion field starts drops out of equilibrium as the universe cools from 10^{12} to about 1 GeV, then relaxes towards the dynamically favored, approximately P and T conserving, value.

tions. I hope I have convinced you that there is abundant fertile territory we can anticipate exploring before we arrive at these limits; as my mother might say, "Such problems you should have."

ACKNOWLEDGMENTS

I thank Keith Dienes for supplying Figs. 6 and 7, and especially Chris Kolda for valuable assistance in the preparation of this manuscript.

REFERENCES

[1] For recent reviews of the Standard Model, see References 4 and 5.

[2] After several partial and tentative proposals, the $SU(2) \times U(1)$ electroweak theory took on in its essentially modern form in: S. Weinberg, Phys. Rev. Lett. **19** 1264 (1967); A. Salam, in *Elementary Particle Physics*, edited by N. Svartholm (Almqvist and Wiksells, Stockholm, 1968), p. 367; S. Glashow, J. Iliopoulos, and L. Maiani, Phys. Rev. D **2** 1285 (1970).

[3] After several partial and tentative proposals, the $SU(3)$ strong interaction theory took on its essentially modern form in: D. Gross and F. Wilczek, Phys. Rev. D **8** 3633 (1973); S. Weinberg, Phys. Rev. Lett. **31** 494 (1973); H. Fritzsch, M. Gell-Mann, and H. Leutwyler, Phys. Lett. **47B** 365 (1973). Not coincidentally, the key discovery that allowed one to connect the abstract gauge theory to experiments, asymptotic freedom, was first demonstrated just prior to these papers, in: D. Gross and F. Wilczek, Phys. Rev. Lett. **30** 1343 (1973); H.D. Politzer, Phys. Rev. Lett. **30** 1346 (1973).

[4] LEP Electroweak Working Group, preprint CERN-PPE/96-183 (Dec. 1996).

[5] M. Schmelling, in *Proceedings of the 28th International Conference on High-energy Physics (ICHEP 96), Warsaw, Poland, 1996*, preprint MPI-H-V39, hep-ex/9701002.

[6] G. 't Hooft, Phys. Rev. Lett. **37** 8 (1976); C. Callan, R. Dashen, and D. Gross, Phys. Lett. **63B** 334 (1976); R. Jackiw, C. Rebbi, Phys. Rev. Lett. **37** 172 (1976).

[7] R. Peccei and H. Quinn, Phys. Rev. Lett. **38** 1440 (1977), Phys. Rev. D **16** 1791 (1977).

[8] S. Weinberg, Phys. Rev. Lett. **40** 223 (1978); F. Wilczek, Phys. Rev. Lett. **40** 279 (1978).

[9] J. Preskill, M. Wise, and F. Wilczek, Phys. Lett. **120B** 127 (1983); L. Abbott and P. Sikivie, Phys. Lett. **120B** 133 (1983); M. Dine and W. Fischler, Phys. Lett. **120B** 137 (1983).

[10] The existence of the infrared fixed point was first discussed in: C. Hill, Phys. Rev. D **24** 691 (1981). More recent examinations including supersymmetry appear in: V. Barger, M. Berger and P. Ohmann, Phys. Rev. D **47** 1093

(1993); P. Langacker and N. Polonsky, Phys. Rev. D **47** 4028 (1993); M. Carena, S. Pokorski and C. Wagner, Nucl. Phys. **B406** 59 (1993).

[11] C. Jarlskog, Phys. Rev. Lett. **55** 1039 (1985).

[12] J. Bahcall, et al., Nature **375** 29 (1995).

[13] G. Fogli, E. Lisi, and D. Montanino, Phys. Rev. D **49** 3626 (1994).

[14] C. Athanassopoulos, et al., Phys. Rev. Lett. **75** 2650 (1995).

[15] M. Gell-Mann, P. Ramond, and R. Slansky, in *Supergravity*, edited P. van Neiuwenhuizen and D. Freedman (North Holland, Amsterdam, 1979), p. 315; T. Yanagida, in *Proceedings of the Workshop on Unified Theory and Baryon Number in the Universe*, edited by O. Sawada and A. Sugamoto (KEK, 1979).

[16] S. Hawking, Phys. Rev. D **14** 2460 (1976).

[17] H. Georgi and S. Glashow, Phys. Rev. Lett. **32** 438 (1974).

[18] H. Georgi, in *Particles and Fields—1974*, edited by C. Carlson (AIP press, New York, 1975).

[19] S. Dimopoulos, S. Raby and F. Wilczek, Phys. Today **44** 25 (1991) and references contained therein.

[20] N. Nielsen, Am. J. Phys. **49** 1171 (1981); R. Hughes, Nucl. Phys. **B186** 376 (1981).

[21] H. Georgi, H. Quinn, and S. Weinberg, Phys. Rev. Lett. **33** 451 (1974).

[22] See for example G. Blewitt, et al., Phys. Rev. Lett. **55** 2114 (1985), and the latest Particle Data Group compilations.

[23] A very useful introduction and collection of basic papers on supersymmetry is S. Ferrara, *Supersymmetry* (2 vols.) (World Scientific, Singapore 1986). Another excellent standard reference is N.-P. Nilles, Phys. Reports **110** 1 (1984). See also [26].

[24] S. Dimopoulos, S. Raby, and F. Wilczek, Phys. Rev. D **24** 1681 (1981).

[25] J. Ellis, S. Kelley, and D. Nanopoulos, Phys. Lett. **B260** 131 (1991); U. Amaldi, W. de Boer, and H. Furstenau, Phys. Lett. **B260** 447 (1991); for more recent analysis see P. Langacker and N. Polonsky, Phys. Rev. D **49** 1454 (1994).

[26] K. Dienes, IASSNS-HEP-95/97, hep-th/9602045 (Feb 1996), Phys. Reports (in press).

[27] Since in this Section many large ideas will be mentioned very briefly, I will not attempt systematic citations.

[28] S. Weinberg, *Dreams of a Final Theory* (Parthenon, New York 1992).

[29] L. Krauss, *The Physics of Star Trek* (Basic Books, New York 1995).

[30] I plan to write about this in a series of Reference Frame articles for Physics Today.

14.6 DISCUSSION

Session Chair: David Gross
Rapporteur: John Brodie

SMITH: Although we were consciously avoiding the subject of searches, experimentally, for CP violation in the near future, I was surprised that there was no comment upon the need, as first stated by Sakharov, for CP violation and baryon non-conservation to get from big bang to where we are today.

WILCZEK: I had to leave out many, many important observations, and that's one of them. It is very attractive to think that the asymmetry between matter and antimatter in the present universe is somehow a secondary effect, and that the initial conditions are more symmetric. This idea is almost mandatory in the context of unified theories that feature baryon-number violating interactions, since in the early universe these interactions could come into equilibrium and thus enforce the symmetry.

SMITH: I know that David (Gross) figures that the CP violation that is seen is simply a set of certain phases, but from my limited understanding that is not what is needed to answer the bigger question.

WILCZEK: The immediate cause of the only CP violation that has actually been observed, in the K meson system, is still not certain. It's very plausible, however, that it just arises as an effect of one of the mixing angles in the weak interactions between quarks being complex, in the context of the Standard Model, as suggested by Kobayashi and Maskawa. It's actually very mysterious why that particular angle is complex and another one, the notorious "θ angle" that generates strong P and T violation is very accurately zero. This puzzle is the strong CP problem, that may be solved by axions. It will be important in coming years to pin down the CP violation in mixing angles, probably through measurements in the $\bar{B} - B$ system, and other sources of CP violation, through measurements of elementary electric dipole moments, and through direct axion searches.

To get a convincing cosmological model of the origin of matter anti-matter asymmetry, one needs more. Sources of CP violation that are exceedingly small and inaccessible under present conditions might have been very important under the cosmological conditions in which the asymmetry formed.

SAMIOS: Would you comment, now that you've unified things in $SU(5)$, on the proton lifetime?

WILCZEK: Yes. An apparent difficulty with unification, which at first sight looks like an absolutely deadly problem, is the fact that when you introduce these other colors, the extra bosons, which change the weak into strong colors, typically they change quarks into leptons. If you work it out more carefully, you see that they inevitably introduce proton decay. If the unification occurred at what you might have thought was a reasonable scale, something not much bigger than the electro-weak scale apriori, in the early days, that would lead to proton decay

14.6 DISCUSSION

Session Chair: David Gross
Rapporteur: John Brodie

SMITH: Although we were consciously avoiding the subject of searches, experimentally, for CP violation in the near future, I was surprised that there was no comment upon the need, as first stated by Sakharov, for CP violation and baryon non-conservation to get from big bang to where we are today.

WILCZEK: I had to leave out many, many important observations, and that's one of them. It is very attractive to think that the asymmetry between matter and antimatter in the present universe is somehow a secondary effect, and that the initial conditions are more symmetric. This idea is almost mandatory in the context of unified theories that feature baryon-number violating interactions, since in the early universe these interactions could come into equilibrium and thus enforce the symmetry.

SMITH: I know that David (Gross) figures that the CP violation that is seen is simply a set of certain phases, but from my limited understanding that is not what is needed to answer the bigger question.

WILCZEK: The immediate cause of the only CP violation that has actually been observed, in the K meson system, is still not certain. It's very plausible, however, that it just arises as an effect of one of the mixing angles in the weak interactions between quarks being complex, in the context of the Standard Model, as suggested by Kobayashi and Maskawa. It's actually very mysterious why that particular angle is complex and another one, the notorious "θ angle" that generates strong P and T violation is very accurately zero. This puzzle is the strong CP problem, that may be solved by axions. It will be important in coming years to pin down the CP violation in mixing angles, probably through measurements in the $\bar{B} - B$ system, and other sources of CP violation, through measurements of elementary electric dipole moments, and through direct axion searches.

To get a convincing cosmological model of the origin of matter anti-matter asymmetry, one needs more. Sources of CP violation that are exceedingly small and inaccessible under present conditions might have been very important under the cosmological conditions in which the asymmetry formed.

SAMIOS: Would you comment, now that you've unified things in $SU(5)$, on the proton lifetime?

WILCZEK: Yes. An apparent difficulty with unification, which at first sight looks like an absolutely deadly problem, is the fact that when you introduce these other colors, the extra bosons, which change the weak into strong colors, typically they change quarks into leptons. If you work it out more carefully, you see that they inevitably introduce proton decay. If the unification occurred at what you might have thought was a reasonable scale, something not much bigger than the electro-weak scale apriori, in the early days, that would lead to proton decay rates of a microsecond or so, contrary to experiment (laughter). But, the result of the unification calculation is that the mass of these bosons, which is governed by the scale at which the unified symmetry breaks, is something like 10^{16} Gev, in supersymmetric unification. So, the most reliable source of proton decay, the most reliably calculable process that gives rise to proton decay in these types of ideas, is governed by the exchange of these very massive particles, and if you calculate, it's comfortably within the experimental limits.

If you didn't have supersymmetry, the scale would be somewhat different, and then straightforward unification would have been ruled out by more recent, heroic experiments, although the lifetime is much larger than a millisecond. In the $SU(5)$ version without supersymmetry, I believe it comes out to be something like 10^{30} years. So, it's a big improvement, but people worked very, very hard as you know, to try to find that decay, and they got limits of something like 10^{31} or 10^{32} years. In supersymmetry models, this particular kind of process, the one that's most fundamental and reliable through exchange of vector bosons, comes out to be something of the order of $10^{35\pm2}$ years. As usual we're teasing the experimentalists.

I should mention that in these supersymmetric theories there are several other potential sources of proton decay, whose strength is a less robust feature of the models. Some of these can be problematic. In building models, one has to arrange that they're sufficiently suppressed. It's possible, but it does constrain the models, and I think it is fair to say that we still do not properly understand why the proton is as stable as it is.